3

571·6 HIN

D1144847

The Facts On File

DICTIONARY
of
CELL
and
MOLECULAR BIOLOGY

The Facts On File

DICTIONARY
of
CELL
and
MOLECULAR BIOLOGY

Edited by
Robert Hine

☑®
Facts On File, Inc.

The Facts On File Dictionary of Cell and Molecular Biology

Copyright © 2003 by Market House Books Ltd

Facts On File, Inc.
132 West 31st Street
New York NY 10001

Library of Congress Cataloging-in-Publication Data

The Facts on File dictionary of cell and molecular biology / edited by Robert Hine.
   p. cm.
  Includes bibliographical references.
  ISBN 0-8160-4912-2
   1. Cytology—Dictionaries.   2. Molecular biology—Dictionaries.   I. Title:
Dictionary of cell and molecular biology. II. Hine, Robert. III. Facts on File, Inc.

QH575.F33 2002
571.6'03—dc21   2002032540

Compiled and typeset by Market House Books Ltd, Aylesbury, UK

Printed in the United States of America

  MP   10 9 8 7 6 5 4 3 2 1

This book is printed on acid-free paper

CONTENTS

PREFACE

This dictionary is one of a series covering the terminology and concepts used in important branches of science. *The Facts On File Dictionary of Cell and Molecular Biology* is planned as an additional source of information for students taking Advanced Placement (AP) Science courses in high schools, but will also be helpful to older students taking introductory college courses.

This volume covers the whole area of modern cell and molecular biology including cell structure, basic molecular biology and biochemistry, molecular genetics, cell metabolism, cell physiology, and laboratory techniques. The definitions are intended to be clear and informative and, where possible, we have provided helpful diagrams and examples. The book also has a selection of short biographical entries for people who have made important contributions to the field. There are a number of useful appendices including a chronology of main advances in the subject and tables of useful information about cell types and sizes. We have also added lists of webpages and an informative bibliography.

The book will be a helpful additional source of information for anyone studying AP Biology, particularly the first part on Molecules and Cells. However, we have not restricted the content to this syllabus. Modern cell biology is one of the most fascinating and fast-moving areas in contemporary science, with relevance in many applied fields including medicine, agriscience, biotechnology, and genetics. We have tried to cover these aspects in the dictionary. We hope that it will prove useful and informative to anyone interested in the subject.

ACKNOWLEDGMENTS

Contributors

Elizabeth Tootill B.Sc.
Sue O'Neill B.Sc.

ABA   *See* abscisic acid.

A band   *See* sarcomere.

ABC transport protein   ATP-binding cassette transport protein; any of a large family of ATP-powered TRANSPORT PROTEINS that move substances across cell membranes. Found in a wide variety of organisms, they have a common basic structure consisting of two transmembrane domains, which form a membrane-spanning channel, and two ATP-binding domains on the cytosolic (inner) face of the membrane. Many bacteria, for example, take up substances such as amino acids from their environment using specific ABC transport proteins called *permeases*. In mammals, cells in the liver, kidney, and intestine can export various natural metabolites or foreign toxins (including certain drugs) using these transport proteins.

abiogenesis   The development of living from nonliving matter, as in the origin of life.

abscisic acid (ABA)   A plant hormone that is involved in seed development and in the closure of leaf pores (stomata) in response to drought stress. Abscisic acid is thought to induce the formation of storage proteins during the later stages of seed maturation, and might also help to prevent precocious germination of seeds. In a quite different role, it is produced in the wilting leaves of plants that lack sufficient water. It diffuses to the GUARD CELLS, signaling them to close the associated stomata and hence reduce further losses of water from the plant. Abscisic acid was first isolated from sycamore and originally thought to be associated with the onset of bud dormancy

(hence its early name 'dormin'). It is also thought to be responsible for the abscission (shedding) of flowers and fruits. However, more recent work has cast doubt on the significance of abscisic acid in these processes.

Abscisic acid

abscission   The shedding of part of a plant, usually a leaf, fruit, or unfertilized flower, as part of the plant's usual life cycle. As the leaf or other organ ages, a distinct *abscission zone* develops at the base of the organ. This zone consists of relatively smaller cells, whose walls weaken and separate. The process is linked to environmental cues, such as colder temperatures and shorter days, and is thought to be controlled by plant hormones called AUXINS. Eventually the connecting tissue is broken mechanically, for example by wind, and the organ is released from the plant.

absorption spectrum   A graph showing the absorbance by a substance of radiation at different wavelengths, usually of ultraviolet, visible, or infrared radiation. It is obtained using a SPECTROPHOTOMETER, and can give information about the identity or

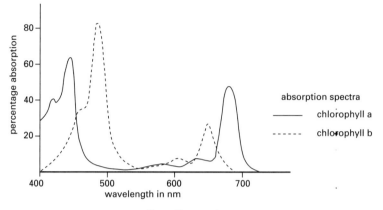

Absorption spectrum

quantity of a substance. Chlorophylls, for example, have absorption peaks in the red and blue (and therefore reflect green light). *See illustration overleaf.*

accessory cell (**subsidiary cell**) Any of several specialized epidermal cells found adjacent to each pair of GUARD CELLS in the leaves and other parts of plants. Accessory cells may help in opening and closing the accompanying pore (stoma).

accessory pigment *See* photosynthetic pigments.

accommodation *See* adaptation.

acellular Describing tissues or organisms that are not composed of discrete cells. Typically, the cytoplasm is continuous throughout the tissue or organism, and contains many nuclei, demonstrating an equivalence to multicellular structures. Examples are aseptate fungal hyphae and muscle fibers. *Compare* unicellular. *See also* syncytium.

acentric Denoting a chromosome or fragment of a chromosome that lacks a CENTROMERE.

acetic acid (**ethanoic acid**) A carboxylic acid, CH_3COOH, obtained by the oxidation of ethanol. Acetic acid is a component of vinegar. This is obtained industrially by the action of acetic acid bacteria, such as *Acetobacter*, on ethanol in wine waste.

acetylcholine (**ACh**) A NEUROTRANS-MITTER found in both vertebrates and invertebrates at synapses in various parts of the nervous system and at NEUROMUSCULAR JUNCTIONS. It can have either excitatory or inhibitory effects, depending on the location. For example, it is excitatory when released by motor nerve endings in skeletal muscle, but has an inhibitory effect on cardiac muscle. It is formed in the presynaptic cell from acetyl coenzyme A and choline, catalyzed by the enzyme choline acetyltransferase. After release, it binds transiently to ACETYLCHOLINE RECEPTORS in the postsynaptic membrane before being broken down by the enzyme acetylcholinesterase to acetate and choline. The choline is then returned to the presynaptic nerve ending to be reused. Nerves that release acetylcholine are called *cholinergic nerves.*

acetylcholine receptors (**cholinoceptors**) Any of various receptors that bind the neurotransmitter acetylcholine at nerve synapses, neuromuscular junctions, or other effector organs. They are described as *cholinergic receptors*, and fall into two broad categories. *Nicotinic (N) acetylcholine receptors* are LIGAND-GATED ION CHANNELS, which open to allow the passage of sodium or potassium ions when activated by binding of acetylcholine (i.e.

the ligand). These occur at neuromuscular junctions, and enable the rapid transmission of a motor nerve impulse, leading to muscle contraction. *Muscarinic (M) acetylcholine receptors* are G-PROTEIN-LINKED RECEPTORS, and have a slower and more prolonged response to the binding of acetylcholine. There are several subtypes, designated M1, M2, M3, etc. They can have either excitatory or inhibitory effects, depending on their location.

acetyl CoA (acetyl coenzyme A) An important intermediate compound in cell metabolism, particularly in the oxidation of sugars, fatty acids, and amino acids, and in certain biosynthetic pathways. It is formed in the mitochondrial matrix by the reaction between pyruvate (from GLYCOLYSIS) and COENZYME A, catalyzed by the enzyme pyruvate dehydrogenase. The acetyl group of acetyl CoA is subsequently oxidized in the reactions of the KREBS CYCLE, to yield reduced coenzymes and carbon dioxide. Acetyl CoA is also produced in the initial oxidation of fatty acids and some amino acids. Other key roles for acetyl CoA include the provision of acetyl groups in biosynthesis of fatty acids, terpenoids, and other substances.

ACh *See* acetylcholine.

acid Any molecule or substance that tends to release hydrogen ions (H+; protons). In aqueous solutions, a free hydrogen ion rapidly combines with a water molecule to form a *hydronium ion* (H_3O^+), also called an *oxonium ion* or *hydroxonium ion*. (However, it is conventional to refer to hydrogen ions in solution, rather than hydronium ions.) An acid solution has a pH below 7. Adding acid to a solution increases the hydrogen ion concentration, and hence lowers the pH. Many organic molecules have several acidic and/or basic groups, and dissociation of particular groups usually depends on the pH of the solution. Hence, the acid (or basic) nature of such molecules varies according to conditions. The pH of cells is stabilized by BUFFERS, such as phosphate ions. *Compare* base.

acid-growth hypothesis A hypothesis proposed by R. Cleland, A. Hager and others in around 1970 to explain cell wall expansion in growing plant cells. According to this, the plant hormone auxin stimulates a proton pump in the cell's plasma membrane, which pumps protons (H+) into the cell wall. The resultant lowering of cell-wall pH activates wall-loosening enzymes, thus enabling extension of the cell wall, and hence cell growth. The wall enzymes, called *expansins*, disrupt hydrogen bonds between the cellulose microfibrils. These bonds re-form when the pH returns to normal, thereby restoring the wall strength when growth is completed.

acidic stain *See* staining.

acquired immune deficiency syndrome *See* AIDS.

acquired immunodeficiency syndrome *See* AIDS.

acrosome A membrane-bound sac in the anterior head region of a spermatozoon. In a mammalian sperm it usually forms a cap over the nucleus. It contains enzymes that are released on contact with the egg at fertilization, as part of the ACROSOME REACTION. The enzymes break down the egg coats, enabling the spermatozoon nucleus to enter the egg.

acrosome reaction The process by which a sperm penetrates the wall of an egg during fertilization. Contact with an egg of the same species triggers a series of events, including the release of lytic enzymes from the ACROSOME. These enzymes break down the outer layer of the egg, and permit fusion of the sperm's plasma membrane with that of the egg. In many invertebrate sperm, actin filaments in the acrosomal region are lengthened by the addition of actin subunits, to form a a fine *acrosomal filament*. This extends from the front end of the sperm and assists penetration.

ACTH *See* corticotropin.

actin A protein that is a major constituent of the cytoskeleton and of muscle. It is the most abundant protein inside eukaryote cells, accounting for 1–5% of total protein. Actin exists as globular monomers (*G-actin*), which assemble into long filamentous polymers (*F-actin*). In conjunction with motor proteins, especially MYOSIN, actin is responsible for many types of cell movement and contraction, including muscle contraction. F-actin forms the MICROFILAMENTS of the cytoskeleton, and thus is responsible for maintaining and changing cell shape. The bundles and networks of actin microfilaments are held together by ACTIN CROSS-LINKING PROTEINS, and are attached to the plasma membrane by other proteins, such as filamin and dystrophin. Changes in cell shape involve extension or shortening of actin microfilaments, brought about by the assembly or dismantling of the G-actin subunits. These processes are regulated by other proteins. For example, profilin promotes assembly of actin filaments, whereas cofilin and gelsolin act as severing proteins, breaking actin filaments into fragments. Actin polymers form the thin filaments that are part of the muscle myofibrils, which consist of alternating and overlapping sets of thick myosin and thin actin filaments. In muscle contraction, overlapping actin and myosin molecules interact to form actomyosin complexes. Actin and myosin also interact in nonmuscle cells. For example, they accomplish the cleavage of cells by CYTOKINESIS, and are responsible for CYTOPLASMIC STREAMING in certain green algae. *See* actomyosin.

actin-binding proteins Proteins that bind to actin filaments – either in muscle to regulate contraction, or as components of the cytoskeleton. In skeletal muscle they include TROPOMYOSIN and TROPONINS, and additionally CALDESMON in smooth muscle. All work essentially by altering the degree to which myosin can bind to actin, under the influence of calcium-ion concentration. In the cytoskeleton, several proteins link actin filaments to each other or to the plasma membrane, including ankyrin, DYSTROPHIN, PROFILIN, and SPECTRIN. *See also* actin cross-linking proteins.

actin cross-linking proteins Proteins that form cross-links between actin filaments in the cytoskeleton. Each cross-linking protein molecule has two actin-binding sites, and can bind two actin filaments. With short linkers, such as fimbrin molecules, the actin filaments form parallel bundles; longer linker proteins, such as spectrin and dystrophin, permit the filaments to form looser networks, as typically occur in the cell cortex just inside the plasma membrane.

actinomycetes (filamentous bacteria) A diverse group of Gram-positive bacteria characterized by a filamentous, often branching growth form that resembles the mycelium of a fungus. They are numerous in the topsoil and are important in soil fertility. Some can cause infections in animals and humans. Many antibiotics (e.g. streptomycin, actinomycin, and tetracycline) are obtained from actinomycetes, especially from members of the genus *Streptomyces*.

action potential The transitory change in electrical potential that occurs across the membrane of a nerve or muscle fiber during the passage of a nervous impulse. The action potential is an ALL-OR-NONE event, and is independent of the strength of the stimulus eliciting the impulse. In the absence of an impulse a RESTING POTENTIAL of about -70 mV exists across the membrane (inside negative) as a result of the unequal distribution of ions between the intracellular and extracellular media. However, during the passage of an impulse the polarity of the membrane reverses due to sequential changes in the permeability of VOLTAGE-GATED ION CHANNELS in the membrane.

Initially, sodium channels open, allowing the influx of sodium ions. This causes a sudden depolarization of the membrane, altering the membrane potential to about $+30$ mV (inside positive) within about 0.5 ms. The sodium channels then close, and are unable to open again (i.e. are refractory) for several milliseconds. As the action potential reaches its peak, potassium channels open, permitting the efflux of potassium ions. This restores the polarity to negative inside the axon and, after a brief

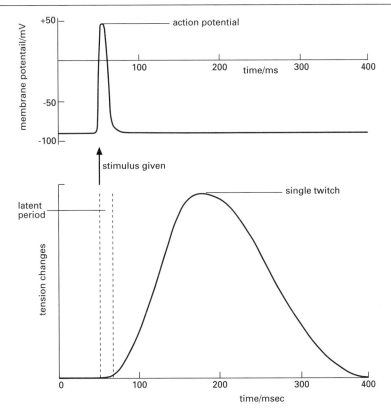

Action potential: the upper diagram shows the action potential stimulating a muscle; the lower diagram shows the resulting change in muscle tension

hyperpolarization, the membrane potential returns to its resting level. Depolarization at any one point spreads to adjoining segments of the axon, thereby triggering opening of the sodium channels and propagating the impulse.

The action potential is unidirectional due to the refractory state of the sodium ion channels following passage of the action potential. This prevents propagation in the reverse direction. In myelinated nerve fibers, the sodium channels are concentrated in the short segments of 'bare' axon called *nodes of Ranvier*. Depolarization of the axon effectively 'jumps' from node to node, thereby greatly increasing the rate of impulse transmission.

action spectrum A graph showing the effect of different wavelengths of radiation,

usually light, on a given process. It is often similar to the ABSORPTION SPECTRUM of the substance that absorbs the radiation and can therefore be helpful in identifying that substance. For example, the action spectrum of photosynthesis is similar to the absorption spectrum of chlorophyll.

activation domain A three-dimensional region (domain) of an ACTIVATOR – a type of transcription factor that can activate gene transcription when the transcription factor is bound to the DNA molecule via another domain (the DNA-binding domain). Activators can have more than one activation domain; these are joined to the DNA-binding domain by highly flexible regions of the protein. Interaction of the activation domains on different transcription factors, and with other co-activators, leads to the assembly of a transcription INITIA-TION COMPLEX.

activation energy The amount of energy required for a reaction to proceed. It is determined by the free energy of any intermediates formed in the reaction, which tends to be greater than that of either reactants or products. Enzymes accelerate the rate of reactions by reducing the free energy of transition states, and hence the activation energy.

activator A type of TRANSCRIPTION FACTOR that activates transcription of a gene or genes. Activators are proteins that bind to DNA via specific molecular regions called *DNA-binding domains*, and interact with other transcription factors by means of one or more ACTIVATION DOMAINS. Their activity is antagonized by repressors. Activators bind cooperatively to enhancer sites on the DNA to form *enhanceosomes*, and stimulate the formation of transcription INITIATION COMPLEXES.

active site The region of an enzyme molecule that combines with and acts on the substrate. It consists of amino acids arranged in a configuration specific to a particular substrate or type of substrate. An active site performs two functions: it binds the substrate, and it catalyses a chemical reaction involving the bound substrate. Binding of an inhibiting compound elsewhere on the enzyme molecule may change the active site's configuration and hence the efficiency of the enzyme activity.

active transport The transport of molecules or ions across a cell membrane against a concentration gradient or electrochemical gradient through the expenditure of energy. Anything that interferes with the provision of energy will interfere with active transport. The mechanism typically involves a transporter protein that spans the membrane and transfers substances across it by changing shape. There are two main categories: transport by ATP-powered pumps, such as the NA+/K+ ATPASE, which derive energy from the hydrolysis of ATP; and transport by COTRANSPORTERS, which derive energy from the coupled movement of another substance (e.g. hydrogen ions) down an electrochemical gradient. Both categories include a range of specific transporters for various ions and small molecules. These actively transport their respective substrates across membranes at diverse sites within cells, including the plasma membrane and organelles such as chloroplasts and mitochondria.

actomyosin A protein complex found in MUSCLE, formed between molecules of actin and myosin present in adjacent thin and thick muscle filaments. Actomyosin complexes are involved in muscle contraction, particularly in the mechanism that pulls the thin filaments past the thick ones. Actomyosin is also formed when a solution

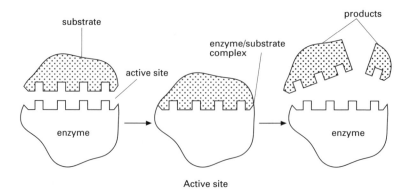

substrate

products

enzyme/substrate complex

active site

enzyme

enzyme

Active site

of actin is mixed with a solution of myosin *in vitro*.

acylglycerol   *See* glyceride.

adaptation   **1.** The extent to which an organism, or a physiological or structural characteristic of an organism, is suited to a particular environment.
2. (**accommodation**) The decrease with time of the frequency of response of sensory receptors subjected to a constant intensity of stimulus. A more intense stimulus is then required to elicit a response. Receptors vary greatly in the rate and degree of adaptation they show depending on their function. Many monitor relatively static stimuli, such as internal temperature, and exhibit no adaptation, while others monitoring rapidly changing stimuli, such as touch sensation in the skin, rapidly adapt to a constant intensity stimulus.

adapter protein   **1.** A protein component of CLATHRIN-COATED PITS and vesicles. The latter arise from invaginations of membranes, and function in transporting materials around the cell. As a pit forms in a membrane, the adapter proteins bind to membrane proteins and form *assembly particles*. These promote the assembly of clathrin molecules to enclose the pit, which is ultimately pinched off as a vesicle. Which proteins enter the vesicle is determined by the type of adapter proteins present.
2. A protein that participates in signal transduction pathways but has no enzymatic activity itself. Adapter proteins contain various binding sites at which other signaling components can dock, and hence facilitate interaction between them.

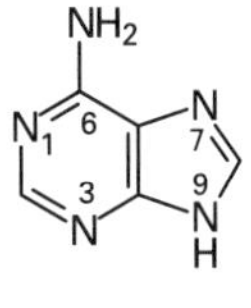

NH$_2$

Adenine

adenine   A nitrogenous base found in DNA and RNA. It is also a constituent of certain coenzymes, for example NAD and FAD, and when combined with the sugar ribose it forms the nucleoside adenosine, found in AMP, ADP, and ATP. Adenine is a PURINE, and contains a pair of fused rings. *See illustration at* DNA.

adenohypophysis   *See* pituitary gland.

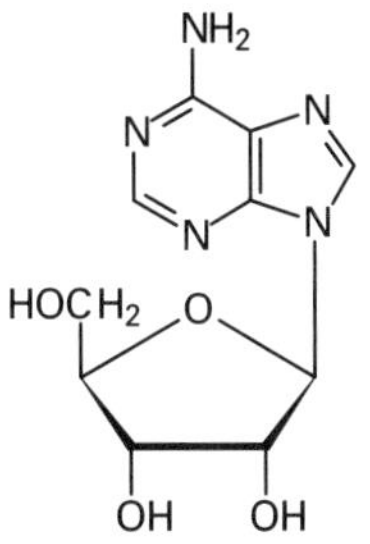

Adenosine

adenosine   A nucleoside formed from adenine linked to D-ribose with a β-glycosidic bond. Adenosine triphosphate (ATP) is a nucleotide derived from adenosine. *See illustration overleaf.*

adenosine diphosphate   *See* ADP.

adenosine monophosphate   *See* AMP.

adenosine triphosphate   *See* ATP.

adenovirus   One of a group of nonenveloped viruses that contain a linear segment of double-stranded DNA. They are about 80 nm in diameter, and cause respiratory diseases in humans and other animals.

ADH (**antidiuretic hormone**)   *See* vasopressin.

adherens junction   A type of cell junction found mainly between epithelial cells that forms a strengthening belt around adjoining cells. It consists of a continuous band of cell adhesion molecules called CADHERINS, situated mainly on the outside of each cell. These bands bind to each other in the intercellular space, connecting neighboring cells together. Inside each cell, the cadherins connect to the cytoskeleton.

adipocyte *See* adipose tissue.

adipose tissue (fatty tissue) A type of CONNECTIVE TISSUE consisting of closely packed cells (*adipocytes*) containing fat. Adipose tissue is found in varying amounts in the dermis of the skin and around the kidneys, heart, and blood vessels. It provides an energy store, heat insulation, and mechanical protection. Most adipose tissue is in the form of *white fat*, whose cells contain a single large fat droplet surrounded by a thin layer of cytoplasm and a nucleus squashed to one side. Hibernating and newborn animals have deposits of darker colored *brown fat*. This is well supplied with blood vessels and consists of smaller cells containing numerous lipid droplets and many mitochondria. Its main function is to generate heat.

ADP **(adenosine diphosphate)** A nucleotide consisting of adenine and ribose with two phosphate groups attached. *See* ATP.

adrenal glands A pair of glands each of which is situated above a kidney. They produce various hormones, notably epinephrine. Secretion is controlled by the nervous system. Each gland comprises a central medulla and an outer cortex, of different embryological origins and behaving independently. The adrenal medulla produces norepinephrine and epinephrine. The adrenal cortex is rich in vitamin C and cholesterol, and produces three types of hormones: aldosterone, cortisol, and sex hormones.

adrenaline *See* epinephrine.

adrenergic **1.** Describing any nerve axon (fiber) that releases epinephrine or norepinephrine from its ending when stimulated by a nerve impulse. Vertebrate sympathetic motor nerve fibers are adrenergic.

2. Describing a neuron, other cell, or cell receptor that is stimulated by epinephrine or norepinephrine. *Compare* cholinergic.

adrenocorticotropic hormone (ACTH) *See* corticotropin.

aerobe An organism that respires aerobically (*see* aerobic respiration) and can live in the presence of atmospheric oxygen (O_2). *Obligate aerobes* must have a supply of O_2 to survive, whereas *facultative aerobes* can live without O_2 by using anaerobic respiration, although they tend to thrive only when O_2 is present. *Microaerophiles* require O_2 but at lower concentrations than that found in air.

aerobic respiration Repiration in which free oxygen is used to oxidize organic substrates to carbon dioxide and water, with a high yield of energy. The free energy released is converted to chemical energy, mainly in the form of ATP. The reaction overall is:
$$C_6H_{12}O_6 + 6O_2 = 6CO_2 + 6H_2O + energy$$
It occurs in a number of stages, the first of which (GLYCOLYSIS) also occurs in anaerobic respiration in the cell cytoplasm. With glucose as the substrate, a sequence of reactions results in the formation of pyruvate. The remaining stages, which do not occur in ANAEROBIC RESPIRATION, take place in the mitochondria in eukaryotes (in prokaryotes, which lack mitochondria, these stages occur on the plasma membrane). Pyruvate is converted to acetyl coenzyme A, which enters a cyclic series of reactions, the KREBS CYCLE, with the production of carbon dioxide and reduced coenzymes (NADH and $FADH_2$). Electrons from the reduced coenzymes pass sequentially along a series of membrane-bound electron carriers (cytochromes and flavoproteins), which make up the ELECTRON-TRANSPORT CHAIN. This movement of electrons is coupled to the passage of hydrogen ions (protons) across a membrane (in eukaryotes the inner membrane of the mitochondrion). This creates a protonmotive force, which is used to drive the formation of ATP from ADP, in the process

called OXIDATIVE PHOSPHORYLATION. The electrons are ultimately transferred to oxygen, with the formation of water. There is a net production of 38 ATPs per molecule of glucose during aerobic respiration, a yield about 19 times that of FERMENTATION alone, which utilizes only SUBSTRATE-LEVEL PHOSPHORYLATION. Aerobic respiration, using oxygen as the terminal electron acceptor, is thus a much more efficient means of energy conversion than fermentation, and the various other forms of anaerobic respiration practised by prokaryotes, which use other substances as electron acceptors. Hence aerobic respiration is the mechanism used by the majority of organisms whenever possible. *Compare* anaerobic respiration.

aerotactic movement   *See* aerotaxis.

aerotaxis (aerotactic movement)   A directed movement (TAXIS) of a cell or organism in response to an oxygen concentration gradient. For instance, motile aerobic bacteria are positively aerotactic, whereas motile obligate anaerobic bacteria are negatively aerotactic.

affinity chromatography   A type of liquid CHROMATOGRAPHY that separates particular proteins or other target molecules according to their ability to bind specifically to another molecule. It uses a column packed with beads to which are attached small molecules capable of binding the target molecule. When a mixture is poured into the column, any target molecules bind to the the small molecules (i.e. the stationary phase), while other components pass through. The bound target molecules are then eluted from the column using, for example, a salt solution. A common variant is *antibody-affinity chromatography*, in which the attached small molecules are antibodies specific to the target molecule.

agamospermy   *See* apomixis.

agar   A gelling agent prepared from seaweed. It is used extensively to set liquid nutrients in growth media, for example for microorganisms. *See also* agarose.

agarose   A component of agar that forms a gel widely used as a matrix in chromatography and electrophoresis.

agglutination   The clumping together of cells (e.g. red blood cells or bacteria) or other ANTIGEN-coated particles as a result of the action of ANTIBODIES. Agglutination may occur in transfusion if blood of the wrong group is given. The surfaces of the donor's red blood cells contain antigen molecules that are attacked by antibody molecules in the serum of the recipient, which causes the donor red cells to clump together. These clumps may block capillaries, causing fatal damage to the heart or brain.

aging   *See* senescence.

agonist   A substance, such as a drug or hormone, that binds to a cell's receptors and elicits a response. *Compare* antagonist.

agranulocyte   A white blood cell (leukocyte) that does not contain granules in its cytoplasm. There are two types: LYMPHOCYTES and MONOCYTES (comprising 25% and 4%, respectively, of all leukocytes). Both have large nuclei and a small amount of clear cytoplasm.

Agrobacterium   A genus of soil bacteria, the species *A. tumefaciens* being the causative agent of crown gall, a type of tumor in plants. A segment of DNA (transferred DNA, T-DNA) from a plasmid in the bacterium is transferred into the host DNA and induces tumor formation. Since the plasmid is capable of independent replication in host cells of many dicotyledonous plants, it has been used as a CLONING VECTOR in genetic engineering. Once the desired segment of DNA, for example a gene, has been spliced into the T-DNA, the plasmid can be introduced into plant cells, which are then cultured to produce entire plants with the desired characteristic. Unfortunately, the bacterium does not infect monocotyledonous plants, which include important cereal crops. *See also* DNA cloning; tissue culture.

AIDS (acquired immunodeficiency syndrome)   A viral disease of humans caused by human immunodeficiency virus (*see* HIV) and characterized by destruction of the lymphocytes responsible for cell-mediated immunity (*see* T-cell). The patient consequently succumbs to opportunistic fatal infections, cancers, etc.

alanine   *See* amino acids.

albinism   *See* melanin.

albuminous cell   A parenchyma cell found in groups within the phloem of gymnosperms. Albuminous cells serve as companion cells to sieve cells, which form the conducting elements.

alcohol   A type of organic compound of the general formula ROH, where R is a hydrocarbon group. Examples of simple alcohols are methanol (CH_3OH) and ethanol (C_2H_5OH).

Alcohols have the –OH group attached to a carbon atom that is not part of an aromatic ring: C_6H_5OH, in which the –OH group is attached to the ring, is thus a phenol. Phenylmethanol ($C_6H_5CH_2OH$) does have the characteristic properties of alcohols.

Alcohols can have more than one –OH group; those containing two, three, or more such groups are described as *dihydric*, *trihydric*, and *polyhydric* respectively (as opposed to those containing one –OH group, which are *monohydric*). Examples are ethane–1,2-diol (ethylene glycol; ($HOCH_2CH_2OH$) and propane-1,2,3-triol (glycerol; $HOCH_2CH(OH)CH_2OH$).

Alcohols are further classified according to the environment of the –C–OH grouping. If the carbon atom is attached to two hydrogen atoms, the compound is a *primary alcohol*. If the carbon atom is attached to one hydrogen atom and two other groups, it is a *secondary alcohol*. If the carbon atom is attached to three other groups, it is a *tertiary alcohol*.

aldaric acid   *See* sugar acid.

aldehyde   A type of organic compound with the general formula RCHO, where the –CHO group (the aldehyde group) consists of a carbonyl group attached to a hydrogen atom. Simple examples of aldehydes are methanal (formaldehyde, HCHO) and ethanal (acetaldehyde, CH_3-CHO). Aldehydes are formed by oxidizing a primary alcohol, and can themselves be oxidized to carboxylic acids. Reduction produces the parent alcohol.

alditol   *See* sugar alcohol.

aldohexose   An aldose SUGAR with six carbon atoms.

aldonic acid   *See* sugar acid.

aldopentose   An aldose SUGAR with five carbon atoms.

aldose   A SUGAR containing an aldehyde (CHO) or potential aldehyde group.

aldosterone   A steroid hormone secreted by the cortex of the adrenal glands. It stimulates the reabsorption of sodium ions from urine by the kidney tubules, and hence causes the reabsorption of water, thereby promoting water retention by the body. This raises blood pressure. It also promotes the excretion of potassium and hydrogen ions by the kidneys.

algae   A large mixed group of photosynthesizing organisms, now usually placed in the kingdom Protista. They often resemble plants and are found mainly in marine or freshwater habitats, although some algae are terrestrial. Algae differ from plants in lacking any real differentiation of leaves, stems, and roots, and in not having an embryo stage in their life cycle. Algae can be unicellular (e.g. *Chlamydomonas*), colonial (e.g. *Volvox*), filamentous (e.g. *Spirogyra*), or thalloid (e.g. *Fucus*). All algae contain chlorophyll but this may be masked by a number of different accessory pigments.

alimentary canal   A tube through which food is passed for digestion and absorption into the bloodstream of an animal. In most animals, it leads from the mouth to the anus, and different parts are modified for digestion and absorption of soluble food. Numerous glands pass secretions, containing enzymes, into the alimentary canal and these digest the food as it is moved along by peristalsis. The inside, bounding the gut space (lumen), is lined by epithelium and underlying connective tissue. Exterior to this are further layers of muscle and connective tissue, with associated blood vessels and lymphatics.

allele (allelomorph)   One of the possible forms of a given gene. The alleles of a particular gene occupy the same positions (*loci*) on homologous chromosomes. A gene is said to be *homozygous* if the two loci have identical alleles and *heterozygous* when the alleles are different. When two different alleles are present, one (the *dominant* allele) usually masks the effect of the other (the *recessive* allele). The allele determining the normal form of the gene is usually dominant while mutant alleles are usually recessive. Thus most mutations only show in the phenotype when they are homozygous. In some cases one allele is incompletely dominant or recessive to another allele, giving rise to an intermediate phenotype in the heterozygote.

allelomorph   *See* allele.

allergen   *See* allergy.

allergy   A type of abnormal immune response in which the body produces antibodies against substances, such as dust or pollen, that are usually not harmful and are normally removed or destroyed in other individuals. The allergy-triggering substances (*allergens*) stimulate certain lymphocytes (B-cells) to secrete antibodies, termed *reagins*, which bind to mast cells. When an allergen encounters and binds to the mast cell the latter degranulates, discharging histamine and other substances. Histamine dilates blood capillaries in the region and increases their permeability, resulting in inflammation and increased mucus secretion. It also stimulates contraction of smooth muscle, leading to bron-

chial constriction. Hence, drugs that block histamine (*antihistamines*) are used to relieve the symptoms of hay fever, asthma, and similar allergies.

allograft   *See* graft.

allopolyploid   *See* polyploid.

all-or-none   Describing a response shown by certain irritable tissues, for example nerve cells and muscle, that occurs either with all possible strength or not at all. A stimulus produces no response until it reaches a threshold level, when it produces a fixed maximum response, independent of stimulus intensity. Individual muscle fibers obey this principle, as do neurons when conducting a nerve impulse.

allosteric site   A part of an enzyme to which a specific effector or modulator can bind reversibly. This creates an enzyme–effector complex, thereby modifying in some way the properties of the enzyme's ACTIVE SITE, to which the substrate binds.

alpha helix   A highly stable SECONDARY STRUCTURE of a protein molecule in which the peptide chain is coiled to form a spiral. Each turn of the spiral contains approximately 3.6 amino-acid residues. The R group (side chain) of each amino acid extends outward from the helix. Hydrogen bonding between successive coils holds the helix together. If the alpha helix is stretched the hydrogen bonds are broken but reform on relaxation. The alpha helix is found in numerous proteins, including muscle protein and keratin.

Altman, Sidney (1939–)   Canadian-born American biochemist who discovered the catalytic properties of RNA. He was awarded the Nobel Prize for chemistry in 1989 jointly with T. R. Cech.

ameba (*pl.* **amebas** *or* **amebae**)   A single-celled, usually microscopic organism with a body that is ever-changing in form and from which extend protrusions of the cytoplasm called pseudopodia (*see* pseudopodium). Amebas are protists, found typically in aquatic and moist habitats; many are free-living in the sea, fresh water, soil, and rotting vegetation, but some are parasitic, living inside other organisms. They belong to several different taxonomic groups. Some are naked, whereas others are enclosed in a 'shell', or test. Amebas move by extending pseudopodia in a certain direction and retracting the trailing edge of the cell while moving the cell contents forward. They feed by engulfing food particles (e.g. bacteria, other protists) using their pseudopodia – a form of PHAGOCYTOSIS. Food and water vacuoles carry out digestion and osmoregulation. Reproduction is by binary fission and spore formation takes place in adverse conditions. Among the best-studied are amebas of the phylum Rhizopoda, which includes the genus *Amoeba*.

amebocyte A cell similar to an ameba found in the walls of sponges. Amebocytes exhibit ameboid movement and can engulf food particles by phagocytosis. As well as feeding and digestion, they perform various other tasks, including the secretion of skeletal material (spongin) and production of gametes.

ameboid movement A type of locomotion shown by amebas and certain other motile cells. It involves the protrusion of the plasma membrane in the forward direction to form finger-like or balloon-like extensions of the cell called pseudopodia, or 'false feet'. Formation of pseudopodia is linked to reversible changes in the cytoplasm, which cycles between a fluid sol state in the cell interior (the endoplasm) and a gel state just beneath the plasma membrane (the ectoplasm). Endoplasm flows to fill a newly forming pseudopodium, and at the leading edge is converted to ectoplasm. At the cell's trailing edge, the reverse occurs, with ectoplasm changing to endoplasm and flowing forward, causing the 'tail' of the cell to retreat. These sol–gel transformations are themselves caused by the assembly and disassembly of actin microfilament networks, which determine the viscosity of the cytosol.

amine A compound containing a nitrogen atom bound to hydrogen atoms or hydrocarbon groups. They have the general formula R_3N, where R can be hydrogen or an alkyl or aryl group. An amine is classified according to the number of organic groups bonded to the nitrogen atom: one, primary; two, secondary; three, tertiary. Since amines are basic in nature they can form the quaternium ion, R_3NH.

amino acids Any of a group of water-soluble organic compounds, found in living organisms principally as the building blocks of proteins. They can be represented by the general formula $R–CH(COOH)-NH_2$, where R (the variable side chain, or R group) is attached to the α carbon atom. Hence, beside the R group, the α carbon atom is bonded to three other different chemical groups: a carboxyl group ($–COOH$); an amino group ($–NH_2$) and a hydrogen atom (H). Amino acids link together via peptide bonds to form short chains (peptides) and longer chains (polypeptides). A protein can contain one or several polypeptides, each comprising tens or hundreds of amino acids. The shape and other properties of any particular protein molecule are determined by the sequence and properties of its constituent amino acids.

Hydrophilic amino acids, such as arginine and serine, have ionized or polar R groups and are water-soluble. These tend to be located on the surface of water-soluble proteins. *Hydrophobic amino acids*, for example valine and tyrosine, have exclusively hydrocarbon side chains and are insoluble or only slightly soluble in water. They pack the interior of soluble proteins, or form the surface of membrane-spanning regions of proteins. The R group of cysteine contains a sulfhydryl group ($–SH$), which can form a disulfide bond with a second cysteine. Such bonds are important in maintaining the correctly folded nature of certain proteins.

Living organisms assemble the polypeptide components of proteins using various combinations of just 20 naturally occurring amino acids. It is important therefore that all the amino acids should be present in sufficient quantities. In humans ten of the twenty amino acids can be synthesized by the body itself. Since these are not required in the diet they are known as *nonessential amino acids*. The remaining ten cannot be synthesized by the body and have to be supplied in the diet. They are known as *essential amino acids. See Appendix for structures.*

Various other amino acids fulfill important roles in metabolic processes other than as constituents of proteins. For example, ornithine
$$(H_2N(CH_2)_3CH(NH_2)COOH)$$
and citrulline
$$(H_2N.CO.NH.(CH_2)_3CH(NH_2)COOH)$$
are intermediates in the production of urea.

aminoacyl tRNA   *See* transfer RNA.

aminoacyl-tRNA ligase   *See* transfer RNA.

amino sugar   A sugar in which a hydroxyl group (OH) has been replaced by an amino group (NH_2). Glucosamine (from glucose) occurs in mucopolysaccharides and is a major component of chitin. Galactosamine or chondrosamine (from galactose) is a major component of cartilage and glycolipids. Amino sugars are important components of bacterial cell walls.

amitosis   Nuclear division in which there is no mitotic spindle, so that the daughter nuclei are likely to have unequal sets of chromosomes. The ordered process of chromosome duplication, dissolution of the nuclear membrane, and production of a spindle as in MITOSIS is apparently absent. Cells produced amitotically inherit variable numbers of chromosomes. However, because many cells that divide amitotically are polyploid, for example the endosperm nucleus in angiosperms and the macronucleus of ciliates, the risk of a daughter cell lacking essential genes is mitigated. *Compare* endomitosis.

Amoeba   *See* ameba.

AMP   (**adenosine monophosphate**)   A nucleotide consisting of adenine, ribose, and phosphate. *See* ATP; cyclic AMP.

amylase   Any of a group of closely related enzymes, found widely in plants, animals, and microorganisms, that hydrolyze starch or glycogen to the sugars maltose or glucose, or to dextrins. Both α- and β-amylases occur in plants, but only α-amylase is found in animals, in the pancreatic juices and in saliva, having an important role in digestion.

amylopectin   The water-insoluble fraction of STARCH.

amyloplast   A membranous organelle (plastid) found in some plant cells that serves as a store for starch grains. Amyloplasts are common in storage organs, for example the potato tuber. In cells of the root cap and elsewhere they are believed to act as gravity detectors, or statoliths. *See* geotropism.

amylose   *See* starch.

anabolism   Metabolic reactions in which molecules are linked together to form more complex compounds. Thus, anabolic reactions are concerned with building up structures, storage compounds, and complex metabolites in the cell. Starch, glycogen, fats, and proteins are all products of anabolic pathways. Anabolic reactions generally require energy, usually provided by ATP. *Compare* catabolism. *See also* metabolism.

anaerobe   An organism that can live and grow in the absence of free oxygen, i.e. it respires anaerobically (*see* anaerobic respiration). *Facultative anaerobes* usually respire aerobically but can switch to anaerobic respiration when free oxygen is in short supply; *obligate anaerobes* never respire aerobically and may even be poisoned by free oxygen. *Compare* aerobe.

anaerobic respiration   Respiration in which oxygen is not involved, found in bacteria, yeasts, some protists, and occasionally in muscle tissue. The food substrate is incompletely oxidized and the energy yield is lower than in AEROBIC RESPIRATION. In the absence of oxygen in animal muscle tissue, glucose is degraded to pyruvate by glycolysis, with the production of a small amount of energy (in the form of ATP) and also lactic acid, which may be oxidized later when oxygen becomes available. FERMENTATION is an example of anaerobic respiration in which ATP is formed utilizing SUBSTRATE-LEVEL PHOSPHORYLATION. In yeasts, for example, alcoholic fermentation produces ethanol and carbon dioxide as end products, but yields only two molecules of ATP per molecule of glucose. Various other types of fermentation are performed by prokaryotes. However, anaerobic respiration in many prokaryotes involves an ELECTRON-TRANSPORT CHAIN and ATP SYNTHASES – similar to aerobic respiration – but with various inorganic or organic substances substituting for molecular oxygen as the terminal electron acceptor. For example, nitrate (NO_3^-) is used by denitrifying bacteria, and sulfate (SO_4^{2-}) by sulfate-reducing bacteria. Some bacteria are capable of both anaerobic and aerobic respiration. *See* anaerobe.

anaphase   The stage in mitosis or meiosis when CHROMATIDS are pulled toward opposite poles of the dividing cell by the SPINDLE. In MITOSIS the chromatids moving towards the poles represent a single complete chromosome. During anaphase I of MEIOSIS a pair of chromatids still connected at their CENTROMERE move to the spindle poles. During anaphase II the centromeres divide and single chromatids are drawn toward the poles.

anatomy   1. The study of the structure and arrangement of the parts of an organism.
2. The organization of the parts of an organism and the structural relationships between them.

androgen   A male sex hormone (a steroid) that controls the development, function, and maintenance of secondary male characteristics (e.g. facial hair and deepening of the voice), male accessory sex organs, and spermatogenesis. Androgens are produced chiefly by the testis (smaller amounts are produced by the ovary, adrenal cortex, and placenta); the most important is testosterone. Androgens are used to treat certain diseases; they also have anabolic activity, promoting growth and formation of new tissue.

aneuploidy   The condition in which one or more chromosomes are missing from or added to the normal somatic chromosome number. It arises during MEIOSIS due to the failure of homologous chromosomes to separate (i.e. NONDISJUNCTION). If both of a pair of homologous chromosomes are missing, *nullisomy* results. *Monosomy* and *trisomy* are the conditions in which one or three homologs occur respectively, instead of the normal two. *Polysomy*, which includes trisomy, is the condition in which one or more chromosomes are represented more than twice in the cell.

Anfinsen, Christian Boehmer (1916–95)   American protein chemist. He was awarded the Nobel Prize for chemistry in 1972 for his work on ribonuclease. The prize was shared with S. Moore and W. H. Stein.

angiogenesis   The formation of the blood vascular system during development. It involves remodeling of the vascular network that forms in the early embryo, with the coalescing of smaller branches into larger ones, and their enclosure by smooth muscle and other support tissue. Growth of blood vessels can also occur in mature individuals following injury or disease, or be induced by tumors.

angstrom   Symbol: Å   A unit of length equal to 10^{-10} meterx. It is still used occasionally for measurements of wavelength or interatomic distance.

aniline stains   *See* staining.

animal   A multicellular diploid eukaryotic organism that feeds on other organisms or on organic matter (i.e. is heterotrophic). Except for sponges, animals develop from a multicellular embryo (blastula), which is formed by fusion of a haploid sperm and haploid egg. Unlike PLANTS, animals are often motile, and react to stimuli quickly. Animal cells are typically highly organized and tightly integrated to form tissues and organs that are specialized to perform particular functions. The constituent cells communicate using an array of intercellular junctions and chemical signals.

Animals comprise the kingdom Animalia. Many single-celled organisms, for example protozoa, that were previously classified as animals are now included in the kingdom Protista, along with other unicellular eukaryotes.

animal pole The part of the surface of an animal egg (ovum) to which the nucleus is closest. It is usually opposite the vegetal, or yolky, pole.

animal starch *See* starch.

anion A negatively charged ion, formed by addition of electrons to atoms or molecules. In electrolysis anions are attracted to the positive electrode (the anode). *Compare* cation.

anisogamete *See* gamete.

anisogamy The sexual fusion of non-identical gametes. Anisogamy grades from situations in which the gametes differ only in size to the extreme of OOGAMY, in which one gamete is a large immotile ovum and the other a small motile sperm. *Compare* isogamy.

antagonism 1. The interaction of two substances, for example drugs or hormones, that have opposing effects in a given system, such that one reduces or completely inhibits the effect of the other. 2. The interaction of two types of organisms, such that the growth of one is partially or completely inhibited by the other. 3. The opposing action of two muscles, such that the contraction of one is accompanied by relaxation of the other. *See* antagonist.

antagonist An agent, such as a drug or other chemical, that opposes the effects of another agent, the agonist (e.g. a hormone or a drug). Many antagonists work by binding to the cell receptors normally occupied by the agonist, and hence reducing or abolishing activation of the cell by the agonist.

antibiotic One of a group of organic compounds, varying in structure, that are produced by microorganisms and can kill or inhibit the activities of other microorganisms. One of the best-known examples is penicillin, which is produced by the fungus *Penicillium chrysogenum* and was discovered by Sir Alexander Fleming. Another example is streptomycin, which is produced by the filamentous bacterium *Streptomyces griseus*. Antibiotics are widely used in medicine to combat bacterial infections. The antibiotic-producing organisms are cultured commercially in large vessels under carefully controlled aerobic conditions, and the antibiotics extracted and purified. Further synthetic chemical modification is often performed to obtain the final therapeutic product.

Antibiotics inhibit bacterial growth or kill bacteria in various ways. For example, the penicillins interfere with synthesis of the bacterial cell wall, resulting in a weakened cell that is ultimately broken down. Others, such as the tetracyclines and streptomycin, disrupt protein synthesis. *Broad-spectrum antibiotics* are effective against a wide range of bacteria, whereas *narrow-spectrum antibiotics* act against relatively few. However, even susceptible strains of bacteria can acquire *antibiotic resistance* – the ability of the cell to resist a particular antibiotic. This is an increasing medical problem, leading to the emergence of pathogenic bacterial strains that are immune to antibiotic treatment. One way in which bacteria acquire antibiotic resistance is via plasmids carrying drug resistance genes, so-called R plasmids. These genes enable the cell to inactivate the drug, or block its uptake.

antibody   A molecule belonging to the IMMUNOGLOBULIN protein family that binds specifically to another substance (called an *antigen*). Antibodies are formed by an animal's immune system to bind to invading pathogens, such as bacteria or viruses, in order to neutralize their effects. They are important in defense against infectious diseases and in developing IMMUNITY. LYMPHOCYTES produce antibodies in response to the presence of antigens. Each antibody has a molecular structure that exactly fits the structure of its particular antigen molecule, like a lock and key. Antibody molecules attach themselves not only to pathogens, but also to transfused red blood cells or grafted tissue of another animal, so bringing about their destruction. Some cause agglutination (clumping) of invading cells, so that they disintegrate; others, called *opsonins*, make bacteria more easily engulfed by phagocytic leukocytes. MONOCLONAL ANTIBODIES are produced commercially using genetically engineered cell cultures. They are used as reagents in a wide range of experimental, analytical, diagnostic, and purification procedures.

anticlinal   Describing a line of cell division at right angles to the surface of the organ. *Compare* periclinal.

anticodon   A nucleotide triplet on TRANSFER RNA that is complementary to and bonds with the corresponding CODON of messenger RNA in the ribosomes.

antidiuretic hormone   *See* ADH.

antigen   A substance that induces the production of ANTIBODIES that are able to bind to the antigen. Large antigens, such as enzymes, typically have several different sites, called EPITOPES, that are each capable of inducing the formation of an antibody that binds specifically to the site.

antigen-presenting cells   *See* T-cell.

antihistamine   *See* allergy.

antioxidant   A substance that slows or inhibits oxidation reactions, especially in biological materials or within cells, thereby reducing spoilage or preventing damage. Natural antioxidants include vitamin E and β-carotene. These work by reacting with peroxides or oxygen free radicals, effectively mopping them up. For example, one of the roles of β-carotene in plants is to protect the photosynthetic pigment chlorophyll from photooxidation by oxygen-free radicals. Antioxidants are added to a range of foods (e.g. margarines) and to industrial materials, such as plastics.

antipodal cells   The three haploid cells found in the embryo sac of seed-bearing plants that migrate to the end of the sac opposite the micropyle. These nuclei arise as a result of the three meiotic divisions that produce the egg cell, synergid cells, and polar nuclei, but they do not themselves take part in the fertilization process. They are eventually absorbed by the developing embryo, and their function is uncertain.

antiporter   A type of COTRANSPORTER that transports a substance across a cell membrane while ions simultaneously flow in the opposite direction. The ions (usually H^+ or Na^+) move through the antiporter protein down an electrochemical gradient, and provide the energy for the transport of the other substance. Antiporters occur in a wide range of cells, and transport various substances. For example, in cardiac muscle cells, a Na^+/Ca^{2+} antiporter is responsible for exporting calcium ions (Ca^{2+}) against a steep concentration gradient; three sodium (Na^+) ions move inward for every one Ca^{2+} ion that is transported outward. *See also* Na^+/H^+ antiporter.

antisense DNA   The DNA strand that is not transcribed. In TRANSCRIPTION the DNA double helix unwinds and only one of the strands acts as the template for messenger RNA synthesis. Short segments of antisense nucleic acid (DNA or RNA), called oligonucleotides (or 'oligos'), can be synthesized in order to bind to specific regions of messenger RNA and prevent translation. They can even bind to some parts of the DNA double helix, forming a triple helix, and so prevent gene transcription. Antisense nucleic acids are being investigated as a possible treatment for certain genetic disorders, or even cancer, as well as having potential as an antiviral therapy. *See also* antisense RNA.

antisense RNA RNA that contains base sequences complementary to the RNA transcript of a gene, and so tends to hybridize with the transcript, thereby interfering with subsequent processing or translation of the RNA transcript. Thus, by introducing a vector that expresses antisense RNA, it can be possible effectively to repress a particular target gene in the host cell. Such 'antisense vectors' are used experimentally, and might have potential as a form of gene therapy to counter the effects of harmful genes.

apical meristem The actively dividing cells constituting the growing point at the tip of the root or stem in vascular plants. They are characteristically small with a dense protoplast and prominent nucleus. As cells of the apical meristem divide, they give rise to distinct files of cells, which constitute the *primary meristems* of the shoot and root. It is these cells that differentiate into the tissues, such as epidermis, xylem, phloem, and root cap. The apical meristems in the lower plants consist of one cell only, as in the ferns, but become more complex and consist of groups of cells in the higher plants. *See also* meristem.

Apicomplexa A phylum of spore-forming protists that are parasites of animals. It includes the malaria parasite, *Plasmodium*, and several important pathogens of domestic livestock, such as *Toxoplasma* (causing toxoplasmosis) and *Eimeria* (responsible for coccidiosis). Apicomplexans are named for a collection of fibrils, vacuoles, and organelles (the 'apical complex') visible at one end of the cell. They reproduce sexually and have complex life cycles, often involving several hosts, and can proliferate rapidly by a series of cell divisions (schizogony). Members of the Apicomplexa were traditionally classified as sporozoans.

apoenzyme The protein part of an enzyme that determines the enzyme's catalytic specificity but is itself catalytically inactive unless it combines with a nonprotein PROSTHETIC GROUP, such as a cofactor or coenzyme. The complete enzyme, comprising apoenzyme and prosthetic group, is called the HOLOENZYME.

apomixis A form of asexual reproduction in plants in which seeds are produced without the fusion of male and female gametes. There are several variations. For example, an embryo can develop from a diploid cell of the nucellus, or from an unfertilized egg cell.

apoplast The system of cell walls and intercellular spaces surrounding the plasma membranes of a plant's cells. It extends through the body of a plant and, where there is no cuticle (as in a primary root), it is continuous with the external environment. The apoplast provides a route – the *apoplastic pathway* – via which water and ions can diffuse passively into root tissues, for subsequent uptake by endodermal cells. *Compare* symplast.

apoptosis (programmed cell death) The orderly destruction and disposal of cells within multicellular organisms. Apoptosis is an essential means of sculpting tissues and remodeling organ systems during development, and of eliminating damaged or superfluous cells during later life. The events of apoptosis follow a characteristic pattern. The cell starts to shrink, the organelles break up, and in the nucleus the chromosomes become condensed and fragmented. Then the nucleus and cytoplasm break up, forming fragments known as *apoptotic bodies*, which are taken up and digested by motile phagocytic cells. Apoptosis involves the activation of enzymes called *caspases*. These cleave certain target proteins and thereby activate other 'executioner' enzymes that chop up the cell's DNA, its cytoskeleton, etc. Apoptosis is controlled by various intracellular and extracellular signals. For example, most cells require continual extracellular survival signals to stay alive. If these signals cease, the cell activates a 'suicide' programme of apoptosis. Apoptosis can also be triggered by specific extracellular signals. This occurs, for instance, in the destruction of nonfunctional lymphocytes, which is prompted by 'murder' signals from other lymphocytes.

apposition The deposition of successive layers of cellulose on the inner face of the secondary cell wall of a plant cell, resulting in an increase in thickness of the wall. *Compare* intussusception.

aquaporin A protein found in certain cells that forms a water channel through the plasma membrane. Aquaporins occur abundantly in erythrocytes and other cells that are highly permeable to water, and accelerate the osmotic flow of water through the otherwise relatively impermeable phospholipid membrane.

Arabidopsis thaliana A flowering plant that is widely used in studies of plant genetics and the genetic control of plant development. It has a relatively small genome, with a haploid complement of five chromosomes, and its genome sequence was fully determined by the end of 2000.

Arber, Werner (1929–) Swiss microbiologist who discovered restriction enzymes. He was awarded the Nobel Prize for physiology or medicine in 1978 jointly with D. Nathans and H. O. Smith.

arbovirus A former term for a large group of RNA-containing viruses that are transmitted from animals to humans by insects (arthropods), hence the name (*arth*ropod-*bo*rne viruses). They include the togaviruses.

archaea (**archaebacteria**) A group of prokaryotic microbial organisms that are distinguished from 'true' bacteria (the eubacteria) principally on the basis of biochemical differences. For example, they differ in the nature of their lipid constituents, and lack a peptidoglycan layer in the cell wall. Most archaeans are anaerobic, and many inhabit harsh environments, such as hot springs, salt flats, or sea vents,

thought to resemble the environment of the Earth soon after the emergence of life. Hence, archaeans may contain the descendants of some of the earliest living cells. They include methanogic (methane producing), themophilic (heat-loving), and halophilic (salt-loving) species. However, they are not restricted to such extreme lifestyles, and are widespread in more congenial settings. *See* extremophile.

archaebacteria *See* archaea.

archenteron The earliest gut cavity of most animal embryos. It is produced by an infolding of part of the outer surface of a blastula to form an internal cavity that is in continuity with the outside via the blastopore.

arginine *See* amino acids.

Arrhenius, Svante August (1859–1927) Swedish chemist and physicist who worked on electrochemistry and on reaction rates. He was awarded the Nobel Prize for chemistry in 1903.

artificial parthenogenesis *See* parthenogenesis.

ascorbic acid *See* vitamin C.

asexual reproduction The formation of new individuals from a single parent without the production of gametes or special reproductive structures. It occurs in many plants, usually by vegetative propagation or spore formation; in unicellular organisms usually by budding, fission, or spore formation; and in multicellular invertebrates by fission, budding, fragmentation, etc.

asparagine *See* amino acids.

aspartic acid *See* amino acids.

assimilation The process by which living cells take in ions or molecules and convert them into the materials required for maintenance, repair and growth. In animals, assimilated materials are derived from the digestion and absorption of food, whereas green plants assimilate carbon dioxide, water, and inorganic ions, using photosynthesis to provide the organic materials required by their cells.

aster A cluster of microtubules found at each end of the SPINDLE during nuclear division, so-called from its starlike appearance in the light microscope. Each aster radiates from the CENTROSOME toward the cell periphery. They are thought to help locate the spindle apparatus in relation to the cell boundaries, and to play a part in cleavage of the cytoplasm (cytokinesis) following nuclear division. One theory is that after the nuclear envelope has formed around each of the daughter nuclei, the astral microtubules send a signal to the cell's equator, initiating assembly of the actin and mysosin molecules that are instrumental in cleavage. Asters are generally more apparent in animal cells, which have CENTRIOLES, than in plant cells, which lack centrioles.

ATP

ATP (**adenosine triphosphate**) The universal energy carrier of living cells. Energy from food (in RESPIRATION) or from sunlight (in PHOTOSYNTHESIS) is used by cells to make ATP from ADP plus inorganic phosphate (P_i). It is then reconverted to ADP in various parts of the cell, the energy released being used to drive numerous cell reactions. ATP supplies the energy for the synthesis of cell components (e.g. proteins and nucleic acids); it powers the active transport of substances into cells; and drives the various steps in cell growth and division.

Muscle contraction and various other forms of cell movement are also fuelled by ATP. Furthermore, the donation of a phosphate group by ATP (i.e. phosphorylation) activates many components of cellular signaling pathways.

ATP is a nucleotide consisting of adenine and ribose with three phosphate groups attached. Hydrolysis of the terminal phosphate bond releases energy (30.6 kJ mol^{-1}) and is coupled to an energy-requiring process. Further hydrolysis of ADP to AMP sometimes occurs, releasing more energy. The pool of ATP is small, but the faster it is used, the faster it is replenished. Aerobic respiration and photosynthesis both use the same type of membrane-bound protein complex called ATP SYNTHASE to generate ATP from ADP and P_i (see oxidative phosphorylation). ATP formation is also accomplished by two enzyme-catalyzed reactions in glycolysis; this is called SUBSTRATE-LEVEL PHOSPHORYLATION, which can take place anaerobically.

ATPase Any enzyme that catalyzes the hydrolysis of ATP to yield ADP and inorganic phosphate (Pi), with the release of energy. ATPases are integral components of numerous cellular processes, including active transport and cell motility.

ATP synthase (F_0F_1 **complex**) An enzyme complex that catalyzes the formation of ATP (adenosine triphosphate) from its precursors ADP (adenosine diphosphate) and inorganic phosphate (Pi). It is the key component responsible for ATP formation by oxidative phosphorylation in mitochondria and bacterial cells, and for photosynthetic ATP formation inside chloroplasts. ATP synthases are found in the inner mitochondrial membrane, the thylakoid membranes of chloroplasts, and the plasma membrane of bacteria, and all have a very similar structure and function. They are driven by the proton gradient across the membrane (see chemiosmotic theory). Each consists of two portions, F_0, a proton (H^+) channel, and F_1, the ATP-synthesizing site. F_0 is embedded in the membrane, whereas F_1 protrudes into the mitochondrial matrix (or equivalent compartment). When protons pass through the proton channel formed by F_0, parts of F_0 and F_1 rotate, causing changes in the conformation of catalytic binding sites in F_1. According to the *binding-change mechanism of ATP synthesis*, these conformational changes are thought to alter the affinity with which the sites in F_1 bind ADP and Pi, leading to the formation of ATP. Three or four protons must pass through the F_0 proton channel for each ATP molecule synthesized.

attenuation 1. A 'back-up' mechanism that helps to regulate gene expression in bacterial OPERONS. An operon has a leader sequence, which lies downstream of the operon's promotor and operator sites, but before the start of the structural genes. An attenuator site is a segment of DNA within the leader sequence. Even if transcription of the leader sequence by RNA polymerase has commenced, transcription can abort when the polymerase complex reaches the attenuator site, according to the cell's requirements for the gene products. For example, if the amino acid tryptophan is readily available in a cell, the *trp* operon, which encodes genes involved in tryptophan synthesis, is inhibited by the *trp* repressor. However, this inhibition is often incomplete, so regulation is augmented by an attenuation mechanism. The high concentration of tryptophan causes translation of the *trp* operon leader sequence by a ribosome to stall. This in turn prompts termination of transcription, and release of the messenger RNA chain from the transcription complex.
2. The loss of virulence of a pathogenic microorganism after several generations of culture *in vitro*. Attenuated microorganisms are commonly used in vaccines.

autogamy 1. (*Zoology*) Reproduction in which the nucleus of an individual cell divides into two and forms two gametes, which reunite to form a zygote. It occurs in some protists, for example *Paramecium*.
2. (*Botany*) Self-fertilization that occurs in plants.

autograft *See* graft.

autolysis The self-destruction of cells following death of the organism. It is due to the action of enzymes released from lysosomes within the cells. *Compare* apoptosis.

autonomic nervous system The division of the vertebrate nervous system that supplies motor nerves to the smooth muscles of the gut and internal organs and to heart muscle. It comprises the SYMPATHETIC NERVOUS SYSTEM, which (when stimulated) increases heart rate, breathing rate, and blood pressure and slows down digestive processes, and the PARASYMPATHETIC NERVOUS SYSTEM, which slows heart rate and promotes digestion. Each organ is innervated by both systems and their relative rates of stimulation determine the net effect on the organ concerned. Many functions of the autonomic nervous system, such as the control of heart rate and blood pressure, are regulated by centers in the medulla oblongata of the brain.

autophagosome *See* autophagy.

autophagy The process whereby redundant, faulty, or aging cell organelles are destroyed. In animal cells the organelle or cell portion is surrounded by a membrane derived from the endoplasmic reticulum, forming an *autophagosome*. This then fuses with a primary LYSOSOME to form a secondary lysosome, in which the materials are degraded by the lysosomal enzymes. In plant cells, the vacuole may perform autophagy in a manner similar to lysosomes. Autophagy is part of the normal turnover of cell constituents, but accelerates during senescence and may be part of a developmental process, such as clearing of cell contents during sieve tube and tracheid formation in plant development.

autopolyploid *See* polyploid.

autoradiography A technique whereby a thin slice of tissue containing a radioactive isotope is placed in contact with a photographic plate. The image obtained on development shows the distribution of the isotope in the tissue. The technique is useful in identifying sites within the cell where particular substances are manufactured, for example by incubating the cell with a radioactively labeled precursor molecule and seeing where the label is subsequently localized.

autosomes   Chromosomes that are not sex chromosomes.

auxin   Any of a group of plant hormones, the most common naturally occurring one being indole-3-acetic acid (IAA). They are synthesized in root and shoot tips and transported throughout the rest of the plant. Auxins induce cell elongation and growth (*see* acid growth hypothesis), and are also involved in various other physiological responses, including root initiation, the suppression of lateral bud growth, tropic responses, and the development of flowers and fruits.

axenic culture (**pure culture**)   A culture containing only one species of microorganism.

axon (**nerve fiber**)   The part of a NEURON (nerve cell) that conveys impulses from the cell body towards one or more terminals, which form synapses with other neurons, muscle cells, glands, etc. It is an extension of the cell body and consists of an axis cylinder (*axoplasm*), surrounded in most vertebrates by a fatty (myelin) sheath, outside which is a thin membrane (*neurilemma*). Microtubules extend through the axoplasm, and enable the transport of materials between the cell body and axon terminals.

axoneme   The central '9+2' core of microtubules found in cilia and undulipodia (flagella) of eukaryotes, consisting of nine pairs of outer microtubules surrounding two single central microtubules. *See* undulipodium.

axoplasm   *See* axon.

bacillus (*pl.* **bacilli**) Any rod-shaped bacterium. Bacilli may occur singly (e.g. *Pseudomonas*), in pairs, or in chains (e.g. *Lactobacillus*). Some are motile.

bacteria A large and diverse group of organisms, which, in terms of numbers and variety of habitats, includes the most successful life forms. In nature, bacteria are important in the nitrogen and carbon cycles, and some are useful in various industrial processes, especially in the food industry, and in techniques of genetic engineering (see also BIOTECHNOLOGY). However, there are also many harmful parasitic bacteria that cause diseases such as botulism and tetanus.

Bacterial cells are simpler than those of animals and plants, and generally relatively small, in the range 1–10 μm. They have a single molecule of DNA – the genophore (bacterial chromosome). This is not contained in a membrane-bound nucleus, but can form a distinct aggregation within the cell called the nucleoid. The cytoplasm contains ribosomes, and possibly some inclusions consisting of storage material, but lacks complex organelles such as chloroplasts and mitochondria. Bounding the cell is a plasma membrane, and external to this is usually a cell wall to provide support and protection. Bacteria reproduce by fission or budding, and can under ideal conditions divide every 20 minutes. They also form resistant spores.

Bacteria include all prokaryotic organisms, and constitute the kingdom Monera (Prokaryotae). They are divided into two main groups: ARCHAEA, which are mostly anaerobic and often occur in extreme conditions, such as hot springs; and the EUBACTERIA, which include the vast majority of bacteria. Many authorities now regard these two groups as so distantly related in evolutionary terms as to merit their separation into different kingdoms.

bacterial artificial chromosome (BAC) A cloning vector used for cloning relatively large DNA fragments (up to about 300 kb). It is based on the naturally occurring F (fertility) plasmid of *E. coli*, and like other PLASMID vectors replicates independently of the host cell genome.

bacteriochlorophyll *See* chlorophylls.

bacteriophage *See* phage.

Balbiani ring *See* puff.

Baltimore, David (1938–) American microbiologist and molecular biologist noted for his work on viral nucleic-acid biosynthesis and his discovery of reverse transcription by the enzyme RNA-directed DNA polymerase in the virions of tumor-producing retroviruses. A classification of animal viruses is named for him. He was awarded the Nobel Prize for physiology or medicine in 1975 jointly with R. Dulbecco and H. M. Temin for their discoveries concerning the interaction between tumor viruses and the genetic material of the cell.

Banting, (Sir) Frederick Grant (1891–1941) Canadian surgeon, physiologist, and endocrinologist distinguished for his discovery (with C. H. BEST) of a way of extracting the hormone insulin from pancreatic tissue. He was awarded the Nobel Prize for physiology or medicine in 1923 jointly with J. J. R. Macleod.

Barton, (Sir) Derek Harold Richard

(1918–98) British organic chemist who developed the concept of conformation and its application in chemistry. He was awarded the Nobel Prize for chemistry in 1969, the prize being shared with O. Hassel.

basal body 1. (kinetosome) A barrel-shaped body found at the base of all eukaryote undulipodia (cilia and flagella) and identical in structure to the CENTRIOLE, comprising a cylindrical array of nine triplet microtubules. Between the end of the basal body and the base of the undulipodium shaft is a transition zone. Two of the three microtubules in each triplet continue through the transition zone into the undulipodium, to form the outer double microtubules of the shaft body (*axoneme*). The basal body provides anchorage for the undulipodium within the cell, and also serves in organizing the assembly of the axoneme microtubules.
2. An assembly of proteins found at the base of a bacterial flagellum, embedded in the plasma membrane and cell wall. It consists of disk-shaped structural proteins, which anchor the shaft of the flagellum; motor proteins, which impart a rotary motion to the base of the shaft; and switch proteins, which can reverse the direction of rotation in response to signals from the cell.

basal lamina (basement membrane) A thin sheetlike mesh of protein fibers that underlies epithelial and endothelial cells, and also surrounds muscle cells, fat cells, and Schwann cells. Typically 60–100 nm thick, it represents a component of the EXTRACELLULAR MATRIX, and is secreted by the adjacent cells themselves. It helps to maintain the structural integrity of the associated tissue cells, as well as regulating the passage of materials to and from the cells into neighboring blood vessels, etc. The kidney glomerulus has a thick basal lamina formed by fusion of endothelial and epithelial laminae. This is the only barrier between the blood and the filtrate. High blood pressure in the glomerulus forces water and all dissolved substances except proteins through the lamina into the capsule. This acts to filter components of the blood to form urine. The main constituents of the basal lamina are COLLAGEN fibrils, which form the structural framework, and LAMININ, a multiadhesive matrix protein that binds the lamina to the adjacent cells via cell adhesion molecules, such as integrins. Other proteins bind these components together, thus stabilizing the laminar construction.

base 1. (*Chemistry*) A compound that releases hydroxyl ions (OH^-) in aqueous solution. Basic solutions have a pH greater than 7. A base can be considered as a substance that tends to accept a proton (H^+). Thus, OH^- is basic because it accepts H^+ to form water. But H_2O is also a base (albeit somewhat weaker) because it can accept a further proton to form H_3O^+. Hence, the ions of mineral acids, such as SO_4^{2-} and NO_3^-, can be regarded as weak bases. *See also* acid; buffer.
2. (*Biochemistry*) A nitrogenous molecule, either a PYRIMIDINE or a PURINE, that combines with a pentose sugar and phosphoric acid to form a nucleotide, the fundamental unit of nucleic acids. The most abundant bases are cytosine, thymine, and uracil (pyrimidines) and adenine and guanine (purines).

basement membrane *See* basal lamina.

base pairing The linking together of two strands of nucleic acids (DNA or RNA) by bonds between complementary bases: adenine pairing with thymine (in DNA) or uracil (in RNA); and guanine pairing with cytosine. Base pairing in DNA produces long double-helical molecules. The specific nature of the base pairing enables accurate replication of the chromosomes and thus maintains the constant composition of the genetic material. Base pairing in RNA produces various structures, including 'hairpins' and 'stem-loops'.

basic stain *See* staining.

basophil A white blood cell (LEUKOCYTE) containing prominent granules that stain with basic dyes. It has a twin-lobed nucleus. Basophils comprise only about 0.5–1% of all leukocytes. They move about in an ameboid fashion and when activated release the contents of their granules – principally HISTAMINE, which promotes inflammation and attracts more phagocytic cells to the infection site, and the anticoagulant heparin. They are equivalent to wandering MAST CELLS, similar to those found in the lining of blood vessels.

B-cell (**B-lymphocyte**) A type of LYMPHOCYTE (white blood cell) that originates in the bone marrow and serves as the main instrument of the specific immune response, particularly the production of ANTIBODIES. (In birds, where antibodies were first studied, the antibody-producing lymphocytes mature in part of the cloaca called the bursa of Fabricius. Hence they were termed bursal cells, or B-cells. This organ has no known counterpart in mammals; mammalian B-cells mature in the bone marrow.) Lymphoid tissue contains vast numbers of B-cells, each with a different type of immunoglobulin receptor on its surface. When an antigen binds to the one or several B-cells carrying its specific receptor, it triggers them to undergo repeated division forming a large clone of cells dedicated to producing antibody specific to that antigen. Hence, according to the CLONAL SELECTION THEORY of antibody specificity, the antigen selects the few appropriate B-cells from among the huge number present in the body. These B-cells then enter the bloodstream as antibody-secreting PLASMA CELLS.

Another type of lymphocyte, known as a T-helper cell, is required to trigger clonal expansion and antibody secretion by B-cells. This recognizes the antigen bound to the surface of the B-cell in association with class II MHC proteins (*see* major histocompatibility complex). The T-helper cell binds to the antigen-MHC complex and releases substances (LYMPHOKINES) that act as the stimulus for B-cell growth.

After an infection has been dealt with and all antigens removed by the antibodies, a small set of B-cells from the clone, called MEMORY CELLS, remain in the circulation. These carry receptors that will bind avidly to the same antigen in any subsequent infection, prompting a much more rapid response by the immune system. *See* immunity; immunoglobulin; T-cell.

Beadle, George Wells (1903–89) American geneticist who formulated the one gene–one reaction hypothesis. He was awarded the Nobel Prize for physiology or medicine in 1958 jointly with E. L. Tatum for their discovery that genes act by regulating definite chemical events. The prize was shared with J. Lederberg.

Benacerraf, Baruj (1920–) Venezuelan-born American immunologist notable for his discovery of histocompatability genes and their role in regulation of the immune response. He was awarded the Nobel Prize for physiology or medicine in 1980 jointly with J. Dausset and G. D. Snell.

Berg, Paul (1926–) American biochemist and molecular geneticist who discovered the method of introducing foreign DNA into bacteria by the use of viruses as vectors. He was awarded the Nobel Prize for chemistry in 1980 for his work on the biochemistry of nucleic acids, with particular regard to recombinant-DNA. The prize was shared with W. Gilbert and F. Sanger.

Bergström, Sune K. (1916–) Swedish biochemist who isolated and purified prostaglandins and certain related substances. He was awarded the Nobel Prize for physiology or medicine in 1982 jointly with B. I. Samuelsson and J. R. Vane.

Best, Charles Herbert (1899–1978) Canadian physician and physiologist. He discovered the enzyme histaminase and the lipotropic action of choline but is best known for the discovery (with F. G. BANTING) of a way of extracting the hormone insulin from pancreatic tissue. Banting divided his share of the Nobel Prize with Best.

beta-sheet A type of SECONDARY STRUCTURE of proteins in which polypeptide chains (or strands) run close to each other and are held together by hydrogen bonds at right angles to the main chain. The structure is folded in regular 'pleats'. The side chains of the constituent amino acids project above and below the plane of the sheet. According to the polarity of the peptide bonds linking the amino acids, the strands of a beta-sheet can run either in the same direction (parallel) or in opposite directions (antiparallel). Protein fibers composed of beta-sheets are tough yet flexible, because the layered beta-sheets can slide over each other.

bilayer *See* membrane.

biochemistry The study of the chemical constituents and chemical reactions occurring in living organisms.

bioinformatics The use of computerized systems for the collection, storage and analysis of DNA and protein sequence data. Such data are held in several large databases, such as SwissProt, maintained by the University of Geneva, and GenBank at the National Institute of Health, Bethesda, Maryland. The data are made available to researchers worldwide via the internet. Newly discovered sequences can therefore be readily compared with existing data and any similarities identified. Such homologous sequences often point to functional similarities in three-dimensional protein structures, and can provide clues about the function of a newly sequenced protein. Moreover, the comparative analysis of sequence data from the genomes of different organisms provides information about the array of proteins any particular organism (e.g. a bacterium or protist) is likely to synthesize from its genome. This sheds light on its biology, even where this has not been investigated directly, and on its taxonomic relationships. *See also* genomics.

biology The study of living organisms, including their structure, function, evolution, interrelationships, behavior, and distribution.

bioluminescence The production of light by living organisms. Bioluminescence is found in many marine organisms, especially deep-sea organisms. It is also a property of some insects, for example fireflies, and certain bacteria. The light is produced as a result of a chemical reaction whereby the compound luciferin is oxidized and gives off photons. An enzyme, LUCIFERASE, catalyzes the reaction, and there several quite different types of luciferin. Most luminescent animals have a light-producing organ (*photophore*). This may contain their own specialized light-producing cells, or house symbiotic luminescent bacteria.

biosynthesis Chemical reactions in which a living cell builds up its necessary molecules from other molecules present. *See* anabolism.

biotechnology The application of technology to biological processes for industrial, agricultural, and medical purposes. For example, bacteria such as *Penicillium* and *Streptomyces* are used to produce antibiotics, and fermenting yeasts produce alcohol in beer and wine manufacture. Recent developments in genetic engineering have enabled the large-scale production of hormones, blood-serum proteins, and other medically important products. Genetic modification of farm crops, and even livestock, offers the prospect of improved protection against pests, or products with novel characteristics, such as new flavors or extended storage properties. *See also* genetic engineering; tissue engineering.

biotin A water-soluble vitamin generally found with other vitamins of the VITAMIN B COMPLEX. It is widely distributed in natural foods, egg yolk, kidney, liver, and yeast being good sources. Biotin is required as a coenzyme for carboxylation reactions in cellular metabolism.

Bishop, John Michael (1936–) American microbiologist. He was awarded the Nobel Prize for physiology or medicine in 1989 jointly with H. E. Varmus for their discovery of the cellular origin of retroviral oncogenes.

bivalent Any homologous chromosomes in a paired configuration during MEIOSIS. Pairing of homologous chromosomes (SYNAPSIS) commences at one or several points on the chromosome during the zygotene stage of meiosis I, and is clearly seen during the pachytene stage.

Black, (Sir) James Whyte (1924–) British pharmacologist who discovered beta blockers in the 1950s. He also discovered the drug cimetidine for the control of gastric ulcers. He was awarded the Nobel Prize for physiology or medicine in 1988 jointly with G. B. Elion and G. H. Hitchings.

blastocoel (**segmentation cavity**) The internal fluid-filled cavity of a BLASTULA, which first appears during cleavage of the egg. It provides space for cell movement during development.

blastocyst A mammalian egg in the later stages of cleavage, before implantation. It is a modified BLASTULA and consists of a hollow fluid-filled ball of cells containing, at one end, the inner cell mass, from which the embryo develops. *See also* trophoblast.

blastoderm The cellular mass that results from cleavage of the superficial disk of cytoplasm (*blastodisc*) of very yolky eggs, such as those of birds, sharks, and cephalopods. The term is also used for the cellular coat of cleaved insect eggs.

blastomere Any of the cells produced during CLEAVAGE of an activated animal zygote, leading to the formation of the early embryo (BLASTULA). There is no growth at this stage, and because the total amount of available cytoplasm remains constant, blastomeres formed by successive divisions of the nuclei become smaller and smaller, until the normal nuclear/cytoplasmic ratio is achieved. *See also* mosaic.

blastopore The opening in a GASTRULA between the archenteron and the outside, through which invagination occurs at gastrulation. In amniotes (reptiles, birds, and mammals) the functional equivalent is the PRIMITIVE STREAK. The dorsal lip of the amphibian blastopore is the primary ORGANIZER of the axial structure of the embryo and corresponds to Hensen's node in a chick or mammal; the future notochord is invaginated over this lip.

blastula The stage in an animal embryo following CLEAVAGE. It is a hollow fluid-filled ball of cells (blastomeres). The modified blastula of mammals is called a blastocyst.

Blöbel, Gunter (1936–) American biochemist who discovered that proteins have intrinsic signals that govern their transport and localization in the cell. He was awarded the 1999 Nobel Prize for physiology or medicine.

Bloch, Konrad Emil (1912–2000) German-born American biochemist who used isotope techniques to study the biosynthesis of cholesterol, creatine, and protoporphyrin. He was awarded the Nobel Prize for physiology or medicine in 1964 jointly with F. Lynen for their work on the mechanism and regulation of the cholesterol and fatty acid metabolism.

blood The transport medium of an animal's body. It is a fluid tissue that circulates by muscular contractions of the heart (in vertebrates) or other blood vessels (in invertebrates). It usually carries oxygen and food to the tissues and carbon dioxide and nitrogenous wastes from the tissues, to be excreted. It also conveys hormones and circulates heat throughout the body. Blood consists of liquid plasma in which are suspended white cells (LEUKOCYTES), which devour bacteria and produce antibodies. In most animals (except insects) the blood carries oxygen combined with a pigment (hemoglobin in vertebrates; hemocyanin in some invertebrates). In some invertebrates the pigment is dissolved in the plasma, but in vertebrates it is contained in the red cells (ERYTHROCYTES). *See also* blood plasma; platelet.

blood cell (**blood corpuscle**) Any of the cells contained within the fluid plasma of blood. In humans, 45% of blood volume is made up of red cells (ERYTHROCYTES) and 1% of white cells (LEUKOCYTES).

blood corpuscle   *See* blood cell.

blood groups   Types into which the blood is classified, based on the occurrence of certain chemical groups on the surface of red blood cells. There are numerous systems, each determined by a particular gene or group of genes. One of the most significant medically is the ABO system. Since 1900 it has been known that human blood can be divided into four groups, A (40% of the population), B (7%), AB (3%), and O (50%), based on the presence or absence of molecular groups, called A and B antigens, on the surface of the red cells. In group AB, for example, both antigens are present, whereas group O has neither antigen. Knowledge of a patient's blood group is essential when a blood transfusion is to be given. If blood from a group A donor is given to a group B recipient, the recipient's anti-A antibodies will attack the donor's A antigens, causing the red cells to clump together. Group O blood, having no antigens, can be given to patients of any blood group since it will not provoke an antibody reaction. Group AB, having both antigens and therefore neither antibody, can receive blood from any group. The ABO antigens are determined by the alleles at a single gene. There are various other blood group systems, including one based on the presence or absence of the RHESUS FACTOR.

blood plasma   The straw-colored liquid that remains when all the cells are removed from blood. It consists of 91% water and 7% proteins, which are albumins, globulins (mainly antibodies), and prothrombin and fibrinogen (concerned with clotting). Plasma also contains the ions of dissolved salts, especially sodium, potassium, chloride, bicarbonate, sulfate, and phosphate. Plasma is slightly alkaline (pH 7.3) and the proteins and bicarbonate act as buffers to keep this constant. It transports dissolved food (as glucose, amino acids, fat, and fatty acids), excretory products, (urea and uric acid), dissolved gases (about 40 mm^3 oxygen, 19 mm^3 carbon dioxide, and 1 mm^3 nitrogen in 100 mm^3 plasma), hormones, and vitamins. Most of the body's physiological activities are concerned with maintaining the correct concentration and pH of all these solutes (i.e. homeostasis), since plasma supplies the extracellular fluid that bathes tissue cells.

blood serum   The pale fluid that remains after blood has clotted. It consists of BLOOD PLASMA without any of the substances involved in clotting.

blue–green algae   *See* cyanobacteria.

blue–green bacteria   *See* cyanobacteria.

B-lymphocyte   *See* B-cell.

Bohr effect   *See* hemoglobin.

Boyer, Paul D.   (1918–)   American biochemist who worked on the enzyme mechanism underlying the synthesis of ATP. He was awarded the 1997 Nobel Prize for chemistry jointly with J. E. Walker. The prize was shared with J. C. Skou.

bp   Symbol for base pair, used as a unit for measuring the length of double-stranded nucleotides, such as DNA. A length of 1000 bp is equivalent to 1 kilobase (1 kb) of a single-stranded nucleotide. *Compare* kilobase; megabase.

brachysclereid   *See* stone cell.

Bragg, (Sir) William Henry (1862–1942) British physicist who, along with his son W. L. Bragg, pioneered x-ray crystallography. He was awarded the Nobel Prize for physics in 1915 jointly with W. L. Bragg for their work on the analysis of crystal structure by means of x-rays.

Bragg, (Sir) William Lawrence (1890–1971) British experimental physicist, born in Australia, noted for his x-ray crystallographic studies in collaboration with his father, W. H. Bragg, and his subsequent application of x-ray crystallography to biological macromolecules. He was awarded the Nobel Prize for physics in 1915 jointly with W. H. Bragg.

broad-spectrum antibiotic   *See* antibiotic.

Brown, Michael Stewart (1941–) American physician and molecular geneticist noted for his discovery of cellular receptors for low-density lipoproteins and their role in the removal of cholesterol from the bloodstream. He was awarded the Nobel Prize for physiology or medicine in 1985 jointly with J. L. Goldstein.

brown fat   *See* adipose tissue.

Brownian motion   The random motion of microscopic particles due to their continuous bombardment by the much smaller and invisible molecules in the surrounding liquid or gas. Particles in Brownian motion can often be seen in colloids under special conditions of illumination.

brush border   The outer surface of columnar epithelial cells lining the intestine, kidney tubules, etc. It consists of fine hairlike projections called MICROVILLI. These greatly increase the surface area of the cell for absorption of dissolved substances. There may be as many as 3000 microvilli on one epithelial cell.

Büchner, Eduard (1860–1917)   German chemist and biochemist who discovered that alcoholic fermentation could be initiated with a cell-free juice obtained from brewers yeast. He suggested that the active principle is a protein (zymase). He was awarded the Nobel Prize for chemistry in 1907 for his discovery of cell-free fermentation.

bud   **1.** An outgrowth of an organism that is capable of vegetative reproduction. In lower animals and yeasts the production of buds that grow into new individuals and then break away from the parent is a common form of asexual reproduction. *See* budding.

2. (*Botany*) A compacted undeveloped shoot consisting of a shortened stem and immature leaves or floral parts. Buds have the potential to develop into a new shoot or flower; sometimes they are a means of asexual reproduction.

budding   **1.** A type of asexual reproduction in which a new individual is produced as an outgrowth (bud) of the parent organism. It is common in certain animal groups, such as cnidarians, sponges, and urochordates, where it is also termed *gemmation*. It also occurs in the unicellular fungi, especially the yeasts. *Compare* fission.
2. The production of buds on plants.

budding yeasts   *See Saccharomyces.*

buffer   A solution that resists any change in pH when small amounts of acid or base are added. Buffers are important in living organisms because they guard against sudden changes in pH. They involve a chemical equilibrium between a weak acid and its salt or a weak base and its salt. In body fluids, the main buffer systems are the phosphate buffer pair ($H_2PO_4^-$/HPO_4^{2-}) and the bicarbonate buffer pair (H_2CO_3/HCO_3^-) systems. Plasma proteins and hemoglobin are also important chemical buffers in blood. Various buffer solutions are used in biochemistry and cell biology to stabilize the pH of solutions, tissue samples, cell-free systems, etc.

buret   A piece of apparatus used for the addition of variable volumes of liquid in a controlled and measurable way. The buret is a long cylindrical graduated tube of uniform bore fitted with a stopcock and a small-bore exit jet, enabling a drop of liquid at a time to be added to a reaction vessel. Burets are widely used for titrations in volumetric analysis. Standard burets permit volume measurement to 0.005 cm^3 and have a total capacity of 50 cm^3; a variety of smaller *microburettes* is available. Similar devices are used to introduce measured volumes of gas at regulated pressure in the investigation of gas reactions.

Burnet, (Sir) Frank Macfarlane (1899–1985) Australian physician, virologist, and immunologist who formulated the clonal selection theory of acquired immunity. He was awarded the Nobel Prize for physiology or medicine in 1960 jointly with P. B. Medawar for discovery of acquired immunological tolerance.

butanedioic acid (succinic acid) A crystalline carboxylic acid, $HOOC(CH_2)_2COOH$, that occurs in amber and certain plants. It forms during the fermentation of sugar (sucrose).

butanoic acid (butyric acid) A colorless liquid carboxylic acid. Esters of butanoic acid are present in butter.

Butenandt, Adolf Friedrich Johann (1903–95) German organic chemist and biochemist who made pioneering studies of steroid hormones and pheromones. He isolated and determined the structures of the first sex hormones (androsterone, estrone, and progesterone) and the first insect hormone (ecdysone), and also investigated the first pheromone (bombykol). He was awarded the Nobel Prize for chemistry in 1939 for his work on sex hormones. The prize was shared with L. Ružička.

cadherin Any of a group of CELL ADHESION MOLECULES that are important components of certain cell junctions, and also play a role in development by mediating interactions between cells. All are glycoproteins that extend from the cytoplasm, through the plasma membrane and into the intercellular space. Here they bind to similar cadherins extending from neighboring cells. Cadherins are found abundantly in junctions between epithelial cells, and require calcium ions for effective binding. During development, cells expressing one class of cadherin will bind to other cells expressing the same class of cadherin, thereby ensuring correct cell–cell interactions.

Caenorhabditis elegans A nematode measuring about 1 mm long and widely used in genetic and developmental studies. It has about 19 000 protein-coding genes, and it was the first multicellular organism to have its genome fully sequenced (completed by 1998). The mature adult worm consists of roughly 1000 cells, and studies of its development have shed light on the genes and proteins involved in the development of more complex organisms, such as humans.

calcium An essential mineral salt for animal and plant growth. It is present between plant cell walls as pectate, and is found in the bones and teeth of animals. Calcium ions, Ca^{2+}, are important in triggering muscle contraction where their rapid release from the cisternae of the sarcoplasmic reticulum triggers the reaction between ATP and the myofilaments. Calcium is important in resting muscles in maintaining the relative impermeability of the cell membranes. If the calcium concentration falls, the potential difference across the membrane also falls so that muscles may spontaneously contract without activation by acetylcholine, giving twitching and spasms. The concentration of calcium ions is also important in influencing the breakdown of glycogen in muscles. Calcium is important in the clotting of blood in the conversion of prothrombin to thrombin. In mammalian stomachs it is also important in precipitating casein from milk.

calcium ion pump Any transport protein that actively carries calcium ions (Ca^{2+}) across a cell membrane against an electrochemical gradient. Such proteins are ATPases, in that they derive energy by the hydrolysis of ATP. *Plasma membrane calcium pumps* are found in animal and yeast cells, and export Ca^{2+} from the cell to maintain low intracellular Ca^{2+} concentrations. The rate of Ca^{2+} export is regulated by the Ca^{2+}-binding protein CALMODULIN. Another type of calcium ion pump occurs in the membrane that bounds the sarcoplasmic reticulum (SR) in skeletal muscle. This *muscle calcium pump* transports Ca^{2+} from the sarcoplasm into the lumen of the SR following MUSCLE CONTRACTION. Two Ca^{2+} are transported for every ATP molecule hydrolyzed.

callus A mass of undifferentiated parenchyma cells formed by the cambium in response to wounding of the vascular tissue. If parenchyma is injured, then the surrounding uninjured parenchyma cells form a cork cambium that produces a layer of suberized cells sealing off the wound. In tissue cultures, callus can be induced to form by various hormone treatments. Adventitious shoots and roots often differentiate from calluses, a phenomenon exploited in the rooting of cuttings. *See also* graft.

calmodulin A small calcium-binding protein found in the cytosol of cells that regulates many cellular activities. Each calmodulin molecule binds four calcium ions (Ca^{2+}) and in so doing undergoes a change in its three-dimensional shape, enabling it to bind to and activate various enzymes.

Calvin, Melvin (1911–97) American biochemist who pioneered the use of radioactive isotopes, especially carbon-14, as tracers in metabolism. He was awarded the Nobel Prize for chemistry in 1961 for his research on carbon dioxide assimilation in plants.

Calvin cycle *See* photosynthesis.

CAM *See* cell adhesion molecule.

cAMP *See* cyclic AMP.

cancer A malignant tumor, or disease caused by it. Malignant tumors, like benign tumors, consist of a clone of cells that no longer respond to the normal controls on cell multiplication, and so proliferate in an unregulated fashion. However, malignant tumors are distinguished from benign ones in that they tend to invade surrounding tissues, and are capable of producing secondary growths (*metastases*) in a part of the body distant from the original tumor. Moreover, typically they are not encapsulated, and their cells are less well differentiated. Cancers are classified into two main groups according to the tissue in which they arise: *carcinomas* arise in epithelial tissue; *sarcomas* in connective tissue.

Loss of the normal control mechanisms in a cancer cell usually arises due to an accumulation of mutations over a long timespan. This is why most cancers develop later in life. These mutations can be caused by carcinogens, such as tobacco smoke, or arise spontaneously. Some result in the loss or inactivation of a TUMOR SUPPRESSOR GENE. Other mutations can convert proto-oncogenes into ONCOGENES. Some viruses, notably the retroviruses, can insert oncogenes into the DNA of the host cell; in some cases this makes the cell cancerous – a process called transformation.

The accumulated oncogenic (cancer-causing) mutations affect the cell in various ways. Proteins involved in the reception and transduction of cell signals may be permanently 'switched on', even in the absence of the normal signals, so leading to overexpression of certain genes. For example, the *ras* oncogene encodes a mutant Ras protein that acts as an abnormal growth-promoting signal within the cell. The normal pattern of cell replication is disrupted by mutations that cause the loss or alteration of key regulatory proteins of the cell cycle. An example is the *APC* tumor suppressor gene, which encodes a crucial cell-cycle protein. Cancer cells also have defects in their DNA repair systems, which lead to a build-up of genetic abnormalities. Such defective cells would normally be destined for programmed cell death (apoptosis), but cancer cells can evade this fate.

A cancer cell is unable to secrete many components of the extracellular matrix; also, its connections with neighboring cells are disrupted. Consequently, the cancer cell can break contact with fellow tissue cells and escape the surrounding matrix to migrate to adjoining tissues, or to blood or lymph vessels, and hence to distant parts of the body. A tumor induces the formation of new blood vessels (*see* angiogenesis) to nourish the tumor mass, otherwise its growth is restricted. Devising drugs to inhibit tumor angiogenesis is one avenue being explored in the perpetual fight against cancer.

cane sugar *See* sucrose.

capillary One of numerous tiny blood vessels (5–20 μm diameter) that branches out from an arteriole to form a dense network (*capillary bed*) amongst the tissues and reunites into a venule. They have thin walls of endothelium through which oxygen, carbon dioxide, inorganic ions, dissolved food, excretory products, etc., are exchanged between the blood and the cells via the tissue fluid.

capsid The protein coat of a VIRUS, surrounding the nucleic acid. A capsid is present only in the inert extracellular stage of the life cycle. The protein subunits (*capsomeres*) are arranged in one of two ways: a helical structure, as in tobacco mosaic virus, or an icosahedral structure, as in the common cold virus. The capsid plus its enclosed nucleic acid is called the *nucleocapsid*. In many viruses it is surrounded by a lipid–protein ENVELOPE.

capsomer *See* capsomere.

capsomere (capsomer) *See* capsid.

capsule **1.** (*Microbiology*) A relatively rigid layer of polysaccharide that coats the surface of some prokaryote cells. It can help a bacterium gain entry to host tissue, and may combat phagocytosis by macrophages or other host defense cells. The capsule is a form of glycocalyx.
2. (*Zoology*) A protective or supportive sheath or envelope that surrounds an organ or part of the body. It is usually composed of connective tissue, as in the capsule of a joint or capsule of a kidney.

carbohydrates A class of compounds occurring widely in nature and having the general formula type $C_x(H_2O)_y$. They are generally divided into two main classes: SUGARS and POLYSACCHARIDES. Carbohydrates are both stores of energy and structural elements in living systems; plants having typically 15% carbohydrate and animals about 1% carbohydrate. The body is able to build up polysaccharides from simple units (anabolism) or break the larger units down to more simple units for releasing energy (catabolism).

carbon An essential element in plant and animal nutrition that occurs in all organic compounds and thus forms the basis of all living matter. It enters plants and other photosynthesizing organisms as carbon dioxide and is assimilated into carbohydrates, proteins, and fats, forming the backbones of such molecules. The element carbon is particularly suited to such a role as it can form stable covalent bonds with other carbon atoms, and with hydrogen, oxygen, nitrogen, and sulfur atoms. It is also capable of forming double and triple bonds as well as single bonds and is thus a particularly versatile building block. Carbon, like hydrogen and nitrogen, is far more abundant in living materials than in the Earth's crust, indicating that it must be particularly suitable to fulfill the requirements of living processes.

carboxylase An enzyme that catalyzes the removal of the carboxyl group (–COOH) from carboxylic acids (i.e. decarboxylation). Carboxylases are found in yeasts, bacteria, plants, and animal tissues. One of the most significant in aerobic respiration is pyruvate carboxylase. As part of the pyruvate dehydrogenase enzyme complex in mitochondria, it brings about the decarboxylation of pyruvic acid to form carbon dioxide and acetyl CoA, the latter then proceeding to enter the Krebs cycle.

Carboxylic acid

carboxylic acid An organic compound of general formula RCOOH, where R is an organic group and –COOH is the carboxyl group. Many carboxylic acids are of biochemical importance. The lower carboxylic acids (such as citric, succinic, fumaric, and malic acids) participate in the Krebs cycle. The higher acids, which are bound in lipids, are also called *fatty acids*, although the term 'fatty acid' is often used to describe any carboxylic acid of moderate-to-long chain length. The fatty acids contain long hydrocarbon chains, which may be saturated (no double bonds) or unsaturated (C=C double bonds). Animal fatty acids are usually saturated, the most common being stearic acid and palmitic acid. Plant fatty acids are often unsaturated: oleic acid is the commonest example. *See also* lipid; phospholipid; triglyceride.

carcinoma *See* cancer.

cardiac muscle The muscle of the vertebrate heart. It consists of individual cells that branch and form strong junctions (intercalated disks) with each other. Each cell contains a single nucleus and numerous striated myofibrils. The cell junctions allow the rapid propagation of contractions through the network of cells, hence the component cells are well integrated and their actions coordinated. Heart muscle beats spontaneously. This inherent rhythmicity of contraction arises from specialized cardiac muscle cells, particularly the sinoatrial node, that initiate and conduct electrical impulses throughout the heart. However, the rate of contraction is regulated by the autonomic nervous system.

Compared to skeletal muscle, the sarcoplasmic reticulum of cardiac muscle is reduced and the transverse tubules (T tubules) are more prominent. This is because much of the calcium (Ca^{2+}) that triggers contraction enters the cell from outside rather than the sarcoplsmic reticulum. Also, the Ca^{2+} ions stay in the sarcoplasm for longer, so that each contraction is prolonged. This means that cardiac muscle does not produce the short 'twitch' contractions of skeletal muscle, nor the powerful sustained tetanic contractions. Neither does it become deficient in ATP and hence fatigued. Instead, contractions are regular, distinct, and continual, consistent with the heart's role as a pump for the vascular system.

cargo protein Any protein that is transported to a particular destination within a cell. Various mechanisms ensure that proteins, whether synthesized in the cell or imported from outside, are directed to their 'target' destination. Many proteins contain an amino acid sequence – the *uptake-targeting sequence* – that binds to specific receptors at the target organelle (e.g. mitochondrion or chloroplast), causing the protein to be taken up by the organelle. Other proteins often accompany the cargo protein, to enable its uptake. During the formation of transport vesicles, specific sets of receptors discriminate between different proteins, so that the vesicle carries only the appropriate cargo proteins (*see* clathrin-coated vesicle).

Carlsson, Arvid (1923–) Swedish biochemist noted for his work on signal transduction in the nervous system. He shared the 2000 Nobel Prize for physiology of medicine with P. Greengard and E. R. Kandel.

cartilage A firm but flexible connective tissue containing cells embedded in a viscous proteoglycan matrix, which may have elastic or collagen fibers in it. In higher vertebrates the embryonic skeleton is formed as cartilage and then replaced by bone; in adults cartilage persists in a few places, such as the end of the nose, the pinna of the ear, the disks between vertebrae, over the ends of bones, and in joints. Cartilage is secreted by cells (*chondroblasts*), which subsequently become enclosed in spaces (lacunae) within the cartilage as *chondrocytes*. The physical properties are determined largely by the proportions of collagen and elastic fibers within the matrix, and there are three types. The most abundant is *hyaline cartilage*, which has a glassy appearance and covers the articular surfaces of bones. *Elastic cartilage*, as found in the external ear, for example, contains large numbers of elastic fibers, whereas *fibrocartilage* (e.g. intervertebral disks) is densely fibrous with relatively less matrix.

Casparian strip *See* endodermis.

caspase *See* apoptosis.

catabolism Metabolic reactions involved in the breakdown of complex molecules to simpler compounds. The main function of catabolic reactions is to provide energy, which is used in the synthesis of new structures, for work (e.g. contraction of muscles), for transmission of nerve impulses, and maintenance of cellular functions. Catabolism also involves the degradation of aged or defective cell components, or harmful materials, including invading pathogenic organisms. *Compare* anabolism. *See also* metabolism.

catalase An enzyme present in both plant and animal tissues that catalyzes the breakdown of hydrogen peroxide, a toxic compound produced during metabolism, into oxygen and water. Large amounts of catalase occur in PEROXISOMES and (in certain seeds) in GLYOXISOMES.

catalyst A substance that increases the rate of a chemical reaction without being used up in the reaction. ENZYMES are highly efficient and specific biochemical catalysts.

catecholamines A group of chemicals (amine derivatives of catechol) that occur in animals, especially vertebrates, and act as neurohormones or neurotransmitters. Examples are norepinephrine, epinephrine, and dopamine.

cathepsin One of a group of enzymes that break down proteins to amino acids within the various mammalian tissues. Several cathepsins have been isolated and are differentiated by their different activities.

cation A positively charged ion, formed by removal of electrons from atoms or molecules. In electrolysis, cations are attracted to the negatively charged electrode (the cathode). *Compare* anion.

caveola (*pl.* **caveolae**) *See* endocytosis.

Cdk *See* cyclin.

CD marker Cluster of differentiation marker: any of a group of antigenic molecules occurring on the surface of leukocytes (white blood cells) and other cells that are used to distinguish subsets of very similar cell populations, for example in immunology. They are identified using specific monoclonal antibodies. Examples include the CD4 antigens that characterize T-helper cells, and the CD8 antigens on cytotoxic T-cells. The CD markers may themselves participate in immunological reactions or other cellular processes. For example, CD4 assists the T-helper cell in recognizing and responding to foreign antigen presented by other immune cells. Certain other CD markers are CELL ADHESION MOLECULES, helping immune cells to bind to target cells. *See also* T-CELL.

cDNA *See* complementary DNA.

cDNA library *See* DNA library.

Cech, Thomas Robert (1947–) American biochemist and molecular biologist. He was awarded the Nobel Prize for chemistry in 1989 jointly with S. Altman for their discovery of catalytic properties of RNA.

cell The basic unit of structure of all living organisms, excluding viruses. Cells were discovered by Robert Hooke in 1665, but Schleiden and Schwann in 1839 were the first to put forward a clear *cell theory*, stating that all living things were cellular. Prokaryotic cells (typically measuring 1–5 μm) are significantly smaller than eukaryotic cells (typical diameters 20–50 μm). The largest cells are egg cells; a human ovum is 300 μm in diameter, and shelled eggs, such as those of reptiles and birds are much larger, although these are surrounded by various extracellular membranes. The smallest are mycoplasmas (about 200–300 nm in diameter).

All cells contain genetic material in the form of DNA, which controls the cell's activities; in eukaryotes this is enclosed in the nucleus. All contain cytoplasm and are surrounded by a plasma membrane. This controls entry and exit of substances. Within the cytoplasm of eukaryotic cells are various organelles (*see diagrams and relevant headwords*), and a cytoskeleton that maintains cell shape. Plant cells and most prokaryotic cells are surrounded by rigid cell walls. Differences between animal and plant cells can be seen in the diagrams; differences between prokaryotic and eukaryotic cells can be seen in the table. In multicellular organisms cells become specialized for different functions, and are organized to form tissues and organs. This is

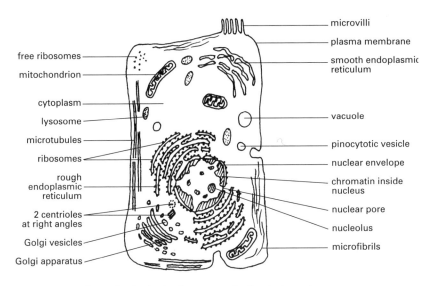

free ribosomes

mitochondrion

cytoplasm

lysosome

microtubules

ribosomes

rough endoplasmic reticulum

2 centrioles at right angles

Golgi vesicles

Golgi apparatus

microvilli

plasma membrane

smooth endoplasmic reticulum

vacuole

pinocytotic vesicle

nuclear envelope

chromatin inside nucleus

nuclear pore

nucleolus

microfibrils

Generalized animal cell as seen under the electron microscope

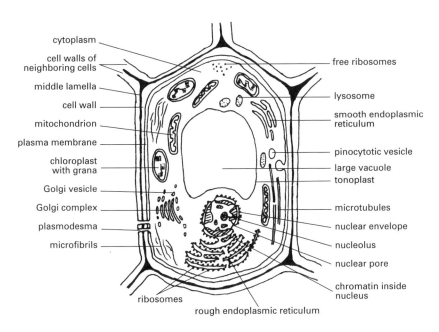

cytoplasm

cell walls of neighboring cells

middle lamella

cell wall

mitochondrion

plasma membrane

chloroplast with grana

Golgi vesicle

Golgi complex

plasmodesma

microfibrils

free ribosomes

lysosome

smooth endoplasmic reticulum

pinocytotic vesicle

large vacuole

tonoplast

microtubules

nuclear envelope

nucleolus

nuclear pore

chromatin inside nucleus

ribosomes

rough endoplasmic reticulum

Generalized plant cell as seen under the microscope

called division of labor. Within the cell, further division of labor occurs between the organelles.

cell adhesion molecule (CAM)   Any of a class of proteins located in the plasma membrane that form various CELL JUNC- TIONS in animal tissues, or mediate cell–cell interactions. CAMs effectively 'glue' cells to each other or to the extracellular matrix (ECM). Part of each CAM resides in the plasma membrane and part protrudes from the membrane to contact another molecule by means of a binding site. There are four main families of CAMs, the largest of which are the CADHERINS. These hold simi- lar cells to each other, for example in ep- ithelial cell layers such as those lining the small intestine and kidney tubules, and also occur in desmosomes. *Integrins* bind cells to components of the ECM, such as collagens and laminins, and can form cell–matrix junctions, such as hemidesmo- somes and focal adhesions. *Nerve-cell ad- hesion molecules* (N-CAMs) are especially important in nervous tissue, and in differ- entiation of muscle and nerve cells during development. *Selectins* participate in inter- actions between endothelial cells lining blood vessels and leukocytes (white blood cells).

cell biology   The study of the structure and function of living cells. Modern cell bi- ology integrates many biological sciences, notably biochemistry, molecular biology and genetics, and can itself be divided into various subdisciplines. Among the many topics studied by cell biologists are how cells grow and replicate, how they are or- ganized internally, how they interact with other cells (e.g. in tissues), how they re- spond to external signals, and how all these activities are controlled ultimately by the cell's genes.

cell body   The part of a nerve cell (*see* neuron) that contains the nucleus and most of its organelles. It has a swollen appear- ance and is a center of synthesis and degra- dation, supplying materials to the rest of the neurons. In many types of neuron, branching dendrites extend from the cell

body to receive incoming signals from synapses with other neurons. The cell body itself may also form synapses with other neurons. When the neuron is stained with basic dyes, fragments of the rough endo- plasmic reticulum are visible as *Nissl gran- ules* within the cell body.

cell culture   The growth and mainte- nance of living cells under artificial condi- tions. Cell biologists commonly study cells that have been grown in culture outside the body of an organism in carefully controlled conditions. This enables the use of more homogeneous cell types, and permits more precise control of experimental conditions. Bacteria and yeasts are widely used as ex- perimental organisms. Common examples are the bacterium *E. coli* and baker's yeast *Saccharomyces cerevisiae*. Such organisms grow naturally as single cells, and have simple nutritional requirements; they are typically grown in petri dishes on an agar- based semi-solid nutrient medium. Under appropriate conditions, each cell will di- vide repeatedly to form a colony of geneti- cally identical cells (clones).

Various types of animal cells can also be cultured artificially, but they are more demanding in their needs than microorgan- isms. For example, they require a more complex nutrient medium, often including growth factors, and they must be grown on a hard surface, such as glass or specially treated plastic. Moreover, animal tissue cells secrete an extracellular matrix of pro- teins that bind the cultured cells together. This matrix must be broken down by en- zymes to release individual cells from the culture. Plant cells can also be grown as isolated cells if kept agitated and sus- pended in liquid culture medium. *See also* tissue culture.

cell cycle   The ordered sequence of phases through which a cell passes leading up to and including cell division (MITOSIS). It is divided into four phases G_1, S, and G_2 (collectively representing interphase), and M-phase, during which mitosis takes place. Synthesis of messenger RNA, transfer RNA, and ribosomes occurs in G_1, and replication of DNA occurs during the

S phase. The materials required for spindle formation are formed in G_2. The time taken to complete the cell cycle varies in different types of cells. For example, in rapidly growing yeast cells the cycle can take only 90 minutes, whereas in cultured human cells it takes about 24 hours.

The many events of the cell cycle must be tightly controlled and coordinated if the cycle is to be successfully completed. This control is accomplished primarily by protein kinases, each consisting of a regulatory subunit called a CYCLIN, and a kinase subunit, called a cyclin-dependent kinase (Cdk). The Cdk activates other key proteins by the addition of a phosphate group (i.e. phosphorylation), as directed by the Cdk's cyclin subunit. The concentrations of these cyclin–Cdk components rise and fall during each phase of the cell cycle, thereby causing the sequential activation of the enzymes, transcription factors, and other components required for each phase. For example, entry into mitosis is initiated by MITOSIS-PROMOTING FACTOR (MPF); this cyclin–Cdk phosphorylates proteins that undertake the tasks needed for mitosis to succeed, such as condensation of chromosomes, formation of the mitotic spindle, and disintegration of the nuclear membrane. For mitosis to be completed, the MPF cyclin subunit is degraded so that its activity declines.

Progress through the cycle is monitored at four main CHECKPOINTS; here the cell ensures that the chromosomes and DNA are undamaged, and that all operations in each stage have been completed. This enables any damage to be repaired before entering the next phase. Non-dividing cells in multicellular organisms leave the cell cycle, often for extended periods, depending on the type of tissue. In vertebrates this non-proliferating phase is called G_0. Cells re-enter the cycle in G_1 when stimulated by external signals, such as growth factors. Late in G_1 they reach a point – the restriction point – when they are committed to enter the S phase and complete the cell cycle.

cell death *See* apoptosis.

cell division The process by which a cell divides into daughter cells. In unicellular organisms it is a method of reproduction. Multicellular organisms grow by cell division and expansion, and division may be very rapid in young tissues. Mature tissues may also divide rapidly when continuous replacement of cells is necessary, as in the epithelial layer of the intestine. Cell division is a part of the CELL CYCLE, and is controlled by a host of external and internal signals. *See* meiosis; mitosis; amitosis.

cell fractionation The separation of the different constituents of the cell into homogenous fractions. This is achieved by suspending the cells in an solution of appropriate pH and salt content, breaking up the cells in a mincer or grinder, and then centrifuging the resultant liquid. The various components settle out at different rates in a centrifuge and are thus separated by altering the speed and/or time of centrifugation. *See* differential centrifugation.

cell fusion The joining together of two or more cells to form a single cell. Cultured animal cells can be induced to fuse in the presence of certain viruses or by treating them with polyethylene glycol (PEG). The fused cells contain two or more nuclei, which eventually fuse producing a single nucleus containing both parental sets of chromosomes. The fusion of cells from different species results in hybrid somatic cells (*see* somatic cell hybridization). These are used in genetic studies to localize genes to particular chromosomes. Plant cells can likewise be induced to form somatic cell hybrids, provided their cell walls are first removed to create PROTOPLASTS.

cell junctions Any of the various types of connections found between individual cells, particularly in animal tissues. TIGHT JUNCTIONS and ADHERENS JUNCTIONS bind neighboring cells tightly together and control cell shape in regularly arranged tissues, such as epithelia. DESMOSOMES also bind cells to each other, and anchor cells to the extracellular matrix. GAP JUNCTIONS permit communication between cells by enabling the diffusion of small molecules and ions between adjacent cells. In plant cells, PLAS-MODESMATA play a similar role.

cell line A culture of cells, usually derived from a tumor, that can divide indefinitely and is effectively immortal. Such cells have lost certain control mechanisms that normally limit the number of replication cycles, and are described as transformed. *See* transformation.

cell membrane *See* plasma membrane.

cell plate A structure that appears in late anaphase in dividing plant cells and is involved in formation of a new cell wall. It is formed by fusion of vesicles from the Golgi complex, resulting in a flattened membrane-bounded sac spreading across and effectively dividing the cell. Cell wall polysaccharides contained in the vesicles contribute to growth of the new wall inside the sac. The process begins at the PHRAG-MOPLAST, a region at the former spindle equator where microtubules remain and the vesicles cluster, possibly oriented by the microtubules. The cell plate membranes eventually form the two new plasma membranes of the daughter cells. *See* cytokinesis.

cell theory The theory that all organisms are composed of cells and cell products and that growth and development results from the division and differentiation of cells. This idea resulted from numerous investigations that started at the beginning of the 19th century, and it was finally given form by Schleiden and Schwann in 1839.

cellulase An enzyme that hydrolyzes 1,4-glycosidic linkages in cellulose, yielding cellobiose (a disaccharide) and glucose. It is important in the degradation of plant cell walls in living plants (e.g. in leaf abscission), and the cellulase enzymes of gut bacteria are essential for digestion in animals, such as ruminants, that consume plant material.

cellulose A polysaccharide forming the framework of the CELL WALLS of all plants, many algae, and some fungi. Each cellulose molecule is an unbranched chain of several thousand β-1,4-linked D-glucose subunits. Bundles of roughly 50–60 cellulose molecules form long parallel arrays, or *cellulose microfibrils*, which form a rigid structure of high tensile strength. In the primary cell wall the microfibrils can be arranged fairly randomly, but in elongating cells they occupy positions transverse to the direction of growth. Cellulose forms an important source of carbohydrate in the diets of herbivores, and is a major constituent of dietary fiber in human diets. It is broken down by the enzyme CELLULASE.

cellulose synthase *See* cell wall.

cell wall A rigid wall surrounding the cells of plants, fungi, some protists (e.g. algae), and bacteria. Plant cell walls are made of cellulose fibers in a cementing matrix of other polysaccharides. Fungi differ, with their walls usually containing chitin. The major groups of algae have cell walls containing cellulose, although this may be augmented with materials such as agar and algin. Bacterial walls are more complex, containing peptidoglycans – complex polymers of amino acids and polysaccharides. Cell walls are freely permeable to gases, water, and solutes. They have a mechanical function, allowing the cell to become turgid by osmosis, but preventing bursting. Plant cell walls get extra strength by addition of lignin (as in xylem and sclerenchyma) or extra cellulose (as in collenchyma). They are an important route for movement of water and mineral salts. Other modifications include the uneven thickening of guard cells, the sieve plates in phloem, and the waterproof coverings of epidermal and cork cells.

At cell division in plants the *primary wall* is laid down on the middle lamella of the CELL PLATE as a loose mesh of cellulose fibers. This gives an elastic structure that allows cell expansion during growth. Later the *secondary wall* grows and acquires greater rigidity and tensile strength. New cellulose fibers are laid down in layers, parallel within each layer, but orientated differently in different layers. Cellulose microfibrils arise on the external face of the plasma membrane by the action of the enzyme, *cellulose synthase*. This consists of a rosette of subunit proteins lying within the plasma membrane; each rosette polymerizes glucose subunits from the cytosol and extrudes the resultant cellulose microfibrils onto the cell surface. Soluble hemicelluloses and pectic substances are formed within the cell and then secreted to form the cell-wall matrix. The primary cell walls of adjacent cells are cemented together by the middle lamella. The secondary cell wall is laid down on the inside of the primary wall. It contains relatively more (up to 45%) cellulose and less hemicellulose and pectic substances, and often has a high content of lignin, making it very strong and chemically resistant.

center A cluster of nerve cells that are concerned with a common function in the nervous system. Spinal cord centers deal with relatively simple reflex actions of the body while centers in the brain regulate such functions as breathing, thirst, hunger, pain, pleasure, etc.

central nervous system (CNS) The part of the NERVOUS SYSTEM that receives and processes sensory information from all parts of the body, and send out messages to muscles and other organs. In vertebrates the CNS consists of the brain and spinal cord. The CNS of invertebrates consists of a connected pair of ganglia in each body segment and a pair of ventral nerve cords running the length of the body. *See also* autonomic nervous system; peripheral nervous system.

centrifuge An apparatus in which suspensions are rotated at very high speeds in order to separate the component solids by centrifugal force. If different components have different sedimentation coefficients then they can be separated by removing pellets of sediment at given intervals. *See also* differential centrifugation; ultracentrifuge.

centriole A cell organelle consisting of two short tubular structures orientated at right angles to each other. It lies within the CENTROSOME, just outside the nucleus, of animal and protist cells, but is absent in plant cells. Each 'barrel' of the centriole consists of a cylinder of nine triplet microtubules. Prior to cell division the centriole and other centrosome components replicate – a process called the *centriole cycle* (or *centrosome cycle*) – and the 'daughter' centrioles move to opposite ends of the cell, each coming to lie within a separate centrosome, and marking a pole of the newly forming mitotic spindle. The centriole is not essential for spindle formation, although an analogous structure, the BASAL BODY, is responsible for organizing the microtubules of undulipodia (cilia and flagella). *See illustration at* cell.

centromere The region of the chromosome that becomes attached to the nuclear spindle during mitosis and meiosis. Following the replication of chromosomes, the resultant sister chromatids remain attached at the centromere. The centromere then becomes attached to the spindle microtubules via a platelike structure called the KINETOCHORE, protein components of which bind to DNA sequences in the centromere region. The microtubules move the chromosomes to the spindle equator and subsequently pull the sister chromatids to opposite poles of the cell.

centrosome A structure found in all eukaryotic cells, except fungi, that serves as a MICROTUBULE-ORGANIZING CENTER (MTOC) and forms the SPINDLE during cell division. In animal cells it also serves as the main MTOC for the assembly of cytoplasmic microtubules, and consists of a matrix of microtubule-associated proteins. The centrosome lies close to the nucleus in nondividing cells, but at the commencement of cell division it divides, and the sister centrosomes move to opposite ends of the cell, trailing the microtubules of the spindle behind them. In animal and protist cells the centrosome contains two short barrel-shaped structures, the CENTRIOLE, but this is not necessary for spindle formation. In fungi, the function of the centrosome is served instead by the SPINDLE POLE BODY. *See also* mitosis.

cerebral cortex The surface layer of the cerebrum of the brain. It contains billions of nerve cell bodies, collectively called gray matter, and is responsible for the senses of vision, hearing, smell, and touch, for stimulating the contraction of voluntary muscles, and for higher brain activities, such as language and memory.

Chain, (Sir) Ernst Boris (1906–79) German-born British biochemist noted his part in the development of penicillin (with H. W. FLOREY) and his work on variants of penicillin as antibacterial agents. He was awarded the Nobel Prize for physiology or medicine in 1945 jointly with A. Fleming and H. W. Florey for the discovery of penicillin.

chain-termination *See* DNA sequencing.

chaperones A group of proteins in living cells that promote the folding of newly synthesized polypeptide chains into their appropriate three-dimensional structures. There are two types, both of which function using energy from the hydrolysis of ATP. *Molecular chaperones* bind and stabilize unfolded or partly folded polypeptides, thus preventing their degradation while they complete the folding process. They bind briefly to specific sites on the target polypeptide. Splitting of bound ATP releases the target polypeptide, which proceeds to fold normally. *Chaperonins* are barrel-shaped protein complexes that directly intervene in the folding of certain polypeptides. Following the initial involvement of a molecular chaperone, a partially folded polypeptide enters the cavity of a chaperonin, binds briefly to the inner wall, and folds to its correct shape. With energy from ATP, the chaperonin then expands and releases the correctly folded polypeptide.

checkpoint A point in the CELL CYCLE at which further progress through the cycle is stalled if certain processes are incomplete or if the cell's DNA is damaged. Such checkpoints ensure that the cell does not enter mitosis prematurely, with potentially disastrous consequences for the cell. For example, if chromosomes enter mitosis while they are still replicating, they become fragmented, leading to death of the cell. There are four main checkpoints, in G_1, S, G_2 and M phases.

chemical cleavage method *See* DNA sequencing.

chemical synapse *See* synapse.

chemiosmotic theory A hypothesis proposed in 1961 by the British biochemist Peter Mitchell (1920–92) to explain ATP synthesis in mitochondria. It asserts that the energy released by the electron transport chain is harnessed to 'pump' protons (H^+) across the mitochondrial inner membrane and so create an electrochemical gradient. As the protons flow back through the membrane down this gradient, they drive the formation of ATP by an integral membrane protein complex (*see* ATP synthase). Although widely dismissed when first published, the hypothesis has since been confirmed by experimental findings, and is now generally accepted. *See* electron-transport chain.

chemoreceptor A receptor that responds to chemical compounds; examples include the taste buds.

chemotactic movement *See* chemotaxis.

chemotaxis (chemotactic movement) A TAXIS in response to a chemical concentration gradient. Motile bacteria can respond to dissolved attractants and repellents by moving up or down the concentration gradient, respectively. They sample the medium at intervals to detect the change in concentration, and alter their pattern of swimming accordingly. This is accomplished by receptors on the bacterial plasma membrane. When chemicals bind to these receptors, they trigger a signal pathway inside the cell that ultimately regulates rotation of the flagellum. Attractants cause counterclockwise rotation, and smooth swimming, whereas repellents cause clockwise rotation, and the cell tumbles. The spermatozoids of primitive plants are often positively chemotactic, swimming toward the female organs in response to a chemical secreted by the latter. For example, the archegonium (female organ) of the moss *Funaria* secretes sucrose.

chiasma (*pl.* **chiasmata**) A cross-shaped connection between homologous chromosomes seen during the prophase I stage of MEIOSIS. Chiasmata are sites where there has been a mutual exchange of material between homologous, nonsister chromatids (crossing over) and provide one mechanism by which RECOMBINATION of genes occurs, through the splitting of LINKAGE GROUPS.

chimera (**mosaic**) An individual or part of an individual in which the tissues are a mixture of two genetically different tissues. It can arise naturally due to mutation in a cell of a developing embryo, producing a line of cells with the mutant gene, and hence different characteristics compared to surrounding cells. Another cause is faulty separation of homologous chromosomes during mitosis at an early developmental stage. Subsequent cell division produces an individual with a mixture of cells containing different chromosome numbers. A chimera may also be induced experimentally. For example, two mouse embryos at the eight-cell stage from different parents can be fused and develop into a mouse of normal size. Analysis of the genotypes of the tissues and organs of such a mouse reveals a random mixture of the two original genotypes.

Many variegated plants are examples of *periclinal chimeras*, in which a mutation has occurred in a sector of tissue derived from the tunica or corpus, resulting in subsequent chlorophyll deficiency. For example, in a white-edged form of *Pelargonium*, the outermost layer is colorless, indicating a lack of chlorophyll, and is the result of a mutation. There is no genetic mixture throughout the plant.

chitin A nitrogen-containing polysaccharide found in the skeleton of some animals, and in the cell walls of most fungi and some algae. The outer covering (cuticle) of arthropods is impregnated with chitin, associated with protein, to give a lightweight, waterproof exoskeleton that is tough but flexible. Chitin is also found in the hard parts of several other groups of animals. It is a polymer of N-acetylglucosamine, consisting of many glucose units, each with a hydroxyl group replaced by an acetylamine group (CH_3CONH).

chloride ion channel A channel through a cell membrane that is selectively permeable to chloride ions (Cl^-). There is typically a much greater concentration of Cl^- outside a cell than inside, but in a resting (nonexcited) cell, only a small proportion of the Cl^- channels are open. Opening of Cl^- channels causes an inflow of Cl^- down this concentration gradient, leading to increased negative charge inside the cell, i.e. hyperpolarization. In the vertebrate brain and spinal cord, inhibition of synapses is the result of such a Cl^--induced hyperpolarization. This involves the activation of ligand-gated Cl^- channels by the inhibitory neurotransmitters glycine and gamma-aminobutyric acid (GABA). These bind to receptor sites in the channel protein, causing the pore to open and allowing Cl^- to flow through the channel.

chlorine An element found in trace amounts in plants and an essential nutrient in animal diets. Common table salt is made up of crystals of sodium chloride. The chloride anion (Cl^-) is an important constituent of body fluids in animals, being present in much higher concentrations in blood and extracellular fluid than inside cells. It helps to determine the electrochemical and osmotic properties of living cells, and in neurons is instrumental in the inhibition of nerve impulses in certain types of synapses (*see* chloride ion channel). Hydrogen chloride secreted in gastric juice lowers stomach pH so that the enzyme pepsin is able to act.

chlorophylls A group of PHOTOSYN-THETIC PIGMENTS. They absorb blue–violet and red light and hence reflect green light, imparting the green color to green plants, algae, and cyanobacteria. The molecule consists of a hydrophilic (water-loving) head, containing magnesium at the center of a porphyrin ring, and a long hydrophobic (water-hating) hydrocarbon tail (the phytol chain), which anchors the molecule in the lipid of the membrane. Different chlorophylls have different chemical groups attached to the head. Some prokaryotes use chlorophylls with slightly different chemical structures, called *bacteriochlorophylls*. These absorb light of different wavelengths. *See* absorption spectrum.

chloroplast A photosynthetic PLASTID containing chlorophyll and other photosynthetic pigments. It is found in all photosynthetic cells of plants and protists but not in photosynthetic prokaryotes. There are two bounding membranes, an outer membrane and an inner membrane, with an intermembrane space between them. Inside is a gel-like ground substance, or STROMA, where the dark reactions of photosynthesis occur, and a membrane system containing the pigments on which the light reactions take place. The typical higher plant chloroplast is lens-shaped and 3–10 μm in diameter. Various other forms exist in the algae; for example spiral in *Spirogyra*, star-shaped in *Zygnema*, and cup-shaped in *Chlamydomonas*. The number per cell varies, from one to several hundred, although a leaf mesophyll cell usually contains 20 to 60.

Chloroplast membranes form elongated flattened fluid-filled sacs called THYLAKOIDS. The sheetlike layers of the thylakoids are called LAMELLAE. In all plants except algae, the thylakoids overlap at intervals to form stacks, like piles of coins, called grana (*see* granum). This arrangement is thought to enhance the efficiency of the light reactions. The stroma may contain storage products of photosynthesis, for example starch grains. The chloroplasts of most algae contain one or more *pyrenoids*. These are dense protein bodies associated with polysaccharide storage. In green algae, for example, starch is deposited in layers around pyrenoids during development.

The stroma also typically contains PLASTOGLOBULI, spherical lipid droplets that become larger and more numerous as the chloroplast senesces, when carotenoid pigments accumulate in them. Apart from enzymes of the dark reactions, the stroma also contains typical protein-synthesizing machinery including circular DNA and ribosomes. There is now strong evidence that chloroplasts and other cell organelles, such as mitochondria, represent prokaryotic organisms that invaded heterotrophic eukaryotic cells early in evolution and are now part of an indispensable symbiotic union (*see* endosymbiont theory). Chloroplast DNA comprises about 120 genes, some of which encode chloroplast proteins; however, most of the latter are encoded by nuclear DNA. *See* photosynthesis. *See illustration at* cell.

choanocyte (**collar cell**) In sponges, a cell bearing an undulipodium (flagellum) surrounded at its base by a raised cylindrical collar. Choanocytes line chambers of the sponge and the beating undulipodia circulate water through the chambers and canals. A very similar cell structure is seen in certain protists, called choanomastigotes (choanoflagellates), and it is thought that these are direct ancestors of sponges.

cholecalciferol *See* vitamin D.

cholesterol A sterol (fat derivative) found in animal cells, and in trace amounts in plants cells. It occurs in cell membranes, blood plasma, and egg yolk, and is a precursor for various steroids, including bile acids and steroid hormones. In the blood it is transported in minute protein/lipid particles called LOW-DENSITY LIPOPROTEINS. It can accumulate in the gall bladder as gallstones, and an elevated level of blood cholesterol is thought to be a contributory cause of hardening and narrowing of the arteries (atherosclerosis).

cholinergic 1. Designating a nerve fiber that releases acetylcholine from its ending when stimulated by a nerve impulse. In vertebrates, motor fibers to striated muscle, parasympathetic fibers to smooth muscle, and preganglionic sympathetic fibers are cholinergic. 2. Designating a receptor that binds acetylcholine. *See* acetylcholine receptors. *Compare* adrenergic.

cholinergic nerve *See* acetylcholine.

Cholodny–Went hypothesis *See* phototropism.

chondroblasts The cells that secrete the matrix of CARTILAGE, and become enclosed as *chondrocytes*.

chondrocytes *See* chondroblasts.

choriogonadotropin *See* chorionic gonadotropin.

chorion 1. One of the three embryonic membranes of reptiles, birds, and mammals. It encloses the amnion (with embryo or fetus inside), yolk sac, and allantois. The trophoblast of mammals is part of the chorion. The allantois is usually fused with it, forming the chorio-allantoic membrane. 2. The tough outer membrane of some eggs, notably those of insects. There is usually a pore, the *micropyle*, to admit spermatozoa.

chorionic gonadotropin (**choriogonadotropin**) A glycoprotein hormone secreted in higher mammals by the chorionic villi of the placenta (fingerlike projections of the chorion into the uterus). It prevents the regression of the corpus luteum in the earlier stages of pregnancy. The detection of human chorionic gonadotropin (HCG) in the urine is often used as a pregnancy test.

chromatid One of any pair of replicated chromosomes found during the prophase and metaphase stages of mitosis and meiosis. During MITOSIS, sister chromatids remain joined by their centromere until anaphase. In MEIOSIS it is not until anaphase II that the centromere divides and the chromatids separate to become daughter chromosomes.

chromatin The chromosomal material of eukaryotic cells, consisting of DNA and proteins (mainly HISTONES). In nondividing cells the DNA molecule is in a condensed form, being wound around histones to form a series of linked globular structures called NUCLEOSOMES. Under the microscope these resemble beads on a string. This form is usually further condensed into a spiral, or solenoid form, visible microscopically as uniformly thicker fibers, about 30 nm in diameter. The genes in this more condensed 30 nm fiber form of chromatin are inaccessible to the enzymes and other components that perform gene transcription, and some unfolding to the 'beads-on-a string' form is necessary to permit transcription of a region of the DNA molecule. During mitosis or meiosis, the chromatin becomes even more condensed, with the 30 nm fiber itself coiling (supercoiling) to form a solenoid of 300 nm diameter.

Chromatin can be distinguished as *euchromatin* or *heterochromatin*, the latter staining much more intensely with basic stains because it is more condensed. Euchromatin is unfolded and actively involved in transcription, whereas heterochromatin is inactive. Euchromatin stains more intensively than heterochromatin during nuclear division. *See also* chromosome.

chromatography A method of analyzing materials involving the separation by selective absorption of various compounds, such as proteins or nucleic acids. In general chromatography involves a test material being carried by a moving phase (liquid or gas) through a stationary phase (solid or liquid). Different substances move at different rates (depending on their absorption–desorption) and therefore separate into bands, the least readily absorbed being carried the farthest. Colorless materials can be used if some means of detecting them is used (electronic detection, radioactive labeling, or ninhydrin developer). *See also* affinity chromatography;

gas–liquid chromatography; gel-filtration chromatography; paper chromatography; thin-layer chromatography.

chromatophore 1. (in prokaryotes) A pigment-bearing membranous vesicle found in photosynthetic bacteria that contains the components of the light reaction of photosynthesis. Numerous chromatophores may be present, each formed by infolding of the plasma membrane.
2. (*Botany*) *See* chromoplast.
3. (*Zoology*) A pigment-containing cell, usually in the skin, whose color can be changed by expansion or contraction, or movement of pigment within the cell. Such changes occur in response to various stimuli, for example light intensity, temperature, fright, or the opposite sex in courtship. They often result in camouflage, as in the chameleon and some fish.

chromomere A region of a chromosome where the chromosomal material is relatively condensed, and consequently stains darker. Clusters of chromomeres produce distinct bands, the pattern of which is characteristic for a particular chromosome and is used to distinguish the chromosomes of a particular organism.

chromoplast (**chromatophore**) A colored PLASTID, i.e. one containing pigment. They include chloroplasts, which contain the green pigment chlorophyll and are therefore photosynthetic, and nonphotosynthetic chromoplasts. The term is sometimes confined to the latter, which are best known in flower petals, fruits (e.g. tomato) and carrot roots. They are yellow, orange, or red owing to the presence of carotenoid pigments.

chromosome A threadlike structure containing nucleic acid and various proteins that carries the genes of a cell. The chromosomes of different organisms differ in length, shape, and number per cell. Bacteria typically have a single circular chromosome, generally aggregated into a region of the cell called the NUCLEOID. Eukaryotes usually have several or many linear chromosomes in each cell, contained within the nucleus. These consist of CHROMATIN – a single molecule of DNA and proteins (mostly HISTONES) – with associated RNA. During nuclear division the chromosomes are tightly coiled and are easily visible through the light microscope. After division, they uncoil and become difficult to see. The number of chromosomes per nucleus is characteristic of the species, for example, humans have 46. Normally one set (haploid) or two sets (diploid) of chromosomes are present in the nucleus. In early prophase of MITOSIS and later prophase of MEIOSIS, the chromosomes split lengthwise into two identical chromatids held together by the centromere. In diploid cells of certain species (e.g. humans), there is a pair of SEX CHROMOSOMES; the remainder are termed AUTOSOMES.

chromosome map A diagram showing the order of genes or other loci along a chromosome. Such maps fall into two main categories: *genetic maps*, which give the relative order of loci, and *physical maps*, which give the distance between loci in terms of the number of intervening bases (or base pairs) along the DNA molecule. The techniques employed differ according to the type of organism being studied. For example, many plants and animals can be crossed experimentally to study inheritance patterns of particular mutations, but this is not possible in humans, where family pedigrees were, until recently, often the only available evidence. However, mapping techniques have been transformed by the introduction of such techniques as DNA cloning, nucleic acid probes, DNA sequencing, and the polymerase chain reaction.

Traditionally, inheritance patterns of mutant characters, especially in laboratory strains of experimental organisms such as mice and fruit flies, have provided a wealth of information about the location of their determining genes. Genes on the same chromomsome are linked, and tend to be inherited together. However, the frequency of recombination between any two linked loci reflects their distance apart; this can be determined by experimental crosses or pedigree analysis. By studying various different character combinations, a LINKAGE MAP can be constructed. In humans, experimental crosses are not possible; however, recombinant DNA technology has provided researchers with a battery of molecular markers to track the inheritance of mutations that cause human diseases. These markers are based on naturally occurring variations (i.e. mutations) in the base sequence of the noncoding DNA (DNA polymorphisms). Non-coding sequences are interspersed throughout the genome, and contain many DNA polymorphisms. The main class of such markers comprises RESTRICTION FRAGMENT-LENGTH POLYMORPHISMS, which are mutations in sites where the DNA is cleaved by specific restriction enzymes. Another useful polymorphism for genetic markers is in the length of repeated short sequences of DNA (*see* variable number tandem repeats). Observations of the banding pattern of stained chromosomes provide information about gross chromosomal abnormalities, which can be correlated with mutations to give a *cytological map*. This technique has been applied especially to polytene (giant) salivary-gland chromosomes of certain insects, for example *Drosophila*.

The mapping of large genomes, such as undertaken by the Human Genome Project, builds on existing linkage and cytological map data. One powerful approach, combining various aspects of recombinant DNA technology, is *positional cloning*. This uses markers for a particular mutant character to locate and clone the normal gene, so that its molecular structure, and hence the protein it encodes, can be determined. The first step is to assign a gene to a particular chromosome. This can be achieved by, for example, using a GENE PROBE for a marker sequence; the probe will bind to its complementary region of chromosomal DNA, somewhere near the gene of interest, using the technique of IN SITU HYBRIDIZATION. An alternative method is to screen hybrid cells that contain only one or a part of one human chromosome (see SOMATIC CELL HYBRIDIZATION). Specific DNA segments in such cell lines can be detected by, for example, Southern blotting.

The next step is to isolate, from a library of the organism's genomic DNA, segments of DNA that extend from either side of the marker, using the technique called CHROMOSOME WALKING. These segments are then assembled to form a contiguous stretch of DNA that includes, presumably, the gene of interest. The gene can then be pinpointed within this stretch by various methods. For example, comparison of the normal DNA with the mutant DNA using Southern blotting can reveal differences that identify at least part of the gene. Further information can be gleaned by examining the expression of messenger RNAs (mRNAs) in normal and mutant individuals, and identifying any that are missing or defective in mutant tissues. Such an mRNA is likely to encode the protein responsible for the mutant character, and can be used to construct a corresponding complementary DNA (cDNA) sequence. This embodies the coding sequences of the gene concerned, and is used to identify the sequence of the protein encoded by the gene, and hence to deduce the likely function of the protein. *See* gene library; restriction map.

chromosome mutation A change in the number or arrangement of genes in a chromosome. If chromosome segments break away during nuclear division they may rejoin the chromosome the wrong way round, giving an *inversion*. Alternatively, they may rejoin a different part of the same chromosome, or another chromosome, giving a *translocation*. A segment that becomes lost causes a *deficiency* or *deletion*; it is often fatal. A part of a chromosome may be duplicated and occur either twice on the same chromosome or on two different nonhomologous chromosomes: this is a *duplication*. Chromosome mutations can occur naturally but their frequency is increased by the effect of x-rays and chemical mutagens. Various cancers are associated with chromosomal MUTATIONS.

chromosome walking A technique for assembling a contiguous stretch of DNA from cloned DNA fragments. It enables the isolation from a cloned DNA library of genomic DNA fragments that extend either side of a starting fragment, for example one containing a marker that is linked to a particular gene. Hence a sequence of cloned fragments can be assembled from scratch that is likely to include part or all of the linked gene. The marker fragment is itself cloned, and a small segment taken from one end. This is used as a DNA probe to isolate a new clone that overlaps with the first. A new probe fragment is taken from the second clone, and the process repeated, and so on, so that overlapping clones are assembled to form a series of steps that 'walk' along the chromosome.

chylomicron   *See* lipoprotein.

ciliated epithelium   A single layer of tightly packed columnar or cubical epithelial cells with numerous cilia projecting from the free surface. The cilia beat in metachronal rhythm (each moves a fraction of a second after the one in front), causing the movement of surrounding fluid or particles. Ciliated epithelium lines the respiratory passages of mammals, for example, where the cilia can exceed a density of $10^7/mm^2$. They sweep away foreign particles, and expel them from the tract. *See also* cilium.

Ciliophora   A phylum of protists that characteristically possess cilia embedded in an outer layer (cortex). The cilia are used for locomotion and feeding (mainly on bacteria). Ciliophorans have two types of nuclei, the *meganucleus* controlling normal cell metabolism, and the smaller *micronucleus* controlling sexual reproduction (conjugation). Binary fission also takes place. Some (e.g. *Paramecium*) are covered with cilia. Others (e.g. *Vorticella*) have cilia only round the mouth, and in some (e.g. *Stentor*) these cilia are specialized for feeding. *See also Paramecium*.

cilium   A whiplike extension of certain eukaryotic cells that beats rapidly, thereby causing locomotion or movement of fluid over the cell. Cilia and flagella represent the two types of eukaryotic UNDULIPODIUM.

Cilia are identical in structure to flagella, though shorter, typically 2–10 µm long and 0.5 µm in diameter and usually arranged in groups. Each cilium has a BASAL BODY at its base. In ciliated protists, sperm, and some marine larvae they allow locomotion. In multicellular animals they may function in respiration and nutrition, wafting water containing respiratory gases and food over cell surfaces, as in filter-feeding mollusks, for example. In mammals the respiratory tract is lined with ciliated cells, which waft mucus, containing trapped dust, bacteria, etc., towards the throat.

Cilia and flagella have a '9 + 2' structure (the axoneme), consisting of nine outer pairs of microtubules with two single central microtubules enclosed in an extension of the plasma membrane. The beat of each cilium comprises an effective downward stroke followed by a gradual straightening (limp recovery). Cilia beat in a coordinated fashion such that each is slightly out of phase with its neighbor (*metachronal rhythm*). This produces a wave of activity passing across a mass of cilia, and hence a constant rather than a jerky flow of fluid.

cisterna   A flattened membrane-bounded sac of the ENDOPLASMIC RETICULUM or the GOLGI COMPLEX, being the basic structural unit of these organelles.

cistron   A genetic unit (i.e. a segment of DNA) that determines a single polypeptide chain of a protein molecule. *See* gene.

citric acid   A six-carbon tricarboxylic acid, occurring in the juice of citrus fruits, particularly lemons, and present in many other fruits. Citric acid is biologically important because it participates in the KREBS CYCLE.

citric acid cycle   *See* Krebs cycle.

clathrin-coated pit   An indentation in a cell membrane that is coated with various proteins, including clathrin. Specific cargo proteins and their receptors accumulate in the pit, which subsequently deepens and pinches off to form a CLATHRIN-COATED VESICLE.

clathrin-coated vesicle A protein-coated vesicle, typically 50–100 nm in diameter, that transports proteins within the cell. It forms by the budding of a CLATHRIN-COATED PIT from the plasma membrane or from membranes of the Golgi complex. The vesicle is bounded by membrane, which is coated by interlocking fibrous clathrin proteins forming a cagelike structure. Between the clathrin coat and the membrane is a space 20 nm wide, which contains various other proteins (assembly particles). These determine which proteins are included for transport in the vesicle during its formation.

Claude, Albert (1899–1983) Belgian-born American cell biologist who developed methods of separating cellular components by centrifugation. He also identified mitochondria as the primary cellular site for biological oxidation. He was awarded the Nobel Prize for physiology or medicine in 1974 jointly with C. R. M. J. de Duve and G. E. Palade for their work concerning the structural and functional organization of the cell.

clearing In the preparation of permanent microscope slides, the stage between dehydration and embedding. Clearing removes the dehydrating agent and replaces it by a substance that is miscible with the embedding substance. It also renders the tissues transparent. Clearing agents include benzene and xylene.

cleavage The series of rapid mitotic divisions that follow fertilization of an animal egg and divide the egg into successively smaller cells (*blastomeres*) with equivalent nuclei. There is no growth, and the result is an early embryo, or BLASTULA, which is no bigger than the original egg. The egg cytoplasm is not usually homogeneous, being divided into special regions that foreshadow the major parts of the future embryo and affect the nuclei of the blastomeres. *See* holoblastic; meroblastic.

cleidoic egg An egg with a tough shell, which permits gaseous exchange but restricts water loss (although it may take up water). Characteristic of reptiles, birds, and insects, it usually has a large food store (yolk and, in birds, albumen), and is largely self-contained.

clonal deletion *See* clonal selection theory.

clonal selection theory A theory, formulated by F.M. Burnet (1899–1985) in 1959, that explains how cells of the immune system produce large amounts of antibodies specific to particular foreign antigens, given the almost infinite number of possible antigens. The mature immune system contains numerous clones of antibody-producing B-CELLS; each clone can produce just a single type of antibody. An invading antigen effectively selects the clone that produces its corresponding antibody, therefore stimulating those cells to proliferate and produce large quantities of the appropriate antibody, at the appropriate time, to deal with an infection, for example. It is now known that B-cells carry surface receptors for one type of antigen. When the matching antigen binds to these receptors, a quiescent B-cell is activated and starts to proliferate, producing a clone of cells with similar receptor specificity. After the infection has been dealt with, a small number of memory cells remain, each bearing the receptor for the antigen. These can quickly mount a response to any subsequent invasion by the same antigen.

Clonal selection is also thought to explain how immune cells acquire tolerance to tissues of the host (self tissues). During their maturation and development, any immune cells that show reactivity to self tissues are eliminated (*clonal deletion*), whereas cells that do not react with self tissues go on to form clones of normally functional cells.

clone 1. A group of organisms or cells that are genetically identical. In nature, clones are derived from a single parental organism or cell by asexual reproduction or parthenogenesis. Such processes are common in bacteria, protists, fungi, plants, and some invertebrates, but identical twins are the only natural clones among vertebrates. Clones of sexually reproducing higher animals have been produced experimentally by embryo-splitting techniques, and by NUCLEAR TRANSFER using early embryo cells and adult body cells.

2. In genetic engineering, a copy of a gene or DNA segment produced in *cloning vectors*, such as plasmids and phages (*see* DNA cloning).

cloning vector   A DNA molecule that can combine with a gene or other DNA fragment and carry it into a host cell to be replicated. This results in numerous replicas, or clones, of both the vector and its inserted DNA fragment. Various types of vector are used in cloning, depending on the size of the DNA fragment to be cloned and its intended use. Most are genetically modified versions of naturally occurring agents, such as bacterial plasmids and bacteriophages. Any vector must have restriction sites, where the vector DNA is cut by a specific type of restriction enzyme and the fragment to be cloned can insert (*see* DNA cloning). It must also contain the sites required for binding the host-cell enzymes used for its replication, and a marker that identifies its presence in a cell. Small DNA fragments, up to about 20 kb, are generally cloned using PLASMID vectors. Various viruses are also used. They infect host cells with high efficiency, and some can be used to insert DNA into the host genome. Lytic viruses, such as LAMBDA PHAGE, replicate and then kill their host cells, so the cloned fragments are packaged with the viral DNA in the virus particles. These are a convenient repository for cloned fragments, as in DNA libraries. COSMIDS are hybrids of lambda phage and plasmids, and are suitable for cloning fragments up to about 45 kb. Bacterial artificial chromosomes are used as vectors for larger fragments (<300 kb), and yeast artificial chromosomes can accommodate fragments up to 1000 kb. *See also* expression vector.

cnidocil   *See* cnidocyte.

cnidocyte   A specialized stinging cell in the ectoderm of cnidarians, mainly on the tentacles. Each contains a *nematocyst*, which consists of a threadlike structure enclosed in a cavity. A fine trigger-like hair (*cnidocil*) projects from the outer end of the thread cell; this is sensitive to substances dissolved in the water; for example, from nearby prey. When the cnidocil is stimulated the thread is discharged from the nematocyst. It penetrates the prey, injecting poisonous substances that paralyze it. As well as this *penetrant* type of thread, other types can coil around hairs or bristles on the prey, or produce sticky substances to prevent the prey from escaping.

CNS   *See* central nervous system.

CoA   *See* coenzyme A.

coated pit   *See* clathrin-coated pit.

coated vesicle   *See* clathrin coated vesicle.

coccus   (*pl.* **cocci**)   A spherical-shaped bacterium. Cocci may be found singly, in pairs (e.g. *Diplococcus*), or chains (e.g. *Streptococcus*), or in regularly or irregularly packed clusters. Different species are characteristically found in certain conformations.

codon   A group of three nucleotide bases (i.e. a nucleotide triplet) in a MESSENGER RNA (mRNA) molecule that codes for a specific amino acid or signals the beginning or end of the message (start and stop codons). Since four different bases are found in nucleic acids there are 64 (4 × 4 × 4) possible triplet combinations. The arrangement of codons along the mRNA molecule constitutes the *genetic code*. When synthesis of a given protein is necessary the segment of DNA with the appropriate base sequences is transcribed into messenger RNA. When the mRNA migrates to the ribosomes, its string of codons is paired with the anticodons of TRANSFER RNA molecules, each of which is carrying one of the amino acids necessary to make up the protein.

coenocyte An area of cytoplasm containing many nuclei, typically found in certain fungi and algae. *Compare* plasmodium; syncytium.

coenzyme A relatively small organic molecule that is required for the catalytically active component of an enzyme (*apoenzyme*) to function. In some cases the coenzyme binds transiently to the apoenzyme. Coenzymes are a type of PROSTHETIC GROUP. *See also* cofactor.

coenzyme A (CoA) A coenzyme that is important in the synthesis and reactions of fatty acids. In the KREBS CYCLE it combines with pyruvic acid, leading to loss of carbon dioxide. It is a complex nucleotide containing an active –SH group. The compound is readily acetylated to CoAS–COCH₃ (acetyl CoA).

coenzyme Q *See* ubiquinone.

cofactor A nonprotein substance that helps an enzyme to carry out its activity. Cofactors may be cations or organic molecules (coenzymes). When a cofactor associates with a catalytically active component of an enzyme (*apoenzyme*) a complex called a *holoenzyme* is produced.

Cohen, Stanley (1922–) American biochemist. He was awarded the Nobel Prize for physiology or medicine in 1986 jointly with R. Levi-Montalcini for their discoveries of growth factors.

colcemid *See* colchicine.

colchicine A drug obtained from the autumn crocus *Colchicum autumnale* that is used to prevent spindle formation in mitosis or meiosis. It binds irreversibly to tubulin subunits and inhibits the assembly of spindle MICROTUBULES. Colchicine treatment can give rise to cells containing double the normal chromosome number. Another related drug, *colcemid*, can act reversibly to stall cells in metaphase. When

colcemid is washed from the cell, mitosis resumes. This technique can be used to assemble a population of cells undergoing synchronous cycles of cell division.

coliform bacteria Gram-negative rod-shaped bacteria able to obtain energy aerobically or by fermenting sugars to produce acid or acid and gas. Most are found in the vertebrate gut (e.g. *Escherichia coli*), and some are pathogenic to humans (e.g. *Salmonella*). The concentration of coliforms in water is used as an indicator of fecal contamination.

collagen The principal insoluble protein of extracellular matrix and fibrous connective tissues, present in bone, skin, and cartilage. It is the most abundant of all the proteins in the higher vertebrates, and gives enormous strength to tissues. There are numerous types of collagens, but all have the same basic molecular structure of three polypeptide chains wound together in a triple helix. These rodlike molecules pack together in a staggered side-by-side arrangement to form *collagen fibrils*, typically 50 nm in diameter. Parallel bundles of fibrils form *collagen fibers*, for example in muscle tendons. Other collagens act as linkers between collagen fibrils, or between other proteins of the extracellular matrix.

collar cell *See* choanocyte.

Collip, James Bertram (1892–1965) Canadian biochemist and endocrinologist. With J. J. R. Macleod he developed a method for the fractionation of pancreatic extracts to containing insulin. These were pure enough to permit the first clinical trials of the hormone to take place. Macleod divided his share of the Nobel Prize with Collip.

colloid A system that consists of a *disperse phase*, comprising large molecules or aggregate particles (typically in the range 10^{-6}–10^{-4} mm), in a continuous medium, or *continuous phase*. Sols, emulsions, and gels are all types of colloid.

colony 1. A group of individuals of the same species, generally attached to each other and dependent on each other to some degree. Colonial organization occurs in certain bacteria, some algae (e.g. *Volvox*), some hydrozoans (e.g. *Obelia* and *Physalia*), some anthozoans (corals and sea fans), and bryozoans.
2. Social colony. A group of individuals of the same species that are not anatomically connected but live together and show some degree of functional and behavioral interdependence. Examples are honey bees and humans.
3. (*Microbiology*) A cluster of cells (especially bacteria) derived from the division of one or a few initial cells in culture.

colony-stimulating factor (CSF) Any of a group of CYTOKINES that control the growth and differentiation of blood cells. Various ones are responsible for different types of cells. For example, *erythropoietin*, produced in the kidney and liver, promotes the formation of red blood cells (erythrocytes); interleukin-3 controls the production of certain white blood cells, namely granulocytes and monocytes/macrophages. *See* interleukin.

companion cell An elongated thin-walled cell that is intimately associated with a SIEVE ELEMENT in the phloem tissue of a plant. Unlike the sieve element, the companion cell has a full complement of cell organelles, and is thought to provide metabolic support for the sieve element via their numerous cytoplasmic connections (plasmodesmata). Companion cells may also play a role in loading and unloading sucrose or other sugars to and from the sieve tube.

competent Describing embryonic tissue that is able to respond to natural (induction) or experimental (evocation) stimuli by becoming or making a specialized tissue. For example, the ectoderm over the optic cup of vertebrate embryos is competent to produce lens tissue.

complement A group of about 20 proteins normally found in vertebrate blood that react in an ordered sequence – the *complement cascade* – when an antigen–antibody complex has formed. The reaction causes lysis of the foreign cells or bacteria, attracts phagocytic cells to the reaction site, and promotes ingestion of antigen-bearing cells by phagocytes. Complement plays a role in inflammation, and is also involved in the tissue damage associated with certain autoimmune disorders.

complementary DNA (cDNA) A form of DNA synthesized from a messenger RNA (mRNA) using a reverse transcriptase enzyme. The mRNA is first isolated from a particular cell or tissue of interest, and a single strand of cDNA is synthesized by reverse transcriptase using the mRNA strand as a template. A double-stranded version of the cDNA is then made prior to cloning with a suitable vector, such as lambda phage. After cloning, the base sequence can be determined. cDNA differs from the original DNA sequence in that it lacks intron and promoter sequences, and so can be used to deduce the amino acid sequence of the encoded protein (reading from the genetic code). Labeled single-stranded cDNA is used as a GENE PROBE to identify common gene sequences in different tissues and species. A collection of cDNAs representing all the mRNAs produced by a particular organism is called a cDNA library. *See* DNA cloning.

compound microscope *See* microscope.

cone One of the two types of light-sensitive cells in the RETINA of the vertebrate eye. They are concerned with vision in bright light and color vision and have a high visual acuity. There are three classes of cones, containing photopigment (iodopsin) sensitive to red, green, or blue light. Iodopsins each consist of a glycoprotein (opsin) combined with retinal, derived from vitamin A.

conformation 1. Any of the possible arrangements of atoms of a molecule resulting from rotation about a single bond.

2. The three-dimensional structure of a protein.This is determined primarily by the protein's amino acid sequence, but is altered by changes in the ionic environment, binding of cofactors, etc. Conformational changes often play a crucial role in the activity of enzymes, transport proteins, and signaling proteins. The most stable conformation is called the *native state*.

conjugation   1. The sexual fusion of gametes, particularly isogametes. *See* isogamy.
2. A type of sexual reproduction found in some bacteria, most ciliates, and certain algae, involving the union of two individuals for the purpose of transferring genetic material. In bacteria (*Escherichia* and related genera), two individuals join by a conjugation bridge and part of the genetic material of one, the donor (or male) cell, is transferred to the recipient (or female) cell. In ciliates (e.g. *Paramecium*) the two individuals unite by a bridge; their macronuclei disintegrate and their micronuclei divide by meiosis to form two gamete nuclei, one of which moves to the other cell and fuses with the remaining gamete nucleus to form a zygote. Each zygote divides and eventually forms four daughter cells. In algae such as *Spirogyra*, which are normally haploid, a conjugation tube forms between cells of two individuals and the gamete formed in one cell (the male gamete) moves through the tube and fuses with the gamete of the other cell (the female gamete).

connective tissue   A tissue in which the cells are isolated from each other by a matrix. It supports, binds, connects, and holds in position the organs of the body and arises from the mesoderm germ layer of the embryo. There are various types of connective tissue, including fibrous, adipose (fatty), bone, and cartilage; blood can also be considered a form of connective tissue. The properties of connective tissue are largely dependent on the composition of the matrix, which contains varying quantities of fibers, fluid, and other material (ground substance). Several kinds of fibers occur, principally tough strong collagenous fibers (made of COLLAGEN) and elastic fibers (ELASTIN). Collagen can also form meshlike reticular tissue. The matrix can vary from relatively soft and gel-like, in the loose areolar connective tissue of the mesenteries and skin, to hard and dense in bone. Cells lying within or surrounding the connective tissue secrete its constituents; for example, FIBROBLASTS in fibrous connective tissue, and OSTEOBLASTS in bone.

constitutive enzyme   An enzyme that is produced continuously by a cell, regardless of the cell's metabolic state. *Compare* inducible enzyme.

continuous phase   *See* colloid.

contractile ring   A belt of proteins that encircles a dividing cell and contracts to pinch the cell into two parts. Actin and myosin proteins assemble in the equatorial region of a cell during mitosis, and organize to form a contractile ring. Following formation of the daughter nuclei within the parental cell, the ring contracts, creating a deepening cleavage furrow around the cell (*see* cytokinesis). This ultimately leads to separation of the two daughter cells.

contractile vacuole   One or more membrane-bound cavities in many protists that act as osmoregulators. They periodically expand as they fill with water by osmosis and contract to discharge their contents to the exterior. This prevents the cell from swelling with water and eventually bursting.

corepressor   *See* repressor.

core promoter   *See* promoter.

Cori, Carl Ferdinand (1896–1984) and **Gerty Theresa Cori** (née Radnitz, 1896–1957)   Czech-born American biochemists distinguished for their studies of glucose and glycogen metabolism, largely carried out in collaboration. They were jointly awarded the Nobel Prize for physiology or medicine (1947) for their discovery of the course of the catalytic conversion of glycogen. The prize was shared with B. A. Houssay.

Cornforth, (Sir) John Warcup
(1917–) Australian-born British organic chemist and biochemist. Cornforth elucidated the steps involved in the biosynthesis of cholesterol from acetate. He was awarded the Nobel Prize for chemistry in 1975 for his work on the stereochemistry of enzyme-catalyzed reactions. The prize was shared with V. Prelog.

cornification *See* keratinization.

corpus luteum *See* ovarian follicle; progesterone.

cortical granules Membrane-bound vesicles in the cortex of many animal eggs, whose contents are extruded at fertilization. Penetration of the egg membrane by a sperm triggers the release of calcium ions from intracellular stores, which in turn causes the release of mucopolysaccharides from the cortical granules. This *cortical reaction* turns the vitelline membrane into the fertilization membrane, preventing further spermatozoa from penetrating the egg.

corticosteroid Any steroid hormone produced by the adrenal cortex. The release of corticosteroids is controlled by corticotropin. They are classified into two groups: mineralocorticoids (i.e. aldosterone) and glucocorticoids (i.e. cortisol, cortisone, and corticosterone).

corticotropin (adrenocorticotropic hormone; ACTH) A polypeptide hormone secreted by the anterior pituitary gland. It acts on the adrenal cortex, stimulating the secretion of corticosteroid hormones. Its release is controlled by the hypothalamus and by circulating corticosteroids, whose production it stimulates.

cortisol *See* hydrocortisone.

cosmid A hybrid CLONING VECTOR formed from a virus and a plasmid, used for cloning relatively large fragments of DNA, up to about 45 kb. Cosmids are constructed from a small *E. coli* plasmid into which is inserted the COS site from the lambda phage genome (hence the name 'cosmid'). Numerous copies of the cosmid are cut with a restriction enzyme, and incubated with restriction fragments of foreign DNA. This produces a long string of restriction fragments, joined together by linear cosmid vectors. By adding appropriate viral packaging enzymes, this string is cut at each COS site, and each fragment is packaged separately into a phage particle. These particles are then used to infect *E. coli* cells, where the recombinant DNA behaves not as a virus, but as a reconstituted plasmid. Hence the foreign DNA is replicated along with the plasmid DNA.

cotransporter A transport protein that couples the movement of an ion or molecule across a cell membrane down its electrochemical gradient, with the 'uphill' transport of another substance against its concentration gradient. The energy for the 'uphill' movement is derived from the 'downhill' movement. There are two types of cotransporter: ANTIPORTERS move the two substances in opposite directions, whereas SYMPORTERS move them in the same direction. *See* active transport.

cotyledon (seed leaf) The first leaf of the embryo of seed plants, which is usually simpler in structure than later formed leaves. Cotyledons play an important part in the early stages of seedling development. For example they act as storage organs in seeds without an endosperm, such as peas and beans, and they form the first photosynthetic organ in seeds showing epigeal germination (e.g. sunflower).

counterstain *See* staining.

C₃ pathway (**Calvin cycle; photosynthetic carbon reduction cycle**) The series of reactions occurring in plants and other photosynthesizing eukaryotes by which carbon (in the form of carbon dioxide) is assimilated by incorporation into organic compounds. It commences with the conversion of carbon dioxide to phosphoglyceric acid (PGA), which is a three-carbon compound, hence the name of the pathway (*see* photosynthesis). This also involves ribulose 1,5-bisphosphate, and is catalyzed by the enzyme RIBULOSE 1,5-BISPHOSPHATE CAR-BOXYLASE (rubisco). The C_3 pathway is the ultimate route of carbon assimilation in all plants, and in most plants (C_3 *plants)* it is also the primary route. *Compare* C_4 pathway.

C_4 pathway The series of reactions occurring in certain plants (C_4 *plants*) by which carbon (in the form of carbon dioxide) is assimilated initially through the formation of oxaloacetic acid and other four-carbon compounds (hence the name of the pathway). The C_4 pathway precedes the C_3 PATHWAY, and has evolved in various tropical and subtropical plants (e.g. sugar cane and corn) as a means of improving the efficiency of CO_2 utilization. This is beneficial to such plants, particularly when leaf pores are closed during daylight to conserve water in hot dry conditions. The pathway is generally accompanied by a characteristic leaf anatomy, called *Kranz anatomy*, in which the chloroplasts are concentrated in thick-walled bundle sheath cells surrounding the leaf vessels. The oxaloacetic acid is formed in spongy mesophyll cells, and converted to other four-carbon compounds, which are transferred to the bundle sheath cells. Here CO_2 is regenerated, and enters the C_3 pathway. The C_4 pathway increases the concentration of CO_2 in the bundle sheath cells, so that carbon assimilation is favored over PHOTORES-PIRATION, even with the relatively low CO_2 concentrations elsewhere in the leaf. However, the C_4 pathway requires energy in the form of ATP, and so the overall efficiency of photosynthesis itself is reduced compared to C_3 plants. But this is outweighed by the virtual absence of wasteful photorespiration in C3 plants.

Crick, Francis Harry Compton (1916–) British physicist and molecular biologist who, with J. D. Watson in 1953, put forward the double-helix structure of DNA, formulated the adaptor hypothesis of protein synthesis and the central dogma of molecular biology, and defined the codon. He was awarded the Nobel Prize for physiology or medicine in 1962 jointly with J. D. Watson and M. H. F. Wilkins.

crista (*pl.* **cristae**) An infolding of the inner mitochondrial membrane, forming a shelf-like projection into the matrix of the MITOCHONDRION. The extent and nature of the folding varies, active cells having complex and closely packed cristae, less active cells having fewer and less complex cristae. Within the membranous cristae are the enzymes and other components responsible for respiration and ATP synthesis.

crossing over (**homologous recombination**) The exchange of material between homologous chromatids by the formation of chiasmata. This takes place during meiosis, and results in the recombining of paternal and maternal alleles of linked genes. A model describing the mechanism of crossing over was proposed in 1964 by Robin Holliday (1932–). This envisaged that the homologous double-stranded DNA molecules become aligned side by side, and nicks appear in two of the strands. The nicked strands then 'swop over', and join up with the equivalent strand in the other DNA molecule, forming an X-shaped intermediate, or *Holliday structure*. Further cuts, in either two or all four strands, then allow the strands to separate and the chromosomes to complete meiosis successfully. Various modifications of the Holliday model have been proposed, but the precise mechanism is still unclear. However, it is now thought to involve a double-strand break in one DNA molecule, and subsequent repair of the gaps by EXCISION REPAIR of the DNA. *See also* chiasma.

cryoelectron microscopy *See* electron microscope.

CSF *See* colony-stimulating factor.

culture A population of microorganisms or tissue cells grown on or within a solid or liquid medium for experimental purposes. This is done by inoculation and incubation of the nutrient medium. *See also* tissue culture.

culture medium A mixture of nutrients used, in liquid form or solidified with agar, to cultivate microorganisms, such as bacteria or fungi, or to support tissue cultures.

Cyanobacteria A phylum of eubacteria containing the blue–green bacteria (formerly called blue–green algae) and the green bacteria (chloroxybacteria). Both groups convert carbon dioxide into organic compounds using photosynthesis, generally using water as a hydrogen donor to yield oxygen, like green plants. However, under certain circumstances they use hydrogen sulfide instead of water, yielding sulfur. Cyanobacteria are an ancient group, and their fossils (STROMATOLITES) have been dated at up to 2500 million years old. Today, most species are found in soil and freshwater. They are spherical (coccoid) or form long microscopic filaments of individual cells. Many species, for example *Nostoc* and *Oscillatoria*, are nitrogen fixers. They reproduce asexually by binary fission, or by releasing sporelike propagules or filament fragments. It is thought that certain ancient cyanobacteria became permanent symbionts of ancestral algae and green plants, taking up residence in their cells as photosynthetic organelles (plastids). This theory would account for the striking similarities between cyanobacteria and plastids.

cyanocobalamin (vitamin B$_{12}$) A water-soluble vitamin having a complex organic ring structure centered on a single cobalt atom. It is required as a coenzyme for various metabolic reactions, and deficiency in humans leads to pernicious anemia. *See* vitamin B complex.

cyclic AMP (cAMP; adenosine-3′,5′-monophosphate) A form of adenosine monophosphate (see AMP) formed from ATP in a reaction catalyzed by the enzyme adenyl cyclase. It has many functions, acting as an enzyme activator, genetic regulator, and chemical attractant. Paramount is its role as a SECOND MESSENGER, especially in animal cells, where it mediates the activity of many hormones, including epinephrine, norepinephrine, vasopressin, corticotropin, and the prostaglandins. Binding of a hormone to its receptor on the outer surface of the plasma membrane activates a membrane G protein, which in turn activates adenylyl cyclase, catalysing the formation of cAMP. This causes a rapid rise in the intracellular concentration of cAMP, which in turn activates various PROTEIN KINASES, responsible for the activation of target enzymes that control various cellular functions. The effects of cAMP vary widely in different types of cells, depending on the nature of the specific protein kinases that are activated.

cyclin Any of a family of proteins found in all eukaryote organisms that control the progress of cells through the CELL CYCLE. The cyclins act as regulatory subunits for a class of enzymes, called *cyclin-dependent kinases* (Cdks). Different cyclins are synthesized during successive phases of the cycle, and form complexes with particular Cdks. The cyclin component regulates the kinase activity of its associated Cdk; i.e. it determines which target proteins the Cdk activates by phosphorylation (the addition of phosphate groups). Each phase of the cycle involves the phosphorylation of a different set of target proteins, which implement the processes required for that particular phase.

cyclosis 1. A circular form of CYTOPLASMIC STREAMING observed in some plant cells, particularly young sieve tube elements.
2. The circulation of cell organelles through the cytoplasm, for example the food vacuoles of *Paramecium*.

cysteine *See* amino acids.

cystine A compound formed by the joining of two cysteine amino acids through a –S–S– linkage (a *cystine link*). Bonds of this type are important in forming the structure of proteins.

cytidine (cytosine nucleoside) A nucleoside formed when cytosine is linked to D-ribose via a β-glycosidic bond.

cytochemistry   *See* histochemistry.

cytochrome P-450   Any of a family of
CYTOCHROME enzymes, found in many
types of organism, that catalyze the oxidation of various lipid-soluble substances,
such as steroids, fatty acids, or certain toxins. The iron atom in the heme prosthetic
group has an associated sulfur atom. For
instance, in animal cells oxidation of toxins, such as hydrocarbons, insecticides, or
drugs, by this enzyme starts the process of
converting these often highly insoluble
compounds into a soluble form that can be
excreted by the body.

cytochromes   Conjugated proteins containing heme, that act as intermediates
in the ELECTRON-TRANSPORT CHAIN. They
transfer electrons by reversible oxidation
and reduction of the iron atom at the center of the heme group. There are four main
classes, designated *a*, *b*, *c*, and *d*. *See also*
cytochrome P-450.

cytogenetics   The study of the relationship between heredity and the cytological
appearance of chromosomes, particularly
their structure and behavior.

cytokeratin   *See* keratin.

cytokine   Any of a large group of small
proteins or peptides that act as highly potent chemical messengers for cells involved
in immune responses. In mammals they
are released by lymphocytes, mast cells,
macrophages, vascular endothelial cells,
and cells in the bone marrow, thymus and
spleen. They may cause a wide variety of
responses, for example, triggering cell differentiation or stimulating secretion. In humans, most cytokines are designated as
INTERLEUKINS or INTERFERONS. A single cytokine can be released by various cell types,
and its effects can also vary depending on
circumstances. Several are secreted by bone
marrow cells, and regulate the formation
of committed progenitor cells that subsequently proliferate to produce B-cells, T-
cells, and other immune cells. Others are
involved in regulating the function of immune cells, and play a role in such phe-

nomena as inflammation, allergies, and
fever. Cytokines released by lymphocytes
are called LYMPHOKINES.

cytokinesis   The final event in cell division, in which the cytoplasm divides to
produce two daughter cells following nuclear division (mitosis or meiosis). The
plane of cleavage corresponds to the equator of the spindle, where the chromosomes
align in the preceding metaphase of nuclear
division. It is thought that the cleavage
plane might be determined by chemical signals sent by the two asters, situated at either pole of the parent cell. In animal cells
cytokinesis involves constriction of the cytoplasm between the daughter nuclei, by
means of a CONTRACTILE RING, and eventual 'pinching off' of the daughter cells. In
dividing plant cells a new plant cell wall
forms (*see* cell plate).

cytokinin   One of a class of plant hormones concerned with the stimulation of
cell division, and root–shoot interactions.
Cytokinins are often purine derivatives: examples include *kinetin* (6-furfuryl aminop-
urine), an artificial cytokinin commonly
used in experiments; and zeatin, found
widely in higher plants.
 Cytokinins are produced in roots,
where they stimulate cell division. They are
also transported from roots to shoots in the
transpiration stream, where they are essential for healthy leaf growth. Subsequent
movement from the leaves to younger
leaves, buds, and other parts may occur in
the phloem and be important in sequential
leaf senescence up the stem. Senescence of
detached leaves can be delayed by adding
cytokinins, which mobilize food from
other leaf parts and preserve green tissue in
their vicinity. Cytokinins promote bud
growth, working antagonistically to auxins
in releasing lateral buds from apical dominance. They work synergistically with auxins and gibberellins in stimulating cambial
activity.
 In cultures of plant callus tissue, high
concentrations of cytokinin relative to
auxin will induce bud and shoot formation, whereas high auxin concentrations
promote root growth. Both hormones are

used widely in agriculture and horticulture as tools in the artificial propagation of plantlets from tissue cultures.

cytological map *See* chromosome.

cytology The study of cells, including their structure, function, biochemistry, and life history, especially in relation to their role in disease. *Compare* cell biology.

cytolysis The destruction of cells, usually by the breakdown of their cell membranes.

cytoplasm The viscous contents of a cell, contained within the plasma membrane but excluding the nucleus and large vacuoles. It comprises a watery solution (cytosol) in which are various ORGANELLES (e.g. mitochondria and chloroplasts), inclusions (e.g. crystals and insoluble food reserves), and components of the CYTOSKELETON. The cytoplasm is about 90% water. It is a true solution of ions (e.g. potassium, sodium, and chloride) and small molecules (e.g. sugars, amino acids, and ATP); and a colloidal solution of large molecules (e.g. proteins, lipids, and nucleic acids). *Compare* protoplasm.

cytoplasmic inheritance The determination of certain characters by genetic material contained in plasmids or organelles other than the nucleus, for example mitochondria and chloroplasts. Characters controlled by the DNA of extranuclear organelles are not inherited according to Mendelian laws and are transmitted only through the female line, since only the female gametes have an appreciable amount of cytoplasm.

cytoplasmic streaming The unidirectional flow of cytoplasm around the inner circumference of a cell. It occurs in certain plant and algal cells, and in amebas, and is a means of distributing metabolites within large cells. The flow is thought to be produced by the interaction between myosin-like motor proteins and actin microfilaments; the latter are arrayed in parallel bundles just inside the cell perimeter. The motor proteins are attached to organelles, such as endoplasmic reticulum, and as these organelles are propelled along, they sweep the cytosol along with them, resulting in bulk cytoplasmic flow.

Cytosine

cytosine A nitrogenous base found in DNA and RNA. Cytosine has the pyrimidine ring structure.

cytoskeleton A network of fibers within the cytoplasm of a cell that maintains its shape, enables movement of the cell, and provides anchorage and movement of its organelles. It comprises various elements, including MICROTUBULES, MICROFILAMENTS, and INTERMEDIATE FILAMENTS. *See also* spindle.

cytosol The soluble fraction of CYTOPLASM remaining after all particles have been removed by centrifugation.

Dale, (Sir) Henry Hallett (1875–1968) British physiologist and pharmacologist who discovered acetylcholine, made fundamental physiological studies of the actions of acetylcholine, epinephrine, and histamine. He was awarded the Nobel Prize for physiology or medicine in 1936 jointly with O. Loewi for their discoveries concerning chemical transmission of nerve impulses.

dalton Symbol: Da   A unit of molecular mass commonly used in measuring the mass of proteins and polypeptides. It is equal to one atomic mass unit, and is numerically equal to the molecular weight. For example, a polypeptide of molecular weight 100 000 has a mass of 100 000 daltons, or 100 kilodaltons (kDA).

Dam, Henrik Carl Peter (1895–1976) Danish biochemist noted for his discovery in green plants of a principle that he named vitamin K (later called vitamin K_1 or phylloquinone). He was awarded the Nobel Prize for physiology or medicine in 1943. The prize was shared with E. A. Doisy.

dark reactions A traditional name for a group of reactions that follow the 'light reaction' in photosynthesis, and form glucose and other reduced products from carbon dioxide. The name is misleading because several key enzymes are inactive without light, and reduction of carbon dioxide does not occur in darkness. *See* photosynthesis.

Dausset, Jean Baptiste Gabriel Joachim (1916–) French physician who first demonstrated the existence in humans of the major histocompatability complex. He was awarded the Nobel Prize for phys-

iology or medicine in 1980 jointly with B. Benacerraf and G. D. Snell.

deamination A type of chemical reaction in which an amino group (NH_2) is removed. It occurs in animals when excess amino acids are to be excreted, by the action of deaminating enzymes (in the liver and kidneys of mammals). Depending on the type of organism, ammonia produced by the reaction may be excreted directly or first converted to urea or to uric acid.

decarboxylase An enzyme that catalyzes the removal of carboxyl groups (–COOH) from carboxylic acids.

dedifferentiation *See* differentiation.

de Duve, Christian René Marie Joseph (1917–) Belgian biochemist and cell biologist noted for his discoveries of the lysosome and the peroxisome. He was awarded the Nobel Prize for physiology or medicine in 1974 jointly with A. Claude and G. E. Palade.

definitive nucleus *See* polar nuclei.

dehydration (*Microscopy*) A process followed when tissues are prepared for permanent microscope slides. Water is removed by immersing the tissue in increasing strengths of ethyl alcohol. The alcohol concentration must be increased gradually as otherwise the cells would dehydrate too quickly and shrink. Dehydration is necessary because water does not mix with the chemicals used in cleaning and mounting sections.

dehydrogenase An enzyme that catalyzes the removal of certain hydrogen atoms from specific substances in biological systems. Hydrogenases are usually called after the name of their substrate, e.g. lactate dehydrogenase. Some dehydrogenases are highly specific, both with respect to their substrate and coenzyme, whilst others catalyze the oxidation of a wide range of substrates. Many require the presence of a coenzyme, which is often involved as a hydrogen acceptor. Dehydrogenases catalyze the transfer of two hydrogen atoms from substrates to NAD and NADP.

Deisenhofer, Johann (1943–) German-born American biochemist who worked on photosynthesis. He was awarded the Nobel Prize for chemistry in 1988 jointly with R. Huber and H. Michel.

Delbrück, Max (1906–81) German-born American physicist and molecular geneticist. He developed the plaque technique for studying viruses and worked on the interaction of bacteriophages with bacteria. Delbrück was awarded the Nobel Prize for physiology or medicine in 1969 jointly with A. D. Hershey and S. E. Luria for their work the replication mechanism and the genetic structure of viruses.

deletion *See* chromosome mutation.

denaturation 1. (of a protein) The unfolding of the peptide chain of a protein, thereby causing loss of the protein's NATIVE STATE, and destroying the molecule's biological activity. Denaturation may be brought about by various physical factors, including changes in pH, extreme temperatures, and radiation. The protein's primary structure remains intact. Denatured proteins generally precipitate out of solution. The changes are often reversible when the denaturing agent is removed, and the protein refolds spontaneously, i.e. it undergoes *renaturation*.
2. (of a nucleic acid) The unwinding and separation of the two strands of DNA, or of double-stranded regions of RNA, due to heating or other factors. The two helically coiled strands of a DNA molecule separate at a characteristic temperature, called the *melting temperature*. When cooled, the

two strands can under certain conditions undergo renaturation and and re-form the double helix. The capacity of nucleic acids to undergo denaturation and renaturation is exploited in the technique of HYBRIDIZATION.

dendrite One of several slender branching projections that arise from or near the cell body of a NEURON (nerve cell) to make contact with other nerve cells. The entire array of dendrites is sometimes called the *dendritic tree*.

dendritic tree *See* dendrite.

denitrification *See* nitrogen cycle.

density-gradient centrifugation A centrifugation method used for separating and purifying particles according to their density. It allows good resolution of different cell organelles from cell samples, especially where these have already been separated into crude fractions by DIFFERENTIAL CENTRIFUGATION. The sample is placed on top of a sucrose or glycerol solution that increases in density along the centrifuge tube. After high-speed centrifugation for several hours, particles migrate to a position in the tube at which their density equals that of the suspending solution. Hence, plasma membrane can be separated from endoplasmic reticulum, for example, because each migrates to a different position in the tube.

deoxyribonuclease *See* DNase.

deoxyribonucleic acid *See* DNA.

deoxy sugar A sugar in which oxygen has been lost by replacement of a hydroxyl group (OH) with hydrogen (H). The most important example is deoxyribose, the sugar component of DNA.

depolarization A reduction in the electrical potential (*see* resting potential) that exists across the plasma membrane of a cell, particularly a neuron (nerve cell) or muscle cell. It occurs, for example, during the passage of a nerve impulse along the axon of a neuron, and when muscle cells of skeletal muscle are stimulated by the arrival of a nerve impulse at a neuromuscular junction. Depolarization is caused by the movement of ions through ion channels in the plasma membrane, notably the influx of sodium ions (Na^{2+}). *See* action potential.

dermatome   *See* somite.

desmosomes   Patchlike CELL JUNCTIONS that serve to attach neighboring cells together, and may help to distribute mechanical stresses through tissues. *Spot desmosomes* act like 'spot welds' between cells. The cytosolic (inner) face of adjacent plasma membranes bears a plaque of proteins (e.g. plakoglobin); the facing plaques are connected by linker proteins, which extend through the plasma membrane into the intercellular space and interlock between the two cells. These linker proteins are CADHERINS, one of the classes of cell adhesion molecules. Inside epithelial cells, the plaques appear to connect to keratin intermediate filaments, which form part of the cytoskeleton. Hence desmosomes may transmit forces through successive tissue layers via the cells' cytoskeletons. *Hemidesmosomes* are structurally similar to spot desmosomes, but anchor the cell to the extracellular matrix rather than to other cells.

desmotubule   *See* plasmodesma.

determined   Describing embryonic tissue whose developmental possibilities are restricted. For example the neural plate is determined to form nervous tissue and can no longer make epidermis. *See* differentiation.

dextrorotatory   Describing compounds that rotate the plane of polarized light to the right (clockwise as viewed facing the oncoming light). *Compare* levorotatory. *See* optical isomer.

diakinesis   The last stage of the prophase in the first division of MEIOSIS. Chiasmata are seen during this stage, and by the end of diakinesis the nucleoli and nuclear membrane have disappeared.

diapedesis   The passage of blood cells or other components of blood through the intact walls of blood vessels. It occurs normally during the migration of leukocytes from the bloodstream to a site of infection in the tissues.

diatoms   Unicellular algae of the phylum Bacillariophyta, found in freshwater, the sea, and soil. Much of plankton is composed of diatoms and they are thus important in aquatic food chains. They have silica cell walls (*frustules*) composed of two parts (valves), often elaborately ornamented. The shape and ornamentation vary with species. For example, some are pill-box shaped (*centric*), others are boat-shaped (*pennate*). The chloroplasts contain chlorophyll *a* and *c*, carotenes, and xanthophylls. Diatoms reproduce asexually by binary fission producing successively smaller generations until size is restored through sexual reproduction by auxospores.

dicaryon   *See* dikaryon.

dictyosome   A stack of membrane-bounded sacs (cisternae) that, together with associated vesicles (Golgi vesicles), forms the GOLGI APPARATUS. The term is usually only applied to plant cells, where many such stacks are found. In contrast, the Golgi apparatus of most animal cells is a continuous network of membranes.

dideoxy method   *See* DNA sequencing.

dideoxy sequencing   *See* DNA sequencing.

differential centrifugation   A technique for separating particles by spinning them in a CENTRIFUGE at successively greater speeds. It is used for separating the components of homogenized tissue samples. Insoluble materials, such as organelles and membranes, sediment out as a pellet, while soluble proteins and other molecules remain in the fluid. This supernatant fraction can then be subjected to further centrifugation at increased speeds and durations, to obtain increasingly homogeneous fractions.

differentially permeable membrane *See* osmosis.

differentiation The process by which cells with generalized form become morphologically and functionally specialized to produce the different cell types that make up the various tissues and organs of the organism. Given that every cell carries the same set of genes, differentiation involves 'switching on' the right genes at the right times to produce the appropriate set of proteins that will specify the form and function of a particular cell. During embryonic development, different structures in the embryo become committed (or determined) to differentiate into particular tissues at specific locations. Determination, and the subsequent differentiation of determined cells, require the intricate interplay of chemical signals, both in time and place, and many of the details are still unknown. Essentially, chemical signals from a cell's surroundings cause the initial determination of a cell, and then trigger differentiation by signaling the synthesis of certain transcription factors within the cell, which orchestrate subsequent transcription of the cell's genes. Determination and differentiation are part of the broader development program of the organism, which also includes growth (usually by cell division) and the rearrangement of cells to form patterns according to the organism's body plan (*see* patterning). A determined cell can undergo proliferation and cell migration necessary for growth and patterning, whereas a differentiated cell cannot.

Many plant cells can, under certain conditions, be stimulated to revert to an undifferentiated form – a process called *dedifferentiation*. Such cells resume cell division and form a cluster of undifferentiated cells called CALLUS. This tissue has the potential to differentiate into various cell types, or even an entire new plant, and is described as TOTIPOTENT. Differentiated animal cells are generally much more difficult to dedifferentiate.

diffraction pattern *See* x-ray crystallography.

dikaryon (dicaryon) A cell containing two different nuclei, arising from the fusion of two compatible cells, each with one nucleus. The nuclei do not fuse immediately, instead dividing independently, but simultaneously. The term is usually applied to fungal mycelia, notably of ascomycetes and basidiomycetes.

dinoflagellate A marine or freshwater protist of the phylum Dinoflagellata (Dinomastigota) that swims in a twirling manner by means of two undulipodia (flagella). These lie at right angles to each other in two grooves within the organism's rigid body wall (*test*). Many possess stinging organelles that they discharge to catch prey; and some produce potent toxins that are capable of killing fish. Roughly half of all known dinoflagellates are capable of photosynthesis.

dinucleotide A compound of two nucleotides linked by their phosphate groups. Important examples are the coenzymes NAD and FAD.

diploid A cell or organism containing twice the haploid number of chromosomes (i.e. 2n). In animals the diploid condition is generally found in all but the reproductive cells and the chromosomes exist as homologous pairs, which separate at meiosis, one of each pair going into each gamete. In plants exhibiting an alternation of generations the sporophyte is diploid, while higher plants are normally always diploid. Exceptions are those species in which polyploidy occurs. *Compare* haploid. *See also* polyploid.

diplotene In MEIOSIS, the stage in late prophase I when the pairs of chromatids begin to separate from the tetrad formed by the association of homologous chromosomes. Chiasmata can often be seen at this stage.

disaccharide   A SUGAR with molecules composed of two monosaccharide units. Sucrose and maltose are examples. These are linked by an –O– linkage (*glycosidic link*).

disperse phase   *See* colloid.

distal   Denoting the part of an organ, limb, etc., that is furthest from the origin or point of attachment. *Compare* proximal.

dizygotic twins   *See* fraternal twins.

DNA (deoxyribonucleic acid)   A nucleic acid, mainly found in the chromosomes, that contains the hereditary information of organisms. The molecule is made up of two helical polynucleotide chains coiled around each other to give a *double helix*. Phosphate molecules alternate with deoxyribose sugar molecules along both chains and each sugar molecule is also joined to one of four nitrogenous bases – adenine, guanine, cytosine, or thymine. The two chains are joined to each other by hydrogen bonds between bases. The sequence of bases along the chain makes up a code – the genetic code – that determines the precise sequence of amino acids in proteins (*see* messenger RNA; protein synthesis; transcription).
   The shape of the DNA molecule is shown in the illustration. The two purine bases (adenine and guanine) always bond with the pyrimidine bases (thymine and cytosine), and the pairing is specific: adenine with thymine and guanine with cytosine. Each chain has a *5′ end*, with a free phosphate group attached to the 5′-carbon atom of the endmost sugar, and a *3′ end*, in which the endmost sugar has a free hydroxyl group attached to its 3′-carbon atom. This directionality is important in relation to how genes are 'read' by the cell. The two chains of a DNA molecule are in an antiparallel orientation, with their 5′→3′ alignments running in opposite directions. It is conventional to write base sequences in the 5′→3′ direction.
   DNA is the hereditary material of all organisms with the exception of RNA viruses. Together with RNA and histones it makes up the CHROMOSOMES of eukaryotic cells. *See* DNA replication; repetitive DNA. *Compare* RNA.

DNA chip (DNA microarray)   A small glass 'chip' onto the surface of which are bound numerous DNA sequences in a regular array. Commonly, the sequences are

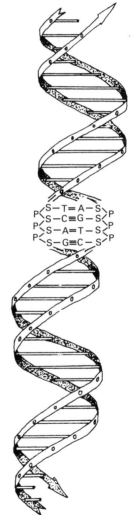

S – P   sugar-phosphate chain
≡   hydrogen bonds linking bases

DNA

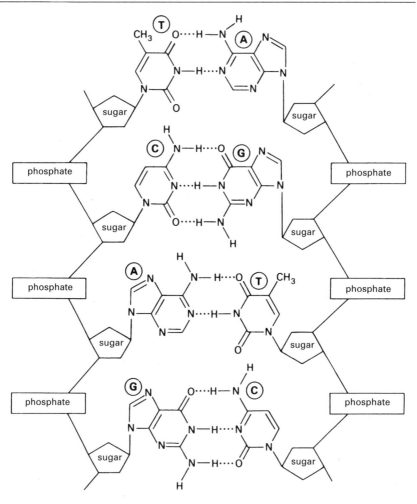

KEY

(A) adenine

(C) cytosine

(T) thymine

(G) guanine

Part of the structure of DNA showing hydrogen bonding (dotted lines) between complementary bases

DNA

COMPLEMENTARY DNAS (cDNAs), about 1 kb long, prepared from known coding sequences of particular genes. Each gene is thus represented on the chip as a tiny spot of DNA. A chip measuring 2 × 2 cm can hold about 6000 such spots, and so all the known genes of an organism can be represented on just a few such chips. DNA chips are a recent invention that are revolutionizing studies of gene expression patterns in cells. They can reveal which genes are 'switched on' and which are 'switched off' during particular metabolic states or stages in development, for example. The total messenger RNAs present at a given time in a given cell or tissue are collected, and used to synthesize cDNAs. These are fluorescently labeled and incubated with the DNA chip. cDNAs in the mixture will bind to their complementary DNA spot on the chip, and the amount of fluorescent label at each spot is detected by a scaning laser microscope with a computer-controlled microscope stage. A computer automatically registers which DNA spots are fluorescing, and the level of brightness. This indicates which genes were being expressed in the original tissue, and what the level of expression was.

DNA cloning A technique whereby a fragment of DNA is replicated, giving many identical copies. The DNA fragment is isolated by using enzymes called restriction endonucleases, or by making a complementary DNA from a messenger RNA template using a REVERSE TRANSCRIPTASE enzyme. It is then inserted into the circular chromosome of a CLONING VECTOR, most commonly a plasmid or a bacteriophage, using enzymes called DNA LIGASES. The resultant hybrid, or recombinant, DNA is used to infect a bacterium, usually *Escherichia coli*, and is replicated within the bacterial cell. A culture of such cells produces many copies of the DNA fragment, which can subsequently be isolated and purified.

Plasmid and phage vectors are suitable for cloning DNA fragments up to about 20–25 kb in length. However, genes of higher eukaryotes often extend over DNA segments of 30–40 kb in length, including introns. To clone DNA fragments of this size, COSMID vectors are used. These enable packaging of the fragments in phage heads, in such a way that the fragments can enter *E. coli* cells to be replicated as plasmids. Very large DNA fragments can be cloned using BACTERIAL ARTIFICIAL CHROMOSOMES (suitable for fragments up to 300 kb long) or YEAST ARTIFICIAL CHROMOSOMES (up to 1000 kb).

Since its introduction in the mid-1970s, DNA cloning has transformed study of the molecular structure of genes. For example, cloning is used to construct DNA libraries, containing sequences of genomic DNA or complementary DNA formed from mRNA molecules. These provide an invaluable resource for geneticists and others involved in many areas, such as sequencing and mapping genes, and studying gene expression. Also, expression of cloned genes enables the host cell, whether prokaryote or eukaryote, to produce novel proteins, which is the basis of GENETIC ENGINEERING.

DNA fingerprinting Any technique for analyzing the DNA content of an individual that can be used to characterize the individual. One technique that is widely used in forensic science is based on analysis of a class of repetitive DNA called VARIABLE NUMBER TANDEM REPEATS (VNTRs). These are tandemly repeated sequences of 15–100 bp, forming short 1–5 kb regions, found dispersed throughout the chromosomes in numbers that vary between individuals. A DNA sample is digested with a restriction enzyme and subjected to SOUTHERN BLOTTING using a probe for a particular repetitive sequence, commonly a 16-base sequence that is found in all humans. This produces an array of bands on the autoradiograph, resembling a bar code, that is unique to the individual, and thus serves like a molecular 'fingerprint'. Each band represents all DNA segments of a particular length that contain the sequence complementary to that of the probe.

Another type of fingerprint is based on *randomly amplified polymorphic DNAs* (RAPDs). This uses a random short primer sequence with the POLYMERASE CHAIN REACTION to find and amplify any sequences in an individual's DNA that happen, by chance, to be bracketed by sequences complementary to the primer. The resulting amplified fragements are separated by electrophoresis according to length, producing a set of bands (a RAPD, or 'rapid') that characterizes the individual.

DNA hybridization *See* nucleic acid hybridization.

DNA library A collection of cloned DNA fragments. A *genomic DNA library* represents the entire genome of a particular species; a *cDNA library* contains complementary DNA (cDNA) copies of messenger RNA (mRNA) molecules, for example from a particular tissue or cell type. To construct a genomic library, the organism's DNA is partially digested with a restriction endonuclease to produce large numbers of overlapping double-stranded fragments, each about 20 kb long. These are then incorporated into lambda phage vectors, and spread onto petri dishes containing *E. coli* cells. Every phage that is replicated by the bacterial cells gives rise to a small plaque of identical progeny phages, each of which contains the same particular DNA insert from the original genome. Different plaques contain different inserts. The number of plaques required to contain all the various genomic DNA fragments varies depending on the size of the organism's genome. For example, the human genome, which contains about 3×10^9 base pairs, is represented by a library of 10^6 recombinant plaques. These can be maintained on just 20–30 petri dishes.

The first step in assembling a cDNA library is to isolate mRNAs from a tissue sample using AFFINITY CHROMATOGRAPHY. The enzyme REVERSE TRANSCRIPTASE then uses the mRNA as a template to synthesize single strands of DNA with base sequences complementary to the mRNAs. The single-stranded cDNAs are converted to double-stranded cDNAs, using DNA POLYMERASE enzyme with a synthetic primer. These double-stranded cDNAs are then treated so they will insert into plasmid or lambda phage vectors, which are added to cultures of *E.coli*, where the cDNA clones are pro-

duced. The clones in a cDNA library contain only base sequences that encode the organism's gene products; they lack the introns and other noncoding sequences found in genomic libraries.

DNA libraries are screened generally using radioactively labeled DNA or RNA probes, in a technique called the MEMBRANE HYBRIDIZATION ASSAY. The cloned DNA is denatured so that base pairs between the strands are broken. This allows the probe to hybridize with any complementary DNA sequence held in the library, and thus enables identification of the relevant viral plaque or bacterial colony within the library. The plaque or colony containing the clone can then be replated and cultured for further analysis.

DNA ligase An enzyme that joins DNA strands together. It catalyzes the formation of $3' \rightarrow 5'$ phosphodieseter bonds between the 3'-hydroxyl end of one nucleotide chain, and the 5'-phosphate end of another (*see* DNA). It joins together the single-stranded segments produced during discontinuous DNA REPLICATION in all cells, and is also widely used in genetic engineering to join restriction fragments of DNA.

DNA microarray *See* DNA chip.

DNA polymerase An enzyme that catalyzes the formation of DNA from its constituent nucelotides. Such enzymes are crucial for the replication and repair of chromosomal DNA. In order to function, they require that the double strands of the DNA molecule are first unwound (by a helicase enzyme), and that a short RNA primer molecule is synthesized (by a primase enzyme). The polymerase then adds nucleotides to the 3' end of the primer, in a sequence complementary to that of an existing strand. Prokaryotes have three different types of DNA polymerase, designated I, II, and III; mammals have five types; designated α, β, γ, δ, and ϵ. *See* DNA replication.

DNA probe (gene probe) A nucleic acid consisting of a single strand of nucleotides whose base sequence is complementary to that of a particular DNA fragment being sought, for example a segment of cloned DNA in a DNA library, a restriction fragment in a DNA digest, or a gene. The probe is labeled (e.g. with a radioisotope or a fluorescent compound) so that when it binds to the target sequence, both it and the target can be identified (by autoradiography or fluorescence microscopy). Such a probe can be used, for example, to detect the gene for a given protein whose amino acid sequence has been at least partially determined. Here, the probe is constructed with a nucleotide sequence that matches part of the protein's amino acid sequence, according to the genetic code. The minimum length of such probes is 20 nucleotides, corresponding to a sequence of seven amino acids.

DNA repair Any of the mechanisms by which cells rectify damage to DNA or errors in the DNA base sequence. Such lesions take a variety of forms, and arise in various ways. For example, a base might be missing or chemically altered due to the effects of a MUTAGEN, or incorrectly inserted because of errors in DNA REPLICATION. Other possible effects of mutagens include breaks in DNA strands, cross-linking of strands, or insertion of extra nucleotides. If these lesions are not repaired, the cell and its progeny will accumulate mutations, which typically disrupt cell function and might affect the viability of offspring.

DNA repair has been studied most extensively in *E. coli*, but many of the mechanisms are also used in eukaryotes. The first line of defense is the *proofreading* ability of the enzyme DNA polymerase, which catalyzes DNA replication. If an incorrect base is incorporated in the newly synthesized DNA strand, the enzyme pauses, removes the incorrect base, and replaces it with the correct one. Any incorrect nucleotides remaining following replication are corrected by a process called mismatch repair, which uses the parental strand as a template to substitute the correct nucleotide. Distortions in the DNA molecule, for example due to chemically modified bases, can be fix by *excision repair*, in which a segment of the strand containing the lesion is removed and a new segment is synthesized to fill the gap. In the case of double-strand breaks, *recombinational repair* uses the homologous chromosome to provide the sequence information needed to repair the damage. Another mechanism can join the ends of nonhomologous double-stranded segments together, although this results in the loss of some nucleotides at the joint site. Extensive DNA damage triggers so-called *inducible DNA-repair systems*, which function when the other repair mechanisms are overwhelmed. An example is the *SOS repair system* of bacteria. This is an error-prone system, introducing many errors as it undertakes repairs.

DNA replication The process, occurring during the S phase of the cell cycle, by which cells make identical copies of the DNA in their chromosomes, allowing them to duplicate their chromosomes in readiness for cell division (mitosis). The two strands of the DNA double helix unwind, and each strand serves as a template for the assembly of a new strand with a complementary base sequence. The result is two DNA molecules, each comprising one 'parent' strand and one new 'daughter' strand. This type of replication is described as *semiconservative*. Assembly of the new strands is catalyzed by the enzyme DNA POLYMERASE, with the assistance of various other enzymes and coenzymes.

In eukaryotes, replication proceeds simultaneously at many points along the DNA molecule. These are the *origins of replication*. At each one, a helicase enzyme unwinds the helically coiled strands, and on each template strand a primase enzyme synthesizes a short length of complementary RNA. This serves as a primer to enable DNA polymerase to begin DNA synthesis. The polymerase starts to assemble the new strands, joining nucleotides together in a sequence complementary to that of each template strand. The two template strands have an antiparallel orientation (*see* DNA): in one strand the nucleotides run in the $3' \rightarrow 5'$ direction, whereas the other strand runs in the $5' \rightarrow 3'$ direction. Each new daughter strand must also run antiparallel to its parent strand, but DNA polymerase can only add nucleotides to the 3′ end of a growing strand. Hence, the two daughter strands are assembled in different ways. The new 5′→3′ strand, called the *leading strand*, is assembled continuously. However, the 3′→5′ strand (the *lagging strand*) is assembled discontinuously, from short segments. For the synthesis of each segment, primase makes a new primer, which is then extended to yield a segment, or Okazaki fragment; the primers are then removed, the gaps filled with DNA, and the fragments are joined together by the enzyme DNA ligase.

Replication proceeds in both directions from each origin of replication, and each complex of DNA polymerase and associated enzymes moves along the parent strands, continually unwinding them and adding nucleotides to the daughter strands at rates of up to 1000 bp/s. Each site of DNA synthesis is marked by a Y-shaped *replication fork*, rather like a zipper, where the DNA strands are unwound. Eventually the replicated regions join up. At the ends of the chromosomes (telomeres), replication requires a telomerase enzyme to add the noncoding repetitive nucleotide sequences that occur at either end of the DNA molecule.

DNase (**deoxyribonuclease**) Any enzyme that hydrolyzes the phosphodiester bonds of DNA. DNases are classified into two groups, according to their site of action in the DNA molecule (*see* endonuclease; exonuclease).

DNA sequencing Determination of the order of bases of a DNA molecule or DNA fragment. This can be extended to the base sequence of an entire gene, in which case multiple cloned copies of the gene are required. Long DNA sequences are cut into more manageable lengths using restriction enzymes. Since these cleave DNA at specific points, it is possible to reconstitute the overall sequence once the constituent fragments have been analyzed individually by the methods outlined below.

There are two methods of sequencing DNA. One is the *chemical cleavage method*, or *Maxam–Gilbert method*. This involves firstly labeling one end of the DNA of the gene or DNA segment with radioactive ^{32}P. The segment is then subjected to a chemical reaction that cleaves the sequence at positions occupied by one of the four bases, say, adenine. Starting with numerous cloned DNA segments, the result is a set of radioactive fragments extending from the ^{32}P label to each successive position of adenine in the segment. This process is repeated for the other three bases, and the four sets of fragments are then separated according to the number of nucleotides they contain by gel electrophoresis, in adjacent lanes on the gel. The sequence can then be deduced directly from the autoradiograph of the gel.

The second method is the *chain-termination* or *dideoxy method* (also called the *Sanger method*). A single-stranded segment of DNA taken from the gene is used as a template to replicate a new DNA strand using the enzyme DNA polymerase. The enzyme is provided with the four normal nucleoside triphosphates (ATP, GTP, CTP, TTP), plus the dideoxy (dd) derivative of one of them, say ddATP. Incorporation of the dideoxy derivative causes replication of the new strand to cease at that point. Hence, the result is a set of new strands of varying length, terminating at all the different positions where adenine, say, normally occurs in the sequence. The process is repeated in turn for each of the three remaining bases, and the set of fragments from each incubation are separated according to size by gel electrophoresis. The sequence of bases in the newly synthesized strand can be deduced directly from the gel.

Automation of both procedures, for example by using laser scanning of fluorescent dye markers instead of autoradiography, has greatly increased the speed with which DNA can be sequenced, and made possible the analysis of entire genomes, including the human genome. *See* chromosome mapping.

DNA walking *See* chromosome walking.

Doherty, Peter C. (1940–) Australian biochemist who was jointly awarded the 1996 Nobel Prize for physiology or medicine with R. M. Zinkernagel for work on the specificity of the cell-mediated immune defense.

Doisy, Edward Adelbert (1893–1986) US biochemist and endocrinologist who isolated the female sex hormones estradiol, estriol, and estrone. Doisy also isolated a principle named vitamin K_2 (now termed menaquinone) and was able to characterize vitamins K_1 and K_2. He was awarded the Nobel Prize for physiology or medicine in 1943. The prize was shared with H. C. P. Dam.

domain A region of a protein with a distinct three-dimensional structure determined by the pattern of folding of the polypeptide chain. Domains have characteristic physical and chemical properties that often relate to their particular function. For example, a domain in an enzyme might be a site for catalytic activity, and a domain in a transcription factor a site that binds to DNA. These properties are determined by the amino acids that occur within the domain. *Compare* motif.

dominant Describing an allele that expresses itself in the phenotype even when it is heterozygous with a different (recessive) allele. Hence, an individual with one dominant and one recessive allele will have a dominant phenotype, and appear identical to an individual with two dominant alleles. *Compare* recessive.

dopamine A catecholamine precursor of epinephrine and norepinephrine. In mammals it is found in highest concentration in the corpus striatum of the brain, where it functions as an inhibitory neurotransmitter. High levels of dopamine are associated with Parkinson's disease in humans.

double helix *See* DNA.

draft sequence *See* Human Genome Project.

Drosophila melanogaster A fruit fly that is extensively used as a model system in genetics and developmental biology.

ductless gland *See* endocrine gland.

Dulbecco, Renato (1914–) Italian-born American virologist who demonstrated how certain viruses are able to transform certain types of cell into a cancerous state. He was awarded the Nobel Prize for physiology or medicine in 1975 jointly with D. Baltimore and H. M. Temin.

duplex Double, or having two distinct parts. The term is particularly used to describe the double helix of the Watson–Crick DNA model.

duplication The occurrence of extra genes or segments of a chromosome in the genome. *See* chromosome mutation.

du Vigneaud, Vincent (1901–78) American biochemist noted for the synthesis of the posterior pituitary hormones oxytocin and vasopressin. He also established the structure of biotin and worked on the synthesis of penicillin. He was awarded the Nobel Prize for chemistry in 1955 for his work on biochemically important sulfur compounds, especially for the first synthesis of a polypeptide hormone.

dynein A large MOTOR PROTEIN, of which there are two main classes. *Cytosolic dyneins* are responsible for conveying transport vesicles along microtubules and for the movement of chromosomes attached to spindle microtubules during cell division. *Axonemal dyneins* are involved in the beating of undulipodia (cilia and flagella). As dynein arms, they project from each of the nine outer doublet microtubules of the axoneme, and generate sliding forces between the microtubules.

E

Eccles, (Sir) John Carew (1903–97) Australian neurophysiologist. He was awarded the Nobel Prize for physiology or medicine in 1963 jointly with A. L. Hodgkin and A. F. Huxley for their discoveries concerning the ionic mechanisms involved in excitation and inhibition in the peripheral and central portions of the nerve cell membrane.

E. coli   See *Escherichia coli.*

ectoderm   The germ layer of metazoans (including vertebrates) that remains on the outside of the embryo and develops into the epidermis and its derivatives (e.g. feathers, hairs, various glands, enamel), the lining of mouth and cloaca, and the major part of the nervous system. *See* germ layers.

ectoplasm (plasmagel)   The gel-like region of cytoplasm located in a thin layer just beneath the plasma membrane of cells that move in an ameboid fashion, such as amebas and macrophages. It consists of a three-dimensional network of actin microfilaments cross-linked by actin-binding proteins, such as filamin, and its semisolid state gives shape to the cell and transmits tension to the substrate. It is thought that the movement of ameboid cells by extension and retraction of cytoplasmic projections (pseudopods) involves reversible changes between ectoplasm and the fluid ENDOPLASM in the cell's interior. The gel–sol transition is brought about by dismantling and assembly of the microfilament network through the action of various other proteins, such as profilin, possibly regulated by calcium ion concentration.

Edelman, Gerald Maurice (1929–   ) American biochemist and molecular biologist. He was awarded the Nobel Prize for physiology or medicine in 1972 jointly with R. R. Porter for their discoveries concerning the chemical structure of antibodies.

effector   An organ or cell that responds in a particular way to a nervous impulse. Effectors include muscles and glands.

egg apparatus   The three haploid nuclei that are situated at the micropylar end of the EMBRYO SAC of a flowering plant. The central nucleus is the female gamete and those to either side of it are called the *synergids. See also* antipodal cells.

egg cell   *See* ovum.

egg membrane   **1.** The vitelline or fertilization membrane: a thin membrane that surrounds the egg cell and is secreted by the oocyte and follicle cells.
2. The tough membrane beneath the shell of a bird's egg, which is secreted by the oviduct wall before it secretes the shell.

Ehrlich, Paul (1854–1915) German medical scientist noted for his early work in chemotherapy, especially his use of arsenic compounds. He also put forward the side-chain theory of immunity. He was awarded the Nobel Prize for physiology or medicine in 1908 jointly with I. I. Mechnikov in recognition of their work on immunity.

Eijkman, Christiaan (1858–1930) Dutch physician, bacteriologist, and nutritionist who first observed that unpolished rice grains contained an antiberiberi principle (later recognized as vitamin B_1 or thiamine). He was awarded the Nobel Prize for physiology or medicine in 1929. The prize was shared with F. G. Hopkins.

elaioplast (lipidoplast; oleoplast) A plastid storing lipids (fats or oils).

elastin A protein component of connective tissue, particularly elastic fibers. It consists of chainlike polymers linked together to form an extensible network, and is found at sites such as the lung and aorta.

electron carrier A chemical compound, group, or atom that transfers electrons between components of an ELECTRON-TRANSPORT CHAIN. For example, electron carriers in mitochondria include flavins, heme, iron-sulfur clusters, and copper, all of which are prosthetic groups bound to protein complexes in the mitochondrial inner membrane.

electron micrograph *See* micrograph.

electron microscope A microscope that uses a high-velocity beam of electrons, instead of light, to form images of an object. Because the effective wavelength of the electrons is much less than light, the resolution of an electron microscope is some 2000 times greater than the best optical light microscopes, typically about 0.1 nm under operating conditions. However, only dead material can be observed because the specimen must be in a vacuum and electrons eventually heat and destroy the material. Electron microscopes are of two main types, the *transmission electron microscope* and the more recent *scanning electron microscope*. The former produces an image by passing electrons through the specimen. With the scanning microscope electrons scan the surfaces of specimens rather as a screen is scanned in a TV tube, allowing surfaces of objects to be seen with greater depth of field and giving a 3D appearance to the image. Scanning microscopes cannot operate at such high magnifications as transmission microscopes.

Transmission electron microscopy can reveal details of thin sections of cells, including the location and distribution of specific components within such sections. It also shows the shapes of subcellular particles, such as filaments and enzymes. However, the image obtained depends on the STAINING and other preparative techniques used. Tissue must be specially fixed and dehydrated before being embedded in plastic and cut into sections using an ULTRAMICROTOME. With *cryoelectron microscopy*, liquid nitrogen is used to cool the sample stage, and the sample can be maintained in a more natural unfixed, hydrated state at the very low temperature (–196°C).

electron-transport chain (respiratory chain) A chain of chemical reactions involving the stepwise transfer of electrons through a series of ELECTRON CARRIERS (enzyme complexes), resulting ultimately in the formation of ATP and the transfer of hydrogen atoms to oxygen to form water. Such reaction sequences play a crucial role in respiration in virtually all living cells; a similar set of reactions forms part of the light-dependent reactions of PHOTOSYNTHESIS. The enzymes and other components of the respiratory electron-transport chain are, in eukaryotic cells, located in the inner membrane of the mitochondria. During aerobic respiration the reduced coenzyme NADH, produced by the KREBS CYCLE in the mitochondrial matrix, gives up two electrons to the first component in the chain, NADH dehydrogenase, and two hydrogen ions (H^+) are discharged from the matrix of the mitochondrion into the intermembrane space. The electrons are transferred along the chain to a carrier molecule (ubiquinone), and then in sequence to a series of cytochromes, finally acting with the enzyme cytochrome oxidase to reduce an oxygen atom, which combines with two H^+ ions to form water. During this electron transfer, a further two pairs of H^+ ions are pumped into the intermembrane space, making a total of six per molecule of NADH. If $FADH_2$ is the electron donor, only four H^+ ions are pumped across.

The function of electron transport in the mitochondrion is to phosphorylate

ADP to ATP. According to the CHEMIOS-MOTIC THEORY, the H^+ ions in the intermembrane space diffuse back to the matrix through the inner mitochondrial membrane down a concentration gradient. As they do so they drive the synthesis of ATP from ADP by an enzyme, called ATP SYNTHASE. Each pair of H^+ ions catalyzes the formation of one molecule of ATP, so for each NADH molecule, three molecules of ATP may be synthesized (two ATP per molecule of $FADH_2$).

electrophoresis The migration of electrically charged particles toward oppositely charged electrodes in solution under an electric field – the positive particles to the cathode and negative particles to the anode. The rate of migration of any particular molecule varies with the molecule's mass, size, shape, and net charge. The technique can be used to separate or analyze mixtures on an inert porous medium such as wetted filter paper or starch gel. However, for separation of proteins and nucleic acids, the best results are obtained using gels, especially polyacrylamide gels. In *SDS-polyacrylamide gel electrophoresis*, proteins are first treated with the detergent SDS (sodium dodecylsulfate); this denatures the proteins and unfolds polypeptide chains, so that the resultant amino acid chains are separated by virtue of their length alone. This enables discrimination between proteins of broadly similar mass, and also allows estimation of the molecular mass of unknown proteins by comparing migration distances with standard proteins on the same gel. Even better discrimination is possible with *two-dimensional electrophoresis*. In this, proteins are separated firstly using a column of gel in which a pH gradient is established. Each protein stops migrating through the gel when it reaches its particular isoelectric point – the pH at which the molecule's net charge becomes zero. Then the column of gel is transferred to a slab of SDS-impregnated gel, in which the proteins migrate in a second dimension, at right angles to the first migration, according to their mass. *See also* immunoelectrophoresis.

electroporation A technique used in genetic engineering for introducing DNA directly into eukaryotic cells. The target cells are subjected to a brief electric shock, which allows large molecules such as foreign DNA to enter the cell. It is thought that the shock triggers the transient opening of pores in the cell plasma membrane.

Elion, Gertrude Belle (1918–99) American biochemist and pharmacologist distinguished for her introduction of a range of widely used synthetic drugs designed as antimetabolites, including the antifolate bactericidal agent co-trimoxazole, the immunosuppressants mercaptopurine and azathioprine, and the antiviral compound acycloguanosine. She was awarded the Nobel Prize for physiology or medicine in 1988 jointly with J. W. Black and G. H. Hitchings.

ELISA (enzyme-linked immunosorbent assay) A sensitive and convenient type of immunoassay used for determining the concentration of proteins or other (potentially) antigenic substances in biological samples. Specific antibodies against the substance being tested are adsorbed onto an insoluble carrier surface, such as a PVC sheet. Then a known amount of the sample is added, so that molecules of the test substance are bound by the antibodies. The carrier is rinsed and a second antibody, specific to a second site on the test substance, is added. Molecules of this also carry an enzyme, which causes a color change in a fourth reagent. The intensity of the color change can then be measured photometrically and compared against known standard solutions of the test substance. ELISA is widely used in medical and veterinary diagnostics, and in research. *See* immunoassay.

elongation factor (EF) Any of various proteins that are required for elongation of a polypeptide during protein synthesis at a ribosome. Different elongation factors play different catalytic roles. For example, in eukaryotes, EF1α, in combination with guanosine triphosphate (GTP), is responsible for bringing in the correct aminoacyl-tRNA and ensuring its binding to the A site of the ribosome (*see* translation). The energy for this is supplied by hydrolysis of the GTP to GDP. Another EF, designated EF2, catalyzes the movement (translocation) of the ribosome along the messenger RNA molecule, from one codon to the next, again with accompanying hydrolysis of GTP.

eluate *See* elution.

eluent *See* elution.

elution The removal of an adsorbed substance in a chromatography column using a solvent (*eluent*), giving a solution called the *eluate*. The chromatography column can selectively adsorb one or more components from the mixture. To ensure efficient recovery of these components graded elution is used. In this technique the eluent is changed sequentially starting with a nonpolar solvent and gradually replacing it by a more polar one. The more polar eluent will wash the strongly polar components from the column.

Embden–Meyerhof pathway *See* glycolysis.

embedding The sealing of tissue prepared for permanent microscope slides in a solid block of paraffin wax prior to sectioning. After CLEARING, tissues are placed in two or three baths of molten paraffin wax. When the tissue is completely infiltrated by the wax it is allowed to harden. As the wax is opaque the block must be marked to insure correct orientation when sectioning. Tissue being prepared for electron microscopy is put in a solution of plastic and heated in an oven to harden, before being sectioned using an ULTRAMICROTOME. *See also* fixing.

embryo 1. (*Zoology*) The organism formed after cleavage of the fertilized ovum and before hatching or birth. In mammals the embryo in its later well-differentiated stages is called a fetus.
2. (*Botany*) The organism that develops from the zygote of bryophytes, pterido-phytes, and seed plants before germination. *See* embryo sac.

embryology The study of the development of organisms, especially animals, usually restricted to the period from fertilization to hatching or birth.

embryonic stem cell A STEM CELL obtained from an early embryo. Such cells have the potential to differentiate into any type of tissue cell, given the appropriate chemical cues and other developmental signals. Hence, it might be possible to use them in human medicine to regenerate diseased or injured body tissues. In genetic studies, embryonic stem cells are used to introduce mutant alleles into mouse embryos to produce gene-targeted knockout mice.

embryo sac A large oval cell in the NUCELLUS of flowering plants in which egg fertilization and subsequent embryo development occurs. It contains a variable number of nuclei derived by division of the megaspore (egg cell) nucleus. There is commonly an *egg apparatus* at the micropylar end, made up of an egg nucleus and two *synergid nuclei*. There may also be three *antipodal cells* at the opposite chalazal end that probably aid embryo nourishment, and two *polar nuclei* in the center that fuse to form the *primary endosperm nucleus*. At fertilization one male nucleus fuses with the egg nucleus to form the zygote, while the second male nucleus fuses with the primary endosperm nucleus to form a triploid cell that later gives rise to the ENDOSPERM. In the gymnosperms the megaspore gives rise to a cell that is termed the embryo sac because of its similarity to the angiosperm structure.

encephalin *See* endorphin.

endocrine gland (**ductless gland**) A gland that has no duct or opening to the exterior. It produces hormones, which pass directly into the bloodstream. The circulatory system then transmits them to other body tissues or organs, where activity is modified. *Compare* exocrine gland.

endocrinology The study of the endocrine glands and their secretions (hormones).

endocytosis The entry of materials into cells by infolding of a small area of the plasma membrane to form a membranous sac, or transport vesicle. This vesicle then pinches off and moves to the cell's interior, where the contents may be sorted and transferred to other vesicles, or to cell organelles, such as lysosomes.

It is described as *pinocytosis* ('cell drinking') when the cell indiscriminately takes in a droplet of extracellular fluid, and any dissolved substances it might contain. More commonly though, cells select certain materials from their environment, by means of *receptor-mediated endocytosis*. This form of endocytosis is used to take up hormones (e.g. insulin), various glycoproteins, the iron-binding protein transferrin, and cholesterol-containing LOW-DENSITY-LIPOPROTEINS. Receptor molecules on the cell surface bind specifically to a particular substance (the bound substance is called the ligand). Receptors with bound ligand form clusters at pits in the membrane. These pits are coated with a protein called clathrin (*see* clathrin-coated pits). The clathrin is thought to polymerize in a controlled manner and cause expansion and deepening of the pit, which separates from the plasma membrane, resulting in a CLATHRIN-COATED VESICLE. Another type of cell-surface invagination involved in receptor-mediated endocytosis is the caveola, which is lined with the protein caveolin. *Compare* phagocytosis.

endoderm (entoderm) The innermost GERM LAYER of most metazoans (including vertebrates) that develops into the gut lining and its derivatives (e.g. liver, pancreas). It also forms the yolk sac and allantois in birds and mammals.

endodermis The innermost part of the cortex of plant tissue, consisting of a single layer of cells that controls the passage of water and solutes between the cortex and the stele. A clearly defined endodermis is seen in all roots and in the stems of the pteridophytes and some dicotyledons. In roots, the radial and transverse walls of endodermal cells contain an impervious thickened band, the *Casparian strip*. This insures that water and solutes pass through the cytoplasm of the endodermal cells, and blocks passage between the cells. *See also* passage cells.

endogenous Produced or originating within an organism. *Compare* exogenous.

endomitosis The duplication of chromosomes without division of the nucleus. Endomitosis may take two forms: the chromatids may separate causing endopolyploidy, for example in the macronucleus of ciliates, or the chromatids may remain joined leading to multistranded chromosomes or *polyteny*, for example during larval development of dipteran flies. Both processes lead to an increase in nuclear and cytoplasmic volume. *Compare* amitosis; mitosis.

endonuclease An enzyme that catalyzes the hydrolysis of internal bonds of polynucleotides such as DNA and RNA, producing short segments of linked nucleotides (oligonucleotides). *See also* DNase; restriction enzyme.

endoplasm (**plasmasol**) The sol-like form of cytoplasm, located inside the EC-TOPLASM. It is free-flowing and contains the cell organelles. Ameboid movement involves sol–gel conversions, i.e. the conversion of endoplasm to ectoplasm, and vice versa.

endoplasmic reticulum (**ER**) A system of membranes forming tubular channels and flattened sacs (cisternae), running through the cytoplasm of all eukaryotic cells and continuous with the nuclear envelope. Although often extensive, it was only discovered with the advent of electron microscopy. Its surface is often covered with ribosomes, forming *rough ER*. The proteins they make can enter the cisternae for transport to other parts of the cell or for secretion via the GOLGI APPARATUS. The synthesis of secretory proteins begins at unattached ribosomes in the cytosol. The first part of the newly formed polypeptide to emerge is a signal peptide. This binds to a SIGNAL RECOGNITION PARTICLE (SRP), which also binds the ribosome to an SRP receptor in the ER membrane. The ribosome is then transferred to a protein complex in the ER membrane called a TRANSLOCON. This acts like a gateway, through which the elongating polypeptide passes, into the lumen of the ER. When its synthesis is complete, the polypeptide separates from the ribosome, the gateway closes, and the ribosome moves away into the cytosol. Inside the ER, the newly synthesized polypeptide might undergo various modifications, including folding to its correct conformation, addition of carbohydrate groups, and assembly with other polypeptide chains into a multimeric protein.

ER lacking ribosomes is called *smooth ER* and is involved with lipid synthesis, including fatty acids, phospholipids, and steroids. It is especially abundant in liver cells, where it is also responsible for detoxifying chemicals such as pesticides and carcinogens. In muscle cells a specialized form of ER called SARCOPLASMIC RETICULUM is present.

endoribonuclease   *See* ribonuclease.

endorphin   One of a group of peptides produced in the brain and other tissues that are released after injury and have pain-relieving effects similar to those of opiate alkaloids, such as morphine. They include the *enkephalins*, which consist of just five amino acids, and are released as inhibitory neurotransmitters to depress pain in certain parts of the central nervous system. Other larger endorphins occur in the pituitary, while some are polypeptides, found mainly in pancreas, adrenal gland, and other tissues. Pain relief from acupuncture may be due to stimulated production of endorphins.

endosome   A membrane-bound sac, or vesicle, involved in the transport and sorting within a cell of materials destined for delivery to lysosomes. Typically, CLATHRIN-COATED VESICLES derived from the plasma membrane during endocytosis, or from membranes of the Golgi complex, lose their protein coats and become transport vesicles, which fuse with existing endosomes. These vesicles might contain, for example, particles taken into the cell from outside, or lysosomal enzymes packaged by the Golgi complex. The materials are further processed within the endosomes and packaged into more transport vesicles, which bud from the endosome and carry the materials to their final destination in the lysosome.

endosperm   The nutritive tissue that surrounds the embryo in angiosperm plants. In *nonendospermic seeds* most of the endosperm is absorbed by the developing embryo and the food is stored in the cotyledons. In *endospermic seeds* the endosperm replaces the nucellus and is often a rich source of hormones. Many endospermic seeds (e.g. cereals and oil seeds) are cultivated for their food reserves. The endosperm develops from the primary endosperm nucleus and is therefore triploid.

endospore   A resting stage produced by certain bacteria under unfavorable conditions. Endospores are formed within the cell and consist of a core surrounded by a cortex of peptidoglycan and several spore coats made of protein. The core is partially dehydrated, and contains high concentrations of calcium ions and dipicolinic acid. On germination the wall is lysed and one vegetative cell is produced. Endospores can remain viable for several centuries and are resistant to heat, desiccation, and x-rays.

endosporium   *See* intine.

endosymbiont theory   The theory, proposed originally in 1970 by US biologist Lynn Margulis (1938–   ), that eukaryotic organisms evolved from symbiotic associations between bacteria. It proposes that integration of photosynthetic bacteria, for example purple bacteria and cyanobacteria, into larger bacterial cells led to their permanent incorporation as forerunners of the mitochondria and plastids (e.g. chloroplasts) seen in modern eukaryotes. There is compelling supporting evidence for the theory, particularly from studies of mitochondrial and plastid DNA and ribosomes, which demonstrate remarkable similarities with those of bacteria.

endothelium The tissue lining the blood vessels and heart. It consists of a single layer of thin flat cells fitting very close together. In capillaries it is the only layer between the blood and the fluid bathing the cells. Water and all dissolved substances of low molecular mass pass through the cells. White blood cells 'squeeze' between the endothelial cells – a movement known as DI-APEDESIS.

endotoxins Toxic cell-wall components of Gram-negative bacteria (e.g. *Salmonella, Escherichia, Shigella*) that are released on disintegration of the cell. They are heat-stable lipopolysaccharide-protein complexes causing nonspecific effects in their hosts, for example, fever, diarrhea, and vomiting. *Compare* exotoxins. *See* toxin.

end plate A flattened nerve ending that occurs at the junction of a motor axon and a muscle cell (*see* neuromuscular junction). It transmits nerve impulses in a way similar to that of other SYNAPSES.

end-plate potential (**EPP**) A brief localized depolarization or potential change across the membrane in the motor endplate region of a muscle fiber, at a neuromuscular junction. A neurotransmitter is released from the presynaptic nerve endings on stimulation by an impulse and increases the permeability of the postsynaptic (muscle) membrane to ions, causing the EPP. The size of the EPP depends on the amount of neurotransmitter released but normally it is sufficiently large to cross the threshold level for response and set off an action potential, which is propagated along the length of the muscle fiber.

enkephalin *See* endorphin.

entoderm *See* endoderm.

envelope 1. A membranous layer that surrounds the protein coat (capsid) of certain viruses, for example the rabies virus. It consists of a lipid bilayer with various proteins, mainly glycoproteins. The latter are encoded by the virus, and play an important role in recognition and penetration of host cells.
2. The structure that bounds a bacterial cell. It consists of the plasma membrane and, generally, a cell wall, plus various other layers according to the type of bacterium.
3. *See* nuclear envelope.

enzyme A protein that catalyzes biochemical reactions. Enzymes act with a given compound (the substrate) to produce a complex, which then forms the products of the reaction. The enzyme itself is unchanged in the reaction; its presence allows the reaction to take place. The names of most enzymes end in -ase, added to the substrate (e.g. lactase) or the reaction (e.g. hydrogenase). To avoid confusion, each enzyme is also given a number by the Enzyme Commission; it consists of the abbreviation EC followed by a set of four numbers. For example, α-amylase, an enzyme found in saliva, is designated as EC 3.2.1.1.
  Enzymes are extremely efficient catalysts for chemical reactions, and very specific to particular reactions. They may have a nonprotein part (cofactor), which may be an inorganic ion or an organic constituent (coenzyme). The mechanism of action of most enzymes appears to be by ACTIVE SITES on the enzyme molecule. The substrate acting with the enzyme changes shape to fit the active site, and the reaction proceeds at a much greater rate than it would without the enzyme (*see* activation energy). Enzymes are very sensitive to their environment – e.g. temperature, pH, and the presence of other substances. *See also* apoenzyme; coenzyme; holoenzyme; ribozyme.

enzyme-linked immunosorbent assay *See* ELISA.

eosin *See* staining.

eosinophil A white blood cell (*see* leukocyte) with a lobed nucleus and cytoplasmic granules that stain orange with acidic dyes such as eosin. Eosinophils comprise 2–5% of all leukocytes, but the number increases in allergic conditions, such as asthma and hay fever, as they have antihistamine properties. They perform phagocytosis of protists and parasitic worms.

epidermis 1. (*Botany*) The outer protective layer of cells in plants. In aerial parts of the plant the outer wall of the epidermis is usually covered by a waxy cuticle that prevents desiccation and protects the underlying cells. Epidermal cells are typically platelike and closely packed together except where they are modified for a particular function, as are GUARD CELLS.
2. (*Zoology*) The outer layer of cells or outer tissue of an animal that generally protects the tissues beneath and insures that the body is waterproof. In vertebrates, the epidermis consists of several layers of cells and forms the outer layer of skin. As it wears away at the surface it is renewed continuously by growth of new cells in the underlying Malpighian layer. The harder cornified cells of the stratum corneum are the chief protective cells. The epidermis of invertebrates is a single layer of cells, often secreting a protective cuticle. In arthropods, this cuticle forms the exoskeleton

epimysium *See* skeletal muscle.

epinephrine (**adrenaline**) A hormone produced by the medulla of the adrenal glands, and a neurotransmitter in certain parts of the central nervous system and autonomic nervous system. Epinephrine secretion from the adrenal medulla into the bloodstream causes the stress response – acceleration of the heart, constriction of arterioles, and dilation of the pupils. In addition, epinephrine produces a marked increase in metabolic rate thus preparing the body for emergency. It triggers its effects by binding to certain types of ADRENERGIC RECEPTORS in various target tissues.

episome A genetic element that exists inside a cell, especially a bacterium, and can replicate either as part of the host cell's chromosome or independently. Homology with the bacterial chromosome is required for integration, therefore a plasmid may behave as an episome in one cell but not in another. Examples of episomes are temperate phages. *See* plasmid.

epistasis The action of one gene (the *epistatic gene*) in preventing the expression of another, nonallelic, gene (the *hypostatic gene*).

epithelium A tissue consisting of a sheet (or sheets) of cells that covers a surface or lines a cavity. The cells are close together, and may be described as *cubical*, *columnar*, *ciliated*, or *squamous* (scalelike), depending on their shape. Epithelial cells typically transport materials from one side to the other. For example, the cells lining the intestine absorb nutrients from the intestinal lumen and transfer these to underlying extracellular fluid, and thence to blood vessels. Hence, the opposite faces of an intestinal epithelial cell have different properties and are said to be *polarized*. The apical surface (facing the lumen) is covered with fine projecting microvilli to increase its surface area, whereas the opposite basal surface is flat and forms junctions with the extracellular matrix or BASAL LAMINA. The two surfaces also have different sets of membrane-bound transport proteins, according to whether their main task is import or export.

Epithelial cells are tightly bound to each other, and to the underlying matrix, by cell junctions, particularly DESMOSOMES. These give strength and integrate the cells into an effective tissue. TIGHT JUNCTIONS form a seal, usually just below the apical surface, that blocks passage of material between the cells. *See also* ciliated epithelium.

epitope The region of an ANTIGEN molecule that is unique to the antigen and therefore responsible for its specificity in an antigen–antibody reaction. The epitope combines with the complementary region on the ANTIBODY molecule.

EPP *See* end-plate potential.

EPSP *See* excitatory postsynaptic potential.

equatorial plate The equator of the nuclear spindle upon which the centromeres of the chromosomes become aligned during metaphase of mitosis and meiosis. It is also the site where the cleavage furrow forms prior to cell division (*see* cytokinesis).

ER *See* endoplasmic reticulum.

Ernst, Richard Robert (1933–) Swiss physical chemist who used pulsed radiofrequency radiation in nuclear magnetic resonance spectroscopy and extended the technique to biological macromolecules. He was awarded the Nobel Prize for chemistry in 1991.

erythroblast One of the cells in the red bone marrow from which ERYTHROCYTES develop. Formed initially by division of HEMATOPOIETIC STEM CELLS, erythroblasts undergo successive divisions to form increasingly mature cells, producing hemoglobin and, in mammals, losing the nucleus to become RETICULOCYTES. These are subsequently released into the blood circulation as erythrocytes. In humans, 200 000 million new erythrocytes are made each day to replace those that are worn out.

erythrocyte (red blood cell) A type of blood cell that contains HEMOGLOBIN and is responsible for the transport of oxygen in the BLOOD. Mammalian erythrocytes are circular biconcave disks without nuclei, whereas those of other vertebrates are oval and nucleated. Human blood contains 5 million red cells per cubic millimeter; each cell lives for about 120 days, after which it is destroyed in the liver and replaced by a new cell from the red bone marrow. In addition to hemoglobin, erythrocytes also contain an enzyme, carbonic anhydrase, and therefore have an important role in transporting carbon dioxide and maintaining a constant pH.

erythropoiesis *See* hematopoiesis.

Escherichia coli A bacterium widely used in genetic and cell biology research, occurring naturally in the intestinal tract of animals and in soil and water. It is Gram-negative and the cells are typically straight round-ended rods, usually occurring singly or in pairs. Some strains are pathogenic, causing diarrhea or more serious gastrointestinal infections. The strains can be distinguished serologically on the basis of their antigens. *E. coli* is killed by pasteurization and many common disinfectants.

essential amino acid *See* amino acids.

essential element An element that is indispensable for the normal growth, development and maintenance of a living organism. The *major elements* are those required in relatively large quantities (*see* carbon; hydrogen; oxygen; nitrogen; sulfur; phosphorus; potassium; magnesium; calcium). Others are required in only small or minute amounts, such as iron, manganese, molybdenum, boron, zinc, copper, cobalt, iodine, and selenium (*see* trace element).

essential fatty acids Fatty acids (*see* carboxylic acid) required for growth and health that cannot be synthesized by the body and therefore must be included in the diet. Linoleic acid and possibly (9,12,15)-linolenic acid are the only essential fatty acids in humans, being required for cell membrane synthesis and fat metabolism. Arachidonic acid is essential in some animals, such as the cat, but in humans it is synthesized from linoleic acid. Essential fatty acids occur mainly in vegetable-seed oils, for example, safflower-seed and linseed oils.

EST *See* expressed sequence tag.

ester A compound formed by reaction of a carboxylic acid with an alcohol:
$$RCOOH + HOR_1 \rightarrow RCOOR_1 + H_2O$$
Glycerides are esters of long-chain fatty acids and glycerol.

estradiol   The most active estrogen produced by the body. It promotes proliferation of the endometrium of the womb during the first half of the estrous cycle to prepare the womb for ovulation, and is also important in female development. It is metabolized to estrone and then estriol. *See* estrogen.

estrogen   Any of various female sex hormones (steroids) involved in the development and maintenance of accessory sex organs and secondary sex characteristics (e.g. growth of the breasts). Estrogens are also released during the estrous cycle (menstrual cycle in humans) to prepare the female sex organs for fertilization of the egg cell and implantation and growth of the embryo. They are used therapeutically and in oral contraceptives. The ovary produces mainly two estrogens: estradiol and estrone. These hormones are produced in smaller amounts by the adrenal cortex, testis, and placenta. *See also* progesterone.

estrone   An ESTROGEN hormone produced by the ovary and by peripheral tissues, with actions similar to ESTRADIOL.

ethanoic acid   *See* acetic acid.

ethene   *See* ethylene.

ethylene (ethene)   A gaseous hydrocarbon (C_2H_4), produced in varying amounts by many plant tissues, that functions as a plant hormone. It is synthesized from the amino acid methionine, and production is usually stimulated by auxins. Almost every aspect of plant growth and development can be affected by ethylene, but it is known particularly to stimulate senescence and fruit ripening, and is produced in response to stress.

etioplast   A modified chloroplast formed from proplastids in leaves grown in total darkness. Instead of the normal chloroplast membrane system, etioplasts contain a highly organized semicrystalline array of tubular membranes (called a *prolamellar body*) showing a hexagonal symmetry. Radiating from the prolamellar body are sin-

gle thylakoids. A normal membrane system develops from this body once the plant is exposed to light.

eubacteria   A large and diverse group of BACTERIA, principally distinguishable from the other major group of bacteria, the ARCHAEA, by the presence of a peptidoglycan layer in their cell walls. There are also differences in the base sequences of RNA subunits of the ribosomes, which reflect wide evolutionary divergence between the two groups. Most are unicells that divide by binary fission. The cells can be spherical, rod-shaped, or helical, and some form assemblages of cells, such as branching filaments. Most are immotile, but some possess flagella. They are a ubiquitous group, some being found in extreme conditions. Many authorities now regard eubacteria as being so distant in evolutionary terms from Archaea that they constitute a separate domain, called Bacteria.

eucaryote   *See* eukaryote.

euchromatin   *See* chromatin.

Euglenophyta   A phylum of aquatic single-celled protists that swim using one or more undulipodia (flagella). It contains both photosynthesizing and nonphotosynthesizing members. For example members of the genus *Euglena* possess a flexible pellicle surrounding the cell, and a single undulipodium. Many species contain chloroplasts, with pigments similar to those found in plants, although these organisms may also consume dissolved or particulate food from their surroundings. Reproduction is asexual, by binary fission. *E. gracilis* is a popular laboratory organism, in which the chloroplasts regress and are effectively turned off in sustained darkness, a process reversed when light is reintroduced. Some individuals may lose their chloroplasts permanently.

eukaryote   Any member of a group (sometimes called Eukaryota or Eukarya) comprising all organisms except bacteria (which comprise the PROKARYOTES). There are several key features that distinguish eukaryotes from prokaryotes. Their genetic material is packaged in chromosomes within a membrane-bound nucleus. Moreover, the cells divide by mitosis, with formation of a mitotic spindle that ensures correct redistribution of the chromosomes to the daughter cells. Unlike prokaryotes, they possess mitochondria and (in plants and other photosynthetic eukaryotes) chloroplasts in the cell cytoplasm. Also they have complex cilia and flagella (undulipodia) composed of arrays of microtubules, whereas the prokaryotic flagellum is a simple shaft of protein.

Euler, Ulf Svante von   *See* von Euler

Euler-Chelpin, Hans Karl August von (1873–1964) German-born Swedish chemist and biochemist, father of U. S. Euler, distinguished for his work on the enzymic mechanism of fermentation. He isolated and studied cozymase (now known as NAD). He was awarded the Nobel Prize for chemistry in 1929, the prize being shared with A. Harden.

euploidy   The normal state in which an organism's chromosome number is an exact multiple of the haploid number characteristic of the species. For example, if the haploid number is 7, the euploid number would be 7, 14, 21, 28, etc., and there would be equal numbers of each different chromosome. *Compare* aneuploidy.

excision repair   *See* DNA repair.

excitatory postsynaptic potential (**EPSP**)   A localized depolarization at an excitatory synapse, due to the release of neurotransmitter from the presynaptic membrane, on stimulation by an impulse. The neurotransmitter acts by binding to receptors in the postsynaptic membrane and opening sodium ion channels, thereby bringing about the depolarization. The size of the EPSP depends on the amount of neurotransmitter released; if it is sufficiently large it will set off an ACTION POTENTIAL in the postsynaptic nerve fiber. It can be raised either by several impulses arriving in quick succession, at one synapse, or by simultaneous impulses arriving at different synapses. *Compare* inhibitory postsynaptic potential.

exocrine gland   A gland that produces a secretion that passes along a duct to an epithelial surface. The ducts may pass to the body surface (e.g. sweat, lacrimal, and mammary glands), or they may be internal (e.g. in the mouth, stomach, and intestines). *Compare* endocrine gland.

exocytosis   The bulk transport of materials out of the cell across the plasma membrane. It involves fusion of vesicles or vacuoles with the plasma membrane in a reversal of endocytosis. The materials thus lost may be secretory, excretory (e.g. from autophagic vacuoles), or may be the undigested remains of materials in food vacuoles. Typical secretions are enzymes and hormones from gland cells. These secretory proteins are packaged in vesicles at the *trans* Golgi reticulum (*see* Golgi complex), from where they travel to the plasma membrane. Docking and fusion of vesicles with the membrane is thought to involve various proteins, including SNAREs and SNAPs, which occur in the vesicle coat and in the target membrane. *See also* lysosome.

exogenous   Produced or originating outside an organism. *Compare* endogenous.

exon   A segment of a gene that is both transcribed and translated and hence carries part of the code for the gene product. Most eukaryotic genes consist of exons interrupted by noncoding sequences (*see* intron). Both exons and introns are transcribed to heterogeneous nuclear RNA (hnRNA), an intermediary form of messenger RNA (mRNA); the introns are then removed leaving mRNA, which has only the essential sequences and is translated into the protein.

exonuclease   An enzyme that catalyzes the hydrolysis of the terminal linkages of polynucleotides such as DNA and RNA, thereby removing terminal nucleotides. *See also* DNase.

exoribonuclease *See* ribonuclease.

exotoxins Toxic substances that are produced by bacteria or other microorganisms and released into the microbial cell's external environment. Exotoxins can therefore travel in the bloodstream from a site of infection to affect host cells elsewhere in the body, and readily cause disease or even death. For example, diphtheria toxin, produced by strains of the bacterium *Corynebacterium diphtheriae*, is a polypeptide that blocks protein synthesis in susceptible cells; only a single molecule is required to kill a cell. Botulinum toxin, produced by *Clostridium botulinum*, is one of the most poisonous substances known. It forms a protein complex that binds to presynaptic membranes at neuromuscular junctions, preventing transmission of nerve impulses to muscles and causing paralysis.

explantation The culture of isolated tissues of adults or embryos in an artificial environment, usually *in vitro*, for maintenance, growth, and/or differentiation. *Compare* implantation; transplantation.

exponential growth A type of growth in which the rate of increase in numbers at a given time is proportional to the number of individuals present. Thus, when the population is small multiplication is slow, but as the population gets larger, the rate of multiplication also increases. Exponential growth starts off slowly and accelerates as time goes by. However, at some point factors such as lack of nutrients, accumulated wastes, etc., limit further increase, and the growth rate starts to decline. When plotted on a graph, as number against time, the curve for such growth is typically sigmoid (S-shaped).

expressed sequence tag (EST) A short sequence of complementary DNA (cDNA) that has been determined from a messenger RNA (mRNA) molecule isolated from a cell. It thus represents a small stretch of the organism's DNA that is expressed in the particular cell. Numerous ESTs for humans and other organisms are held on databases, and the data are made available via the internet. These provide researchers with a means of matching a particular amino acid sequence of a newly discovered protein, say, to an actual stretch of that organism's DNA. If a matching EST is found, its sequence can be used to assemble a DNA probe for screening a DNA library for segments of cloned DNA that contain the protein-coding sequence.

expression vector A type of CLONING VECTOR in which the foreign DNA is not only cloned but also expressed by the host cell as messenger RNA (mRNA), which is subsequently used to synthesize the foreign protein. Protein fragments can be expressed by vectors based on lambda phage. Plaques containing the cloned DNA can thus be identified using antibody probes specific to the protein fragment of interest. Other types of expression vector are used to produce full-length proteins (e.g. therapeutic proteins) at high levels in bacterial cells, such as *E. coli*. These vectors are typically engineered PLASMIDS that, apart from the usual features of such vectors, also contain a promoter site, which binds the RNA polymerase enzyme required for transcription. The foreign gene, in the form of complementary DNA (i.e. without the noncoding introns found in eukaryotic genomic DNA), is therefore transcribed by the bacterial enzymes. Some eukaryote proteins that normally undergo extensive modification after synthesis cannot be satisfactorily produced in bacterial cells, which lack the necessary enzymes. For these, various eukaryotic expression vectors have been devised to work in eukaryotic host cells.

extracellular Occurring or situated outside a cell.

extracellular matrix (ECM) A viscous watery mixture of proteins, polysaccharides, and other substances that surrounds cells in the tissues of animals. It provides a framework that helps bind cells together, giving strength and integrity to tissues. It is also the environment in which tissues develop and cells can move. Hormones and other signal molecules contained in the

ECM are the means by which tissue cells communicate with each other, and by which their activities are regulated from elsewhere in the body. The components of the ECM are secreted by the cells themselves, and belong mainly to three classes of proteins. COLLAGENS are insoluble fibrous proteins that form various chains, bundles, or other structures. *Multiadhesive matrix proteins* bind to CELL ADHESION MOLECULES in cell membranes and to other ECM components. *Proteoglycans* are proteins with attached polysaccharide chains; they contain water and form the bulk of the matrix or attach to cell surfaces. CONNECTIVE TISSUE, such as cartilage, bone, tendon, and adipose (fat tissue), consists of cells within a relatively large amount of extracellular matrix (sometimes called 'ground substance'). *See also* basal lamina.

extrachromosomal DNA   In eukaryotes, DNA found outside the nucleus of the cell and replicating independently of the chromosomal DNA. It is contained within self-perpetuating organelles in the cytoplasm, notably mitochondria, chloroplasts, and plastids, and is responsible for cytoplasmic inheritance. *See* mitochondrial DNA.

extremophile   A prokaryote organism that thrives under extreme environmental conditions, such as very high or very low temperatures or high salt concentrations. Extremophiles can function because their enzymes and other cellular components are adapted to the particular extreme conditions. Many extremophiles are archaeans.

eyespot (**stigma**)   A light-sensitive structure of certain protists and invertebrate animals. The eyespot of unicellular and colonial algae and their gametes and zoospores contains globules of orange or red carotenoid pigments. It controls locomotion, ensuring optimum light conditions for photosynthesis. Its location varies. In *Chlamydomonas* it is just inside the chloroplast; in *Euglena* it is near the base of the undulipodium (flagellum). A light-sensitive pigmented spot is also found in the cells of some jellyfish and flatworms, for example, the miracidium larva of liver fluke.

facilitated diffusion (facilitated transport) The transport of molecules through a cell membrane down a concentration gradient, mediated by membrane transport proteins called UNIPORTERS. Cells take up various small molecules in this manner, including amino acids, sugars, and nucleosides. The transport proteins are specific for particular types of molecule, and greatly increase the rate of diffusion compared to passive diffusion. No energy is expended in this process, but it depends on the presence of a concentration gradient to make transport thermodynamically favorable.

facilitation The phenomenon in which passage of an impulse across a synapse renders the synapse more sensitive to successive impulses so increasing the post-synaptic response. Eventually one stimulus will evoke a response large enough to trigger an impulse. *Compare* summation.

FACS *See* fluorescence-activated cell sorter.

F-actin *See* actin.

FAD (flavin adenine dinucleotide) A derivative of riboflavin (*see* flavoprotein) that is a coenzyme in electron-transfer reactions. The reduced form, $FADH_2$, is generated by the Krebs cycle during aerobic respiration, and subsequently is oxidized to FAD by transferring electrons to components of the ELECTRON-TRANSPORT CHAIN located on the inner membrane of mitochondria. *See also* FMN.

fast green *See* staining.

fat A type of LIPID comprising one or a mixture of triacylglycerols (triglycerides) of long-chain carboxylic acids (fatty acids) that are solid below 20°C. They commonly serve as energy storage material in higher animals (*see* adipose tissue) and some plants.

fate map A map of the fates of the various regions of a zygote, embryo or embryonic structure. Different regions are distinguished by color, shading, or numbering to indicate the various tissues or structures they will subsequently give rise to during development. In animals with mosaic development, fate maps can be drawn for the zygote. Here, the different areas of the egg-cell cytoplasm control the developmental fate of embryonic cells from the outset. In animals showing regulative development, the fate of cells can change depending on their surroundings.

fatty acid *See* carboxylic acid.

fatty tissue *See* adipose tissue.

feedback inhibition The inhibition of the activity of an enzyme (often the first) in a reaction sequence by the product of that sequence. When the product accumulates beyond an optimal concentration it binds to a site (allosteric site) on the enzyme, changing the shape so that it can no longer react with its substrate. However, once the product is utilized and its concentration drops again, the enzyme is no longer inhibited and further formation of product results. The mechanism is used to regulate the concentration of certain substances within a cell.

fermentation The breakdown of organic substances, particularly carbohydrates, in the absence of an externally supplied electron acceptor (e.g. oxygen) to yield usable energy in the form of ATP. It is a form of ANAEROBIC RESPIRATION and occurs in certain bacteria, yeasts, some invertebrate animals, and in skeletal muscle cells during prolonged contraction. The most common fermentations involve the partial oxidation of glucose to pyruvic acid by GLYCOLYSIS, thereby generating two molecules (net) of ATP per glucose molecule. In *alcoholic fermentation*, the pyruvic acid is subsequently converted to acetaldehyde (yielding carbon dioxide) and then to ethanol. This reaction, performed by yeasts, is the basis of the brewing and baking industries. In *lactic acid fermentation*, the pyruvic acid is converted directly to lactic acid. This is performed by LACTIC ACID BACTERIA (e.g. in souring of milk) and also by skeletal muscle under the anaerobic conditions found during strenuous exercise. In each case, the conversion of pyruvate is a means of transferring electrons from NADH to an internal electron acceptor (acetaldehyde or pyruvic acid, respectively) to regenerate NAD, which is needed for glycolysis.

ferredoxins   A group of red–brown proteins found in green plants, many bacteria and certain animal tissues. They contain nonheme iron in association with sulfur at the active site. They are strong reducing agents (very negative redox potentials) and function as electron carriers, for example in photosynthesis and nitrogen fixation. They have also been isolated from mitochondria.

fertilization (**syngamy**)   The fusion of a male GAMETE with a female gamete to form a zygote; the essential process of sexual reproduction. Fertilization in animals typically involves penetration of a relatively large nonmotile egg cell (ovum) – the female gamete – by a much smaller motile sperm – the male gamete. In many species this is achieved by means of an ACROSOME REACTION. After penetration the sperm's nucleus swells to form a male pronucleus inside the egg cell. In some cases this then fuses with the female pronucleus, to form the zygote nucleus. In other cases, the

pronuclei dissolve and their chromosomes take part in the first mitotic division of the zygote. Several mechanisms prevent the further entry of sperm (which could lead to abnormal numbers of chromosomes in the zyggote). Immediately after a sperm enters, the egg cell membrane becomes permeable to sodium ions and undergoes depolarization, preventing binding of other sperm. A longer-lasting barrier is created by the release of mucopolysaccharides from CORTICAL GRANULES, leading to the cortical reaction and formation of the fertilization envelope.

In flowering plants there is a *double fertilization*. Following pollination, the pollen tube, with its two sperm cells, penetrates the ovule, and the EMBRYO SAC within. One sperm nucleus fuses with the egg cell to form a diplod zygote. The other sperm nucleus fuses with the two POLAR NUCLEI to form the triploid primary endosperm nucleus, which subsequently divides to form the endosperm tissue.

External fertilization occurs when gametes are expelled from the parental bodies before fusion; it is typical of aquatic animals and lower plants. *Internal fertilization* takes place within the body of the female and complex mechanisms exist to place the male gametes into position. Internal fertilization is necessary for terrestrial animals because the male gametes are typically very small and require external water for swimming towards the female gametes. In addition, the propagules produced on land require waterproof integuments, which would be impenetrable to male gametes, so they must be fertilized before being discharged from the female's body. Internal fertilization also allows a considerable degree of nutrition and protection of the early embryo, which is seen in both mammals and seed plants. As plants are relatively immotile, they are dependent on other agents such as wind or insects to carry the male gamete to the female plant.

fetus (**foetus**)   The EMBRYO of a mammal, especially a human embryo, when its external features resemble those of the mammal after birth, i.e. after it has developed limbs, eyelids, etc. Technically, the term should be restricted to those embryos with an umbilical cord (not a short stalk). In humans it usually refers to the unborn child from after the seventh week of the pregnancy.

Feulgen's stain   *See* staining.

fiber   **1.** (*Botany*) A form of sclerenchyma cell often associated with vascular tissue, primarily to give support. Fibers are long narrow cells, with thickened walls and finely tapered ends, and are dead at maturity. The fibers of many plants (e.g. flax) are economically important.
2. (*Zoology*) A narrow thread of material, usually flexible and having high tensile strength. Examples include the strengthening COLLAGEN fibers in such tissues as skin, cartilage, and tendons; the silk of a spider's web; the fibroin fibers of the horny sponges; and the fibrin fibers formed from fibrinogen at the site of a wound. The elongated cells of muscles and the axons of neurons are also called fibers.

fiber-tracheid   An elongated cell found in wood, intermediate in form between a fiber and a tracheid.

fibroblast   A cell that produces the fibers and other components of the extracellular matrix (ground substance) of CONNECTIVE TISSUE. Usually fibroblasts are long flat cells found alongside the fibers. Fibroblasts are among the most readily cultured of vertebrate cell types, and are commonly used for experimental work in cell biolgy. *See also* cell culture.

fibrocartilage   *See* cartilage.

filamentous bacteria   *See* actinomycetes.

filopodium (*pl.* **filopodia**)   A fine fingerlike extension of a cell. Filopodia occur at the leading edge of motile or spreading cells, such as activated fibroblasts migrating to a wound site. Each one is reinforced internally by longitudinal bundles of actin filaments.

fimbria (*pl.* **fimbriae**)   *See* pilus.

fine structure   *See* ultrastructure.

Fischer, Edmond Henri (1920–) Swiss-born biochemist. He was awarded the Nobel Prize for physiology or medicine in 1992 jointly with E. G. Krebs for their discoveries concerning reversible protein phosphorylation as a biological regulatory mechanism.

Fischer, Emil Hermann (1852–1919) German organic, father of H. O. L. Fischer, noted for his work on a wide range of natural products (including sugars, purines, amino acids, and polypeptides). Fischer determined the configurations of all the aldohexoses and aldopentoses. He also recognized the stereochemical specificity of enzyme action (lock-and-key model). One of the greatest chemists of the nineteenth century, Fischer was awarded the Nobel Prize for chemistry in 1902.

Fischer, Hans (1881–1945) German organic chemist noted who worked on tetrapyrroles. He was awarded the Nobel Prize for chemistry in 1930 for his research on hemin and chlorophyll, especially his synthesis of hemin.

Fischer, Hermann Otto Laurenz (1888–1960) German organic chemist and biochemist, son of Emil H. Fischer, noted for his synthesis of glyceraldehyde 3-phosphate and of glycerone phosphate.

fission   A type of asexual reproduction in which a parent cell divides into two (binary fission) or more (multiple fission) similar daughter cells. Binary fission occurs in many unicellular organisms, including certain protists, bacteria, and yeasts (e.g. the fission yeast *Schizosaccharomyces pombe*). Multiple fission occurs in apicomplexans, such as the malaria parasite *Plasmodium*. Fission begins with division of the nucleus by mitosis, followed by cytoplasmic division and sometimes sporulation.

fixing   In the preparation of microscope slides, the process by which tissue components are stabilized and rendered insoluble to preserve their architecture. For light microscopy, common fixing agents (fixatives) are alcohol (especially ethanol), formaldehyde (methanal), and acetic acid. Formaldehyde, for example, creates cross-links between amino groups and thus binds adjacent proteins and nucleic acids to each other. These reactions also destroy the molecules' biological properties, and thus kill the tissue. As well as preventing deterioration, fixation should also render cell organelles and inclusions more visible and harden the tissue to prevent shrinkage and distortion during DEHYDRATION, EMBEDDING, sectioning, and STAINING. Fixatives used in electron microscopy include glutaraldehyde and osmium tetroxide (which also stains certain cell components).

flaccid Lacking turgor. *See* plasmolysis.

flagellate *See* mastigote.

flagellum (*pl.* **flagella**) A whiplike extension of prokaryote cells with a basal body at its base, whose beat causes locomotion of the cell. Strictly the term is now reserved for the bacterial flagellum. The flagella and cilia of eukaryote cells have a quite different structure and are called undulipodia. Bacterial flagella are much simpler than undulipodia, being hollow cylinders about 15 nm in diameter, consisting of subunits of a protein (flagellin) arranged in helical spirals. Unlike eukaryote flagella they are not membrane-bounded, are rigid, and function by a complex rotation of their bases. *Compare* undulipodium.

flavin A derivative of riboflavin occurring in the flavoproteins; i.e. FAD or FMN.

flavin adenine dinucleotide *See* FAD.

flavin mononucleotide *See* FMN.

flavone *See* flavonoid.

flavonoid One of a common group of plant compounds having the C_6–C_3–C_6 chemical skeleton in which C_6 is a benzene ring. They are an important source of nonphotosynthetic pigments in plants, including the yellow chalcones and aurones; the pale yellow and ivory flavones and flavonols and their glycosides; the red, blue, and purple anthocyanins and anthocyanidins; and the colorless isoflavones, catechins, and leukoanthocyanidins. They are water soluble and usually located in the cell vacuole.

flavonol *See* flavonoid.

flavoprotein A conjugated protein in which a flavin nucleotide (FAD or FMN) is joined to a protein component. Flavoproteins are found in virtually all cells, acting as enzymes or as ELECTRON CARRIERS in the electron-transport chain.

Fleming, (Sir) Alexander (1881–1955) British bacteriologist famous for his discovery of the bacteriolytic agents lysozyme and penicillin. He was awarded the Nobel Prize for physiology or medicine in 1945 jointly with E. B. Chain and H. W. Florey.

Florey, (Sir) Howard Walter, Lord Florey (1898–1968) Australian physiologist noted for his role in the development of penicillin (with E. B. CHAIN) and cephalosporin C. He was awarded the Nobel Prize for physiology or medicine in 1945 jointly with E. B. Chain and A. Fleming for the discovery of penicillin.

flow cytometry Any automated technique for counting or sorting cells or cell components. The underlying principle of such techniques is that the items to be counted are forced to pass through a narrow aperture in single file, where they are detected according to the presence or absence of some label, such as a fluorescent marker, and counted automatically (e.g. by a computer). *See* fluorescence-activated cell sorter.

fluid mosaic model The generally accepted model of the structure of cell membranes. It consists essentially of a double layer (bilayer) of phospholipids, within which are various proteins. The lipids and many of the proteins diffuse laterally through the membrane, exchanging places with other molecules. Not all membrane proteins are freely mobile. For example, in the PLASMA MEMBRANE, some proteins are fixed to the microscopic fibers of the underlying cytoskeleton.

fluorescence-activated cell sorter (FACS) A form of flow cytometry in which cells are sorted and counted according to whether or not they have a specific fluorescent label attached. It is commonly used to separate different types of lymphocytes in blood samples. The cells pass in single file through a beam of laser light, and any fluorescence is measured by a detector. Tiny droplets, each containing a single cell, form at a nozzle, and each droplet is given an electric charge proportionate to the amount of fluorescence detected. Deflector plates sort the droplets, and hence cells, according to their charge.

fluorescence microscopy A form of microscopy that highlights certain cells, structures, or molecules by means of fluorescent dyes and a fluorescence microscope. It can be applied to living cells or prepared specimens, and the dyes can be linked to antibodies that bind specifically with particular cell components, thus revealing their location in the specimen. A fluorescent substance absorbs light at one particular wavelength (the excitation wavelength) and emits light at another wavelength. In the fluorescence microscope the illuminating light is filtered to obtain the desired excitation wavelength. This light is deflected by a special mirror and focused by the objective lens onto the specimen. When illuminated, any fluorescent dye in the specimen emits light of a particular color – i.e. it fluoresces (e.g. rhodamine dye emits red light). This emitted light is focused by the objective lens and passes through a filter that blocks all wavelengths except that of the fluorescence, before the final image is formed at the eyepiece.

fluorine *See* trace element.

FMN (flavin mononucleotide) A derivative of riboflavin that is a coenzyme for various electron carriers in electron-transfer reactions. It is related to FAD, but lacks an adenosine group, and has only one instead of two phosphate groups. *See also* flavoprotein.

foetus *See* fetus.

follicle A ball of cells surrounding a small cavity or sac within an organ or tissue. Follicles within the ovary, for example, contain developing ova. *See* ovarian follicle.

follicle-stimulating hormone (FSH) A gonadotropin, also called follitropin, produced by the anterior pituitary gland. It acts on the ovary to stimulate the growth and maturation of the tissues forming follicles and ova, which, under the action of luteinizing hormone, mature and are released from the ovary. It also stimulates spermatogenesis in males. It has been used in the treatment of female sterility.

frameshift mutation *See* mutation.

Franklin, Rosalind Elsie (1920–58) British biophysicist who worked on the structure of DNA using x-ray crystallography. Her results contributed to the ideas of J. D. Watson and F. H. C. Crick concerning their double-helix structure for DNA.

fraternal twins (dizygotic twins) Two offspring born to the same mother at the same birth, resulting from the fertilization of two eggs at the same time. They may be of unlike sex and are no more genetically similar than any two siblings. *Compare* identical twins.

freeze etching *See* freeze fracturing.

freeze fracturing A method of preparing material for electron microscopy, particularly useful for studying membranes. Material is frozen rapidly (e.g. by immersion in liquid nitrogen) thus preserving it in lifelike form. It is then fractured, usually with a sharp knife. The fracture plane tends to follow lines of weakness, such as between the two lipid layers of membranes, revealing their internal surfaces. The specimen then undergoes SHADOWING before being examined in the electron microscope. Freeze fracturing is especially useful for revealing the size, shape, and location of integral membrane proteins, which tend to stick to one or other of the lipid layers. In *freeze etching* the fractured surface is etched, i.e. some of the surface ice is allowed to sublime away in a vacuum, before shadowing. This exposes further structure, such as the outer surface of the membrane.

freezing microtome *See* microtome.

fructose A sugar ($C_6H_{12}O_6$) found in fruit juices, honey, and cane sugar. It is a ketohexose, existing in a pyranose form when free. In combination (e.g. in sucrose) it exists in the furanose form.

FSH *See* follicle-stimulating hormone.

fucoxanthin A xanthophyll pigment of diatoms, brown algae, and golden brown algae. The light absorbed is used with high efficiency in photosynthesis, the energy first being transferred to chlorophyll *a*. It has three absorption peaks covering the blue and green parts of the spectrum.

Fungi A kingdom of nonphotosynthetic mainly terrestrial organisms that includes molds, mushrooms, and yeasts. Fungi are now regarded as quite distinct from plants or other living kingdoms. They are characterized by having cell walls made chiefly of chitin, not the cellulose of plant cell walls, and they all develop directly from spores without an embryo stage. Moreover, undulipodia (cilia or flagella) are never found in any stage of their life cycles. Fungi are generally saprophytic or parasitic, and may be unicellular (e.g. yeasts) or composed of filaments (hyphae) that together comprise the fungal body or MYCELIUM. *See* hypha.

furanose A SUGAR that has a five-membered ring (four carbon atoms and one oxygen atom). *Compare* pyranose.

Furchgott, Robert F. (1916–) American biochemist who was awarded the 1998 Nobel Prize for physiology or medicine jointly with L. J. Ignarro and F. Murad for work on nitric oxide acting as a signaling molecule in the cardiobascular system.

G

G-actin   *See* actin.

GAG   *See* glycosaminoglycan.

galactose   A SUGAR found in lactose and many polysaccharides. It is an ALDOHEXOSE, isomeric with glucose.

gamete   A cell capable of fusing with another cell to produce a ZYGOTE, from which a new individual organism can develop. The female and male gametes of a particular species can have similar structure and behavior (*isogametes*), as in many simple organisms, but they are usually dissimilar in appearance and behavior (*anisogametes*). The typical female gamete (*see* ovum) is large because of the food reserves it contains, immotile, and is produced in small numbers. The typical male gamete (e.g. a SPERMATOZOON) is small, motile, and produced in large numbers. Fusion of gametes (*see* fertilization) results in the nucleus of the zygote having exactly twice the number of chromosomes present in the nucleus of each gamete.

gametogenesis   The formation of sex cells or gametes, i.e. ova and spermatozoa. *See* oogenesis; spermatogenesis.

ganglion   A collection of nerve cell bodies, usually bound by a sheath or capsule. In vertebrates the ganglia are located chiefly outside the central nervous system; in invertebrates ganglia occur along the major nerve cords and are the centers of nervous integration.

gap junctions   A type of cell junction found in closely packed tissue cells of animals. The narrow intercellular gap, some 2–3 nm wide, is traversed by arrays of proteins called *connexons*. These enclose cylindrical water-filled channels that connect the cytosol of adjacent cells, permitting the passage of ions and small molecules (with molecular weights up to about 2000). These channels are an important means of communication and coordination between tissue cells. For example, chemical signals such as cyclic AMP and calcium ions produced in one cell can diffuse to neighboring cells, thereby transmitting the signal throughout the tissue. This is important in, for example, coordinating the contractions of smooth muscle cells.

gas chromatography   A technique widely used for the separation and analysis of mixtures whose components are either gaseous or vaporize relatively easily. Gas chromatography employs a column packed with either a solid stationary phase (*gas–solid chromatography* or *GSC*) or a solid coated with a nonvolatile liquid (*gas–liquid chromatography* or *GLC*). The whole column is placed in a thermostatically controlled heating jacket, usually in the range 50–300°C. A volatile sample is introduced into the column using a syringe, and an unreactive carrier gas, such as nitrogen, passed through it. The components of the sample will be carried along in this mobile phase at different times. The emergent sample is passed through a detector, which registers the presence of the different components in the carrier gas. Once separated, the components can be identified by means of a mass spectrometer.

gas–liquid chromatography (GLC)   *See* gas chromatography.

gas–solid chromatography (GSC)   *See* gas chromatography.

gastrula *See* gastrulation.

gastrulation The process by which the cells of an early animal embryo (blastula) are rearranged into the positions appropriate for continuing development. The resultant embryological stage, called a *gastrula*, has bilateral symmetry, and is enclosed by distinct layers of cells, called GERM LAYERS. These layers consist of ectoderm on the outside and endoderm on the inside, with usually mesoderm between them. During gastrulation, cell division and migration (morphogenetic movements) bring about remodeling of the blastula, and enable differentiation of tissues and formation of organ systems to proceed.

The nature of gastrulation depends on the type of egg and blastula. In birds, reptiles, and mammals, for example, the germ layers develop from an embryonic disk (or epiblast) at the animal pole of the blastula. Cells migrate from all sides within the dorsal surface of the disk towards the midline, where they sink inward (ingress) and form a longitudinal groove – the PRIMITIVE STREAK – in the disk's surface. Within the disk, the cells become organized into mesoderm and endoderm; cells remaining on the dorsal surface become the ectoderm. In other animals, part of the blastula wall folds inward, eventually forming a cavity (the ARCHENTERON) that obliterates the existing cavity of the blastula (the BLASTO-COEL) and communicates with the outside via a pore (the BLASTOPORE). In amphioxus, the blastopore ultimately becomes the anus, whereas in annelids, mollusks, and many arthropods, it forms the mouth. During gastrulation in amphibians, cells in the surface of the animal half, or animal hemisphere, migrate inside the hollow blastula to form mesoderm. This inward migration occurs at the dorsal lip of the blastopore. Endoderm develops from yolky cells of the vegetal hemisphere. Later the blastopore closes, and the anus develops from another invagination nearby. The primitive streak (see above) is considered to be functionally equivalent to the blastopore. *See also* organizer.

gel A semi-solid, easily deformable jelly-like mass, such as that formed when an aqueous solution of gelatin is allowed to cool. Other substances that form gels with water include agar, starch, and acrylamide (which forms polyacrylamide gel). Gels may be subdivided into elastic gels (e.g. gelatin) and rigid gels (e.g. silica gel). Various laboratory techniques employ gels as media for separating and characterizing biomolecules, including gel electrophoresis and gel filtration chromatography.

gel filtration chromatography (gel-permeation chromatography) A chromatographic method using a column packed with porous gel particles. It is a standard technique used for separating and identifying macromolecules of various sizes, including proteins or nucleic acids. A solution of the mixture of macromolecules is added to the top of the column and allowed to flow through by gravity. The smaller molecules are hindered in their passage down the column because they are better able to penetrate the hydrated pores within the particles of the gel. Molecules too large to penetrate the pores are excluded, and thus flow more rapidly through the column. By analyzing the liquid that drips from the bottom of the column (the eluate) at set intervals and comparing it with a standard (obtained by running a known macromolecule through the column) information about the sizes and molecular weights of the components of the mixture is gathered. The most frequently used commercial gel is Sephadex.

gel-permeation chromatography *See* gel filtration.

gemmation *See* budding.

gene In classical genetics, a unit of hereditary material located on a chromosome that, by itself or with other genes, determines a characteristic in an organism. It corresponds to a segment of the genetic material, usually DNA (although the genes of some viruses consist of RNA). Genes can exist in a number of alternative forms, termed *alleles*. In a normal diploid cell only two alleles can be present together, one on each of a pair of homologous chromosomes: the alleles may both be of the same type, or they may be different. The segregation of alleles at meiosis and their dominance relationships are responsible for the particulate nature of inheritance. Genes can occasionally undergo changes, called MUTATIONS, to new allelic forms.

A gene can be defined in several different ways. Geneticists have classically investigated genes by observing inheritance patterns in sexually reproducing organisms, and so genes were defined in terms of their effects on phenotypic characteristics. Hence, a gene could be defined as the smallest hereditary unit capable either of recombination or of mutation or of controlling a specific function. These three definitions do not necessarily describe the same thing. In molecular terms, a gene is the entire nucleic acid sequence that is required for the cell to synthesize a given polypeptide. This definition includes not only the coding region, or *cistron*, which is transcribed into RNA, but other transcription-control regions, such as enhancers. These may lie a considerable distance from the coding region, but are nonetheless essential for gene function. A single function can be determined by one or more cistrons, depending on the number of polypeptides involved.

Although the DNA molecules of the chromosomes account for the great majority of genes, genes are also found in certain cell organelles in eukaryotes, particularly mitochondria (*see* mitochondrial DNA) and plastids.

There are essentially three types of gene: (i) *structural genes*, which code for polypeptides of enzymes and other proteins; (ii) *RNA genes*, which code for ribosomal RNA and transfer RNA molecules used in polypeptide assembly; and (iii) *regulator genes*, which regulate the expression of the other two types. *See also* operon.

gene family A set of genes with similar structures that encode proteins with similar but nonidentical amino-acid sequences. Such genes often lie close to each other on the same chromosome, probably having arisen by duplication of a single ancestral gene. Over time the duplicate genes acquire mutations that alter the amino-acid sequence and structure of the ancestral protein, so changing its functional capabilities. This results typically in a cluster of closely linked genes whose proteins have evolved distinct but related functions. The classic example is the human β-globin gene family, whose proteins are components of the blood pigment hemoglobin. This family comprises a cluster of five functional genes and two nonfunctional PSEUDOGENES on chromosome 11. The five different β-globins encoded by this gene family have distinct properties that suit them for different physiological roles.

gene knockout A technique for selectively inactivating a certain gene within a living cell or organism, by replacing the normal alleles with mutant nonfunctional alleles. It is applied to experimental organisms such as yeasts and mice, in order to study how normal function is disrupted by loss (i.e. 'knockout') of the gene. A specific gene is cloned, and then inactivated by making precise changes in its sequence using a technique called *in vitro mutagenesis*. The inactivated gene is incorporated into a suitable targeting vector containing marker genes, and introduced to normal cells, for example, by ELECTROPORATION or direct injection into the nucleus. In a small fraction of diploid cells, the gene becomes incorporated into the chromosome at its target site by a process called homologous recombination, producing a heterozygote with one normal allele and one mutant allele. The marker genes enable selection of cells containing the inactivated gene at the correct target site.

Gene knockout is relatively easy to perform in single-celled organisms, such as yeasts, and the effects of the knockout can be readily assessed by inducing the diploid heterozygotes to undergo meiosis. Half the haploid spores from such a cell will contain the knockout allele alone, and the viability of these spores reflects the functional importance of the inactivated gene. In multicellular organisms, like the mouse, gene knockout requires a more complex procedure, but nonetheless provides a powerful tool for studying the roles of specific genes in development, physiology, disease, etc. In the mouse, the knockout gene is incorporated into its target site in embryonic stem cells. Successfully targeted cells are introduced to a second embryo from a different mouse strain. This embryo then develops inside a surrogate mother, to produce a chimera – i.e. a newborn mouse comprising cells from both strains. When mature, the chimeric mouse is used to breed generations of progeny mice, some of which are homozygous for the knockout gene. Knockout mice are especially valuable as model systems for studying various human genetic diseases, such as cystic fibrosis.

gene library *See* DNA library.

gene mutation *See* mutation.

gene probe *See* DNA probe.

generation time The average time between the cell division of parent and daughter cells within a population of cells.

gene splicing **1.** The joining of exons after the intron sequences have been removed, to produce functional messenger RNA. In the nucleus this is performed by a special assemblage of RNA and proteins called a SPLICEOSOME.
2. (DNA splicing) In genetic engineering, the joining of DNA fragments by the action of the enzyme DNA LIGASE.

genetic code The sequence of bases in either DNA or messenger RNA (mRNA) that conveys genetic instructions to the cell. The basic unit of the code consists of a group of three consecutive bases, the base triplet or CODON, which specifies instructions for a particular amino acid in a polypeptide, or acts as a start or stop signal for translation of the message into polypeptide assembly. For example, the DNA triplet CAA (which is transcribed as GUU in mRNA) codes for the amino acid valine. There are 64 different triplet combinations but only 20 amino acids; thus many amino acids can be coded for by two

or more triplets. The code is said to be *degenerate*, since it appears that only the first two bases, and in certain cases only one base, are necessary to insure the coding of a specific amino acid. Three triplets, termed 'nonsense triplets', do not code for any amino acid and have other functions, for example, marking the beginning and end of a polypeptide chain.

genetic engineering (recombinant DNA technology) The direct introduction of foreign genes into an organism's genetic material by micromanipulation at the cell level. Genetic engineering techniques bypass crossbreeding barriers between species to enable gene transfer between widely differing organisms. Gene transfer can be achieved by various methods, many of which employ a replicating infective agent, such as a virus or plasmid, as a vector (*see* DNA cloning). Other methods include microinjection of DNA into cell nuclei and direct uptake of DNA through the cell membrane. Recognizing whether or not transfer has occurred may be difficult unless the new gene confers an obvious visual or physiological characteristic. Consequently the desirable gene may be linked to a MARKER GENE, for example, a gene conferring resistance to an antibiotic in the growth medium. The transferred gene must also be linked to appropriate regulatory DNA sequences to insure that it works in its new environment and is regulated correctly and predictably.

Initial successes in DNA transfer were achieved with bacteria and yeast. Human genes coding for medically useful proteins have been transferred to bacteria. Human insulin, growth hormone, and interferon are now among a wide range of therapeutic substances produced commercially from genetically engineered bacteria. Genetically engineered vaccines have also been produced by transfer of antigen-coding genes to bacteria.

Modified microorganisms are grown in large culture vessels and the gene product harvested from the culture medium. However, the problems associated with scaling up laboratory systems are still limiting the exploitation of genetic engineering. Genetic manipulation of higher animals and plants has been achieved more recently. Transgenic mammals, including mice, sheep, and pigs, have been produced by microinjection of genes into the early embryo, and it is also now possible to clone certain mammals from adult body cells (*see* clone). Such technology may have considerable impact on livestock production, for instance, by injection of growth hormone genes. Dicotyledonous plants, including tobacco and potato, have been transfected using the natural plasmid vector of the soil bacterium *Agrobacterium tumefaciens* (*see* *Agrobacterium*). Genes have been introduced to crop plants for various reasons, for instance to reduce damage during harvest or to make them resistant to the herbicides used in controlling weeds. Tomatoes, soya beans, corn, and cotton are among the genetically modified crops now widely grown. There is also hope that in future many genetic diseases will be treatable by manipulating the faulty genes responsible. However, genetic engineering raises many legal and ethical issues, and the introduction of genetically modified organisms into the environment requires strict controls and monitoring. *See also* recombinant DNA.

genetic map   *See* chromosome map.

genetic marker   A character or chromosomal site (locus) that can be used as a signpost to track closely linked genes or DNA sequences. Any marker must exist in at least one form (preferably more) that is distinct from the normal (wild-type) condition. Markers are used in establishing inheritance patterns or in chromosome mapping, for example, and there are several types. *Phenotypic markers* are variants of genetically determined traits, such as curly wing shape or white eye color in fruit flies (*Drosophila*), whose inheritance can be easily assessed. They serve as labels to flag the inheritance of the determining alleles. Since the early 1900s, numerous examples have been described in various experimental organisms, and used in recombinational analysis to yield linkage maps of chromosomes (*see* chromosome map). The advent of modern analytical techniques in the 1980s led to the discovery of a vast collection of *molecular markers*. These are variations in DNA sequence that have no apparent effect on the phenotype of the organism concerned, but can be detected by biochemical analysis. The main examples are RESTRICTION FRAGMENT LENGTH POLYMORPHISMS, which result from silent mutations at cleavage sites for restriction enzymes; and VARIABLE NUMBER TANDEM REPEATS – variations in the number of certain tandemly repeated DNA sequences. Molecular markers have filled in the large gaps between existing phenotypic markers, and provide the signposts needed for fine-structure mapping.

genetics   The term coined by Bateson to describe the study of inheritance and variation and the factors controlling them. Today the subject has three main subdivisions – Mendelian (classical) genetics, population genetics, and molecular or biochemical genetics.

genome   All the genetic information contained within a cell or organism. It occurs as DNA – except in certain viruses that have RNA genomes – and is packaged into chromosomes.

genomic DNA library   *See* DNA library.

genomics   The branch of molecular genetics concerned with the study of genomes. It has emerged since the 1980s as one of the fastest-growing fields in all biology, following the development of automated techniques for nucleic acid and protein sequencing, and computerized systems for handling and analyzing the resultant data (*see* bioinformatics). It can be divided into three main areas. *Structural genomics* deals with determining the DNA sequence of all the genetic material of a particular organism. Such a definitive physical map of the genome is the ultimate objective of all genome projects. *Functional genomics* focuses on characterizing all the messenger RNAs produced by transcription of an organism's genome, and the polypeptides that they encode. Complementary DNAs can be constructed from such transcripts, and compared with the genome sequence data to identify corresponding coding regions. *Comparative genomics* is concerned with comparing the genomic sequences of different species, to see how genomes change in the course of evolution, and which parts remain unchanged. The presence of such highly conserved sequences sheds light on their functional significance, and helps in making predictions about similar sequences in other organisms.

genotype The genetic make-up of an organism. The actual appearance of an individual (the phenotype) depends on the dominance relationships between alleles in the genotype and the interaction between genotype and environment.

geotropism *See* gravitropism.

germ cell Any of the cells in animals that give rise to the gametes.

germination The first outward sign of growth of a reproductive body, such as a spore or pollen grain. The term is most commonly applied to seeds, in which germination involves the emergence of the radicle or coleoptile through the testa. Both external conditions (e.g. water availability, temperature, and light) and internal biochemical status must be appropriate before germination can occur.

germ layers The three major body layers – ECTODERM, MESODERM, and ENDODERM – that develop in the embryos of most animals during GASTRULATION. These layers do not include special cells or groups of cells that may be migratory (e.g. neural crest cells of vertebrates) or perform special functions (e.g. germ cells).

germ line The lineage of cells from which gametes arise, continuous through generations.

germ plasm 1. The hereditary material that, according to Weissmann in the 1880s, is transmitted unchanged from parents to offspring, to form the germ cells of successive generations. It is now known that this material is DNA in the chromosomes. Plant breeders have established germ plasm collections, mainly in the form of seeds, to conserve the genetic material of rare plant species and varieties.
2. The special cytoplasm of the eggs of most animals that becomes the germ cells when provided with nuclei. It lies at one end of the eggs of insects, under the gray crescent of amphibian eggs, and in the endodermal area of amniotes. In insect eggs, for example, the germ plasm at the posterior pole contains substances that determine the fate of nuclei migrating to that region of the egg; such nuclei subsequently develop into germ cells.

giant chromosome *See* polytene.

giant fiber (**giant axon**) A nerve fiber (axon) that has a relatively large diameter, enabling the rapid conduction of a nerve impulse. Giant fibers occur in many invertebrate groups and usually supply the muscles used in a protective response, such as the end-to-end contraction in earthworms.

gibberellic acid (GA_3) A common gibberellin and one of the first to be discovered. Together with GA_1 and GA_2 it was isolated from *Gibberella fujikuroi*, a fungus that infects rice seedlings causing abnormally tall growth. *See* gibberellin.

gibberellin Any of a group of plant hormones involved chiefly in shoot extension. Gibberellins are diterpenoids; their molecules are based on the *ent*-gibberellane skeleton. More than 30 have been isolated, the most common being GIBBERELLIC ACID (GA_3). Gibberellins stimulate elongation of shoots of various plants, especially the extension to normal size of the short internodes of genetically dwarf pea or maize plants. Increased gibberellin levels can mimic or mediate the effect of long days. Thus they stimulate internode extension and flowering in long-day plants such as lettuce and spinach. Synthesis of α-amylase and certain other hydrolytic enzymes in barley aleurone layers is regulated by gibberellin produced by the embryo. This initiates germination by mobilizing endosperm food reserves. Gibberellins may be produced in both shoots and roots and travel in both xylem and phloem.

Gilbert, Walter (1932–) American molecular biologist. He was awarded the Nobel Prize for chemistry in 1980 jointly with F. Sanger for their contributions concerning the determination of base sequences in nucleic acids. The prize was shared with P. Berg.

Gilman, Alfred Goodman (1941–) American pharmacologist. He was awarded the Nobel Prize for physiology or medicine in 1994 jointly with M. Rodbell for their discovery of G-proteins.

gland 1. (*Zoology*) An organ that synthesizes a specific chemical substance and secretes this either through a duct into a tubular organ or onto the surface of the body or directly into the bloodstream.
2. (*Botany*) A specialized cell or group of cells concerned with the secretion of various substances produced as by-products of plant metabolism. The secretions may pass to the exterior or be contained in the plant body. For example, the hydathodes of leaves exude a watery solution onto the surface of the leaf in the process termed guttation. Glandular hairs develop from the epidermis of many plants, for example, stinging nettle and geranium.

GLC *See* gas chromatography.

glia *See* neuroglia.

globulin One of a group of globular proteins that are insoluble in water but will dissolve in neutral solutions of certain salts. Three types of globulin are found in blood – *alpha* (α), *beta* (β), and *gamma* (γ) – which can be separated by electrophoresis. α- and β-globulins are made in the liver and are used to transport nonprotein material. γ-globulins are made in reticuloendothelial tissues, lymphocytes, and plasma cells and most of them have antibody activity (*see* immunoglobulin).

glucagon A polypeptide hormone produced by the A-cells of the islets of Langerhans of the pancreas in response to somatotropin. Its action opposes that of INSULIN, causing an increase in blood glucose by promoting the breakdown of glycogen to glucose in the liver.

glucans *See* glycan.

glucocorticoid A type of steroid hormone produced by the adrenal cortex. Glucocorticoids (e.g. corticosterone and hydrocortisone) accelerate the formation of glucose from protein (gluconeogenesis) and the breakdown of glycogen. They also inhibit inflammation, for example by depressing T-cell activity. *See also* corticosteroid.

gluconeogenesis A metabolic process by which glucose or other carbohydrates can be manufactured from lactic acid, glycerol, fatty acids, or noncarbohydrate precursors, such as amino acids. In animals it takes place in the liver. During prolonged fasting it is the route by which amino acids derived from the breakdown of muscle proteins are converted into usable energy. The pathway is stimulated by the hormone glucagon. Forms of gluconeogenesis also occur in plants and microorganisms; for example, during the mobilization of fat reserves in germinating seeds (*see* glyoxylate cycle). The reactions are essentially those of GLYCOLYSIS in reverse, starting from phosphoenolpyruvate.

glucose (**dextrose**) A crystalline six-carbon (hexose) sugar found widely in nature both as the free monosaccharide D-glucose, and combined in various polysaccharides, notably in glycogen, cellulose, and starch. Glucose is one of the main sources of energy for living cells, being metabolized initially via GLYCOLYSIS, and is transported in blood. D-glucose can exist in three forms: two different ring forms (the abundant D-glucopyranose and the rare D-glucofuranose) and a linear form.

glutamic acid *See* amino acids.

glutamine *See* amino acids.

glutathione A tripeptide of cysteine, glutamic acid, and glycine, widely distributed in living tissues. It takes part in many oxidation–reduction reactions, due to the reactive thiol group (–SH) being easily oxidized to the disulfide (–S–S–), and acts as an antioxidant, as well as a coenzyme to several enzymes.

glycan A polysaccharide; any polymer made up of more than about 10 monosaccharide (sugar) units joined by glycosidic bonds (*see* glycoside). *Homoglycans* consist of a single type of sugar (e.g. glucose) whereas *heteroglycans* comprise more than one sugar. Glycans serve both as structural units (e.g. cellulose in plants and chitin in invertebrates) and energy stores (e.g. starch in plants and glycogen in animals). The most common homoglycans are made up of D-glucose units and called *glucans*. *See* polysaccharide.

glyceride (**acylglycerol**) Any ester of glycerol and one or more fatty acids. They may be mono-, di-, or triglycerides according to the number of –OH groups esterified. The fat stores of the body consist mainly of triglycerides. These can form a source of energy when carbohydrate levels are low, being broken down by lipases into fatty acids, which can enter metabolic pathways. *See also* lipid.

glycerin *See* glycerol.

glycerol (**glycerin**) 1,2,3-propanetriol; an alcohol with three OH groups. Glycerol is biologically important as the alcohol involved in lipid formation (these particular lipids being called GLYCERIDES).

glycine *See* amino acids.

glycogen (**animal starch**) A polysaccharide that is the main carbohydrate store of animals. It is composed of many glucose units linked in a similar way to starch, and is readily hydrolyzed to glucose. It is widely distributed in cells but is stored largely in the liver and in muscle. After a meal, most of the glucose contained in food is absorbed via the intestine and blood and converted to glycogen in the liver (*glycogenesis*). The concentration of glucose in the blood is then normally regulated by conversion of glycogen back to glucose (*glycogenolysis*).

glycogenesis *See* glycogen.

glycogenolysis *See* glycogen.

glycolate cycle *See* photorespiration.

glycolysis (**Embden–Meyerhof pathway**) The conversion of glucose into pyruvate, with the release of some energy in the form of ATP. Glycolysis occurs in the cell cytosol. In ANAEROBIC RESPIRATION, breakdown proceeds no further and pyruvate is converted into ethanol or lactic acid for storage or elimination. In AEROBIC RESPIRATION, glycolysis is followed by the Krebs cycle. Glycolysis alone yields only two molecules of ATP (net) per molecule of glucose in anaerobic respiration. In aerobic respiration there is a net yield of six ATP (the conversion of NADH back to NAD yields a further four ATP molecules, and can occur only when oxygen is present to function as electron acceptor). *See also* fermentation.

glycoprotein A protein that is covalently bound to one or more monosaccharide units. The bound monosaccharide can be single, but is more commonly the end unit of an oligosaccharide or small polysaccharide. Certain antigens, enzymes, and hormones are glycoproteins, as are some components of the extracellular matrix. *Compare* proteoglycan.

glycosaminoglycan (**GAG**) One of a group of compounds, sometimes called *mucopolysaccharides*, consisting of long unbranched chains of repeating disaccharide sugars; one of the two sugar residues is an amino sugar – either N-acetylglucosamine or N-acetylgalactosamine – and the other sugar is a uronic acid. Sulfate groups occur in one or both sugars. These compounds are present in connective tissue; they include heparin and hyaluronic acid. Most glycosaminoglycans are linked to protein to form PROTEOGLYCANS (sometimes called *mucoproteins*).

glycoside A derivative of a pyranose sugar (e.g. glucose) in which there is a group attached to the carbon atom that is joined to the –CHO group. In a glycoside the C–OH is replaced by C–OR. The linkage –O– is a *glycosidic link*; it is the link joining monosaccharides in GLYCANS (polysaccharides).

glycosidic link *See* disaccharide; glycoside.

glyoxylate cycle A series of reactions, involving the formation of glyoxylate and certain steps of the KREBS CYCLE, that occurs in some microorganisms, algae, and plants. It is found especially in regions where fats are being rapidly metabolized, for example in germinating fat-rich seeds. In plants, part of the cycle is performed by enzymes contained in cell organelles called glyoxysomes (*see* microbody). The cycle essentially converts the insoluble fatty acids into sucrose, which is soluble and can be easily transported from storage cells to actively metabolizing cells in the seedling. Within a glyoxysome, acetyl groups formed from the fatty acids, are combined with oxaloacetate (from the Krebs cycle in the mitochondrion) to form citrate. The following reactions generate malate (via glyoxylate), and succinate. The latter returns to the mitochondrion and the Krebs cycle, to regenerate oxaloacetate. The malate enters the cytosol and is converted (via oxaloacetate) to phosphoenolpyruvate, which undergoes reverse glycolysis (gluconeogenesis) to yield, ultimately, sucrose.

glyoxysome *See* microbody.

goblet cell A cell that secretes mucus onto a surface or into a cavity. In columnar epithelium some cells produce mucus as a droplet, which enlarges until it distends the upper part of the cell, giving the cell the appearance of a wine glass or goblet. The mucus is discharged from the apical surface of the cell to lubricate and protect the epithelial surface. Goblet cells are found in the lining of the alimentary, reproductive, and respiratory tracts, and in the skin of such animals as earthworms.

Goldstein, Joseph Leonard (1940–) American physician and molecular geneticist notable for his discovery of cellular receptors for low-density lipoproteins and their role in the removal of cholesterol from the bloodstream. He was awarded the Nobel Prize for physiology or medicine in 1985 jointly with M. S. Brown.

Golgi, Camillo (1843–1926) Italian anatomist, cytologist, neurologist, and pathologist noted for his studies of the central nervous system and, in particular, for his discovery of the cellular organelle named for him. He was awarded the Nobel Prize for physiology or medicine in 1906 jointly with S. Ramón y Cajal for their work on the structure of the nervous system.

Golgi apparatus *See* Golgi complex.

Golgi complex (**Golgi apparatus**) An organelle of eukaryotic cells discovered by Camillo Golgi in 1898, and sometimes called a dictyosome in plant cells. It consists of stacks of flattened membrane-bounded sacs (cisternae) associated with vesicles, and is concerned with the modification of newly synthesized proteins, and packaging of proteins and other materials for consignment to their destinations in the cell, especially to the plasma membrane for secretion. The cisternae can occupy any of three functionally distinct positions in the Golgi stack, called *cis-*, *medial-* and *trans-*Golgi. Proteins synthesized in the ENDOPLASMIC RETICULUM enter into transport vesicles, which move to the *cis-*Golgi and fuse either with an existing cisterna or with each other to form a new cisterna. Each cisterna migrates through the complex in the *cis-* to *trans-*Golgi direction. As it does so, the proteins are modified by enzymes within the cisterna, especially by addition of sugar groups to form glycoproteins. On reaching the *trans*-Golgi, they are sorted into vesicles (*Golgi vesicles*) for either immediate secretion, for storage, or for transport to lysosomes. Processes involving the Golgi apparatus include formation of zymogen granules; synthesis and transport of secretory polysaccharides (e.g. cellulose in cell plate formation or secondary wall formation and mucus in goblet cells); and packaging of hormones in nerve cells carrying out neurosecretion. *See illustration at* cell.

Golgi vesicle   *See* Golgi complex.

gonad   The reproductive organ of animals. It produces the sex cells (gametes) and sometimes hormones. The female gonad, the ovary, produces ova; the male gonad, the testis, produces spermatozoa.

gonadotropic hormone   *See* gonadotropin.

gonadotropin (**gonadotropic hormone**) A hormone that acts on the gonads (ovary and testis). The *pituitary gonadotrophins* are follicle-stimulating hormone, luteinizing hormone, and prolactin. They are involved in the initiation of puberty, regulation of the menstrual cycle, and lactation in females, and in the control of spermatogenesis in males. They are used in the treatment of infertility. *See also* chorionic gonadotrophin.

G protein   Any of a class of proteins that play a crucial role in relaying signals from certain types of cell-surface receptors (*G protein-coupled receptors*) to activate signal pathways inside the cell. G proteins are involved in signal transduction from receptors for various hormones and neurotransmitters, as well as light-activated receptors in the eye and odor receptors in the nose. Essentially, binding of the ligand (hormone, neurotransmitter, etc.) to the exterior region of the receptor changes its interior (cytosolic) region, which activates an associated G protein. The activated G protein in turn might activate or inhibit a

further enzyme that produces a SECOND MESSENGER, or change the permeability of certain ion channels and alter membrane potential. For example, in epinephrine and glucagon receptors, the activated G protein binds to and activates the enzyme adenylyl cyclase, which catalyzes the synthesis of the second messenger cyclic AMP (cAMP).

Graafian follicle   *See* ovarian follicle.

graft   The transplantation of an organ or tissue in plants and animals. In plants, grafting is an important horticultural technique in which part (the scion) of one individual is united with another of the same or different species. Usually the shoot or bud of the scion is grafted onto the lower part of the stock. Incompatibility between species is much less likely to occur in plants than in animals.

In animals, a graft is a transplantation of an organ or tissue, either on the same individual or on different individuals (i.e. from a donor to a recipient). Immune defenses of the recipient recognize a graft of nonself ('foreign') tissue and tend to cause its rejection. A successful graft requires matching of the antigenic markers carried by cells of donor and recipient (*see* HLA system); close relatives are most likely to be compatible. Grafts may be from one place to another on the same individual (*autograft*) or between different individuals. A graft between individuals of the same species is a *homograft* (or *allograft*); between individuals that are genetically identical (as between identical twins) it is an *isograft* (or *syngraft*); and between different species it is a *heterograft* (or *xenograft*).

Gram negative   *See* Gram's stain.

Gram positive   *See* Gram's stain.

Gram's stain   A stain containing crystal violet and safranin used for bacteria. It is the basis for the division into *Gram positive* and *Gram negative* bacteria. The former retain the deep purple color of crystal violet; the latter are counterstained red with safranin. This differential staining is due to differences in the composition of the cell walls between the two groups. Gram-positive bacteria, such as *Staphylococcus aureus*, have a thick cell wall composed mainly of peptidoglycan, a sheet-like molecule comprising sugar and amino acid subunits. The cell wall of Gram-negative bacteria is more complex, with an additional outer layer of lipopolysaccharide – the *LPS layer* – comprising sugars and lipids. The LPS layer often causes toxicity in animals infected by Gram-negative bacteria, such as *Salmonella*.

Granit, Ragnar Arthur (1900–91) Finnish-born Swedish neurophysiologist. He was awarded the Nobel Prize for physiology or medicine in 1967 jointly with H. K. Hartline and G. Wald for their discoveries concerning the physiological and chemical visual processes in the eye.

granulocyte A white blood cell (leukocyte) that has granules in the cytoplasm. Granulocytes are sometimes called *polymorphonuclear leukocytes* (*polymorphs*) because the nucleus is lobed. There are three types: NEUTROPHILS (70% of all leukocytes), EOSINOPHILS (1.5%), and BASOPHILS (0.5%).

granum (*pl.* **grana**) A stack of membranes (resembling a pile of coins) in a CHLOROPLAST. With the light microscope these stacks are just visible as grains (grana).

graticule *See* micrometer.

gravitropism (**geotropism**) A directional growth movement of part of a plant in response to gravity. For example, primary roots (tap roots) grow vertically towards gravity (*positive gravitropism*) whereas primary shoots grow vertically away from gravity (*negative gravitropism*). Gravitropic responses are believed to be mediated by the distribution of the plant hormone auxin within cells and tissues. If a shoot or coleoptile is lying on its side, the auxin moves to the lower surface, stimulates growth, and causes upward growth of the organ. In a horizontally placed root, the same high level of auxins inhibits growth of the lower surface, resulting in

downward curvature. These gradients are thought to arise owing to the detection of gravity by organelles (called *statoliths* or *amyloplasts*), which contain starch grains. The statoliths sink toward the lower wall of the cell exerting a pressure that in some way increases the concentration of auxin. This is termed the *statolith hypothesis*. How the statoliths influence auxin concentration remains uncertain. *See also* tropism.

gray Symbol: Gy The SI unit of absorbed energy dose per unit mass resulting from the passage of ionizing radiation through living tissue. One gray is an energy absorption of one joule per kilogram of mass.

gray matter Nerve tissue that consists mainly of nerve cell bodies and their connections, giving it a grayish color. It occurs in the core of the spinal cord and in many parts of the brain, especially the cerebral cortex.

Greengard, Paul (1925–) American biochemist noted for his work on signal transduction in the nervous system. He shared the 2000 Nobel Prize for physiology or medicine with A. Carlsson and E. R. Kandel.

ground tissues The tissues in any region of a plant not occupied by the specialized tissue of vascular bundles, cambium, epidermis, etc. The pith and cortex of the root and stem are ground tissue, as are the mesophyll layers of the leaf. Ground tissue generally consists of PARENCHYMA cells, but other cell types, for example, collenchyma and sclerenchyma, are also often present.

growth hormone (**somatotropin**) A polypeptide hormone that is produced by the anterior pituitary gland (adenohypophysis) and acts on the cells of the body, particularly those of bone, to stimulate metabolism and growth. Its actions *in vivo* are thought to be largely mediated by *somatomedins*, notably polypeptides resembling insulin and called INSULIN-LIKE GROWTH FACTORS. Deficiency during development leads to a form of dwarfism; excessive secretion leads to gigantism.

GSC (gas–solid chromatography) *See* gas chromatography.

GTP (guanosine triphosphate) A nucleoside triphosphate occurring in all cells as a coenzyme for various key processes. Often it provides energy by undergoing hydrolysis to GDP (guanosine diphosphate) and a phosphate group, a reaction catalyzed by an enzyme or other component having *GT-Pase* activity. In protein synthesis, GTP is essential for the assembly of ribosomes and elongation of the polypeptide chain. It is also required for the assembly of microtubules, for protein transport within cells, and for the relaying of messages to various cell components in signal transduction.

Guanine

guanine A nitrogenous base found in DNA and RNA. Guanine has a purine ring structure. *See illustration at* DNA.

guanine nucleoside *See* guanosine.

Guanosine

guanosine (guanine nucleoside) A nucleoside present in DNA and RNA and consisting of guanine linked to D-ribose via a β-glycosidic bond.

guard cell A specialized kidney-shaped epidermal cell, a pair of which encircle each epidermal pore (STOMA) in plants, and control the opening and closing of the stomatal aperture. Adjoining the guard cells, and assisting them in their actions, are SUBSIDIARY CELLS. Changes in the aperture are effected through changes in turgidity of the guard cells. The wall of the guard cell bordering the pore is heavily thickened whereas the opposite wall is comparatively thin. Thus when the guard cell is turgid the thin wall becomes distended, bulging out away from the pore, and causing the thickened wall, which cannot distend, to be drawn outwards with it. This results in opening of the aperture between the guard cells. When osmotic pressure of the guard cells drops, the guard cells shrink, and the pore closes.

The osmotic pressure inside the guard cells is regulated by light-dependent changes in the cytosolic concentration of dissolved solutes. During daylight, starch stored in the guard cell chloroplasts is converted to malate (the anion of malic acid) and hydrogen ions (H^+). ATP-powered proton pumps in the plasma membrane remove H^+ from the cell, creating an H^+ gradient across the plasma membrane. The consequent increased negative charge inside the cell triggers the opening of potassium channels in the plasma membrane and the influx of potassium ions (K^+), followed by the entry of chloride ions (Cl^-) via chloride channels. Hence the cytosolic ion concentration rises, drawing water into the cell by osmosis. During darkness, these events are reversed. The H^+ gradient across the plasma membrane falls, and channels open to transport K^+ and Cl^- from the cell. Hence water also leaves the cell, and the elastic walls of the guard cells cause them to shrink.

Guillemin, Roger Charles Louis (1924–) French-born American physician who worked on peptide hormone production in the brain. He was awarded the Nobel Prize for physiology or medicine in 1977 jointly with A. V. Schally. The prize was shared with R. S. Yalow.

H

hairpin A region of secondary structure of a linear molecule (e.g. a nucleotide) in which the chain of constituent groups is folded back on itself like a hairpin and held in place by chemical bonds (e.g. hydrogen bonds). Hairpins in nucleic acids occur as a result of bonding between quite widely separated stretches of complementary base sequence within single-stranded DNA and RNA molecules.

haploid (**monoploid**) Describing a cell or organism containing only one representative of each pair of homologous chromosomes found in the normal diploid cell. Haploid chromosomes are thus unpaired and the haploid chromosome number (n) is half the diploid number ($2n$). MEIOSIS, which usually precedes gamete formation, halves the chromosome number to produce haploid gametes. The diploid condition is restored when the nuclei of two gametes fuse to give the zygote. In humans there are 46 chromosomes in 23 pairs and thus the haploid egg and sperm each contain 23 chromosomes. Gametes may develop without fertilization, or meiosis may substantially precede gamete formation. This is especially true in plants, and leads to the formation of haploid organisms, or haploid phases in the life cycles of organisms. Various multiples of the haploid number, for example, the tetraploid ($4n$), hexaploid ($6n$), and octaploid ($8n$) conditions, are common in some plant groups, especially in certain cultivated plants. *Compare* diploid. *See also* polyploid.

Harden, (Sir) Arthur (1865–1940) British biochemist who discovered (with W. J. Young) that fermentation of sugar by yeast juice involved a separable substance (later called cozymase), required the presence of phosphate, and resulted in the intermediate formation of two hexose phosphates. He was awarded the Nobel Prize for chemistry in 1929, the prize being shared with H. von Euler.

Hartline, Haldan Keffer (1903–83) American neurophysiologist. He was awarded the Nobel Prize for physiology or medicine in 1967 jointly with R. Granit and G. Wald for their discoveries concerning the physiological and chemical visual processes in the eye.

Hartwell, Leyland H. (1939–) American cell biologist who was awarded the 2001 Nobel Prize for physiology or medicine jointly with R. T. Hunt and P. M. Nurse for discoveries of the key regulators of the cell cycle.

Hassel, Odd (1897–1981) Norwegian physical chemist who worked on the concept of conformation and its application in chemistry. He was awarded the Nobel Prize for chemistry in 1969, the prize being shared with D. H. R. Barton.

haustorium (*pl.* **haustoria**) **1.** An outgrowth of a fungal hypha that penetrates between or inside the cells of a host, for example, a plant or alga, to absorb nutrients. **2.** A specialized outgrowth of certain parasitic plants, for example, dodder, that infiltrates the tissues of a host plant to withdraw food and water.

Haworth, (Sir) Walter Norman (1883–1950) British organic chemist noted for his structural studies of sugars and polysaccharides. Haworth made the first chemical synthesis of ascorbic acid. He was awarded the Nobel Prize for chemistry in 1937. The prize was shared with P. Karrer.

H band *See* sarcomere.

hedgehog protein Any of a family of proteins that play key roles in the induction of patterning and tissue formation during embryonic development, in both invertebrates and vertebrates. In the fruit fly *Drosophila*, for example, Hedgehog protein is important in establishing the polarity (anterior–posterior orientation) of segments in the early embryo, and in patterning of adult body appendages, such as the wings. In vertebrates homologous proteins, such as *Sonic Hedgehog* in birds and mammals, are involved in the development of various embryonic and fetal structures, including the notochord, neural tube, gut, limbs, and heart. Hedgehog proteins are generally tethered to the exterior of the cell producing them. They bind to receptors on neighboring cells to activate or regulate the expression of other developmental genes, thus exerting a very localized influence.

helicase An enzyme that moves along double-stranded DNA to unwind the strands ahead of DNA REPLICATION. It melts the hydrogen bonds that link complementary bases between the two strands.

heliotropism *See* phototropism.

helper T-cell *See* T-cell.

hematopoiesis (hemopoiesis) The formation of blood cells. In the fetus hematopoiesis occurs in the spleen and liver; in the adult it occurs in the bone marrow (erythrocytes and polymorphonuclear white cells) and lymphoid tissue (lymphocytes and monocytes). All blood cells are ultimately derived from a population of *hematopoietic stem cells* in red bone marrow. Division of these gives rise to various cell lines, each of which undergoes proliferation and maturation via a series of cell divisions. For example, the formation of erythrocytes – a process called *erythropoiesis* – starts from stem cells called

proerythroblasts. Similarly, lymphocytes arise from precursors called lymphoblasts.

hematopoietic stem cell (hemopoietic stem cell) A stem cell, found in bone marrow or lymphoid tissue, that gives rise to any of the various types of blood cells. All ultimately originate from pluripotent stem cells in bone marrow. These undergo repeated division to give two lines of stem cells: *myeloid stem cells*, which largely remain in bone marrrow and give rise (via intermediate progenitor cells) to erythrocytes, granulocytes, monocytes, and platelets; and *lymphoid stem cells*, which migrate into lymphatic tissue and give rise to lymphocytes.

hematoxylin *See* staining.

Heme

heme An iron-containing porphyrin that is the prosthetic group in various proteins, including HEMOGLOBIN, MYOGLOBIN, and CYTOCHROMES.

hemicellulose Any of a group of substances that make up the matrix of plant cell walls (together with pectic substances, and occasionally in mature cells, with lignin, gums, and mucilages). They vary greatly in composition between species, and consist of chains of $\beta(1-4)$-linked sugar units, mainly the hexoses (mannose and galactose) and the pentoses (xylose and arabinose). Hemicelluloses form cross-links between the cellulose microfibrils, and help bind them together. In conjunction with the pectins, they contribute to the strength and elasticity of the cell wall. In some seeds (e.g. the endosperm of dates) hemicelluloses are a food reserve.

hemidesmosome   *See* desmosomes.

hemizygous   Describing genetic material that has no homologous counterpart and is thus unpaired in the diploid state. Both single genes and chromosome segments may be hemizygous; for example, the X chromosome in the heterogametic sex, and whole chromosomes in aneuploids.

hemoglobin   The pigment of the red blood cells (erythrocytes) that is responsible for the transport of oxygen from the lungs to the tissues. Adult human hemoglobin consists of four polypeptide chains – two alpha and two beta chains – each of which is associated with an iron-containing heme prosthetic group. Its molecular weight is 64 500. Variations in the polypeptide chains give rise to different types of hemoglobins in different species. The most important property of hemoglobin is its ability to combine reversibly with one molecule of oxygen per iron atom to form *oxyhemoglobin*, which has a bright red color. The iron is present in the divalent state (iron(II)) and this remains unchanged with the binding of oxygen. Oxygen molecules, diffusing across the red cell membrane, are very readily attached to hemoglobin in the lungs and equally readily detached in the tissues. This is the mechanism by which blood transports oxygen through the body.

The binding of oxygen depends on the oxygen partial pressure; high pressure favors formation of oxyhemoglobin and low pressure favors release of oxygen. It also depends on pH. The affinity of oxygen decreases as the pH is lowered (more acid, as a result of dissolved carbon dioxide). This dependence is known as the *Bohr effect*. Carbon dioxide can also combine with hemoglobin at amino groups along its polypeptide chains, producing carbaminohemoglobin. By this mechanism, red cells transport up to a quarter of all carbon dioxide carried in the bloodstream.

hemolysis   The release of hemoglobin from red blood cells (erythrocytes) due to rupture of the plasma membrane. It may be caused by such factors as toxins and incompatible blood transfusion.

hemopoiesis   *See* hematopoiesis.

Hench, Philip Showalter (1896–1965) American physician who showed (with E. C. Kendall) that rheumatoid arthritis could be treated with an adrenal hormone (later known as cortisone). He was awarded the Nobel Prize for physiology or medicine in 1950 jointly with E. C. Kendall and T. Reichstein for their discoveries relating to the hormones of the adrenal cortex.

Hensen's node   *See* primitive streak.

herpesvirus   One of a group of DNA-containing viruses, about 100 nm in diameter, that cause such diseases as cold sores, shingles, and chickenpox. These viruses can remain latent in their host cells for long periods until triggered to produce symptoms of disease. *See* latent virus.

Hershey, Alfred Day (1908–97)   American geneticist. He was awarded the Nobel Prize for physiology or medicine in 1969 jointly with M. Delbrück and S. E. Luria for their discoveries concerning the replication mechanism and the genetic structure of viruses.

heterochromatin   *See* chromatin.

heterogeneous nuclear ribonucleoprotein (hnRNP)   *See* heterogeneous nuclear RNA.

heterogeneous nuclear RNA (hnRNA) An assortment of RNA molecules found associated with proteins to form *heterogeneous ribonucleoprotein* (hnRNP) particles in the nuclei of eukaryotic cells. hnRNA consists of newly transcribed pre-messenger RNA (pre-mRNA) plus other nuclear RNAs. The proteins bind to different regions of the pre-mRNA, and to each other. They are thought to facilitate subsequent processing of the pre-mRNA by the enzymes of the spliceosome to produce the mature mRNA transcript, for example, by preventing base pairing between complementary regions of the RNA. They might also function in transporting the mRNA out of the nucleus.

heteroglycan *See* glycan.

heterograft (xenograft) A type of graft from one organism to another of a different species. *See* graft.

heterokaryon 1. A fungus whose mycelium contains paired nuclei of different genotype. Typically this results from the mating of individuals of different mating types.
2. Any cell containing two or more nuclei, or sets of chromosomes, with differing genotypes. Induced fusion of different animal or plant cells is effective in creating heterokaryons (*see* cell fusion).

heterozygous Describing an organism that has two different alleles at a given genetic locus. Usually only one of these, the dominant allele, is expressed in the phenotype. *Compare* homozygous; hemizygous.

Hevesy, György (Károly) (1885–1966) Hungarian-born Swedish chemist and biochemist who was the first to recognize the possibility of using natural radioactive elements as indicators or tracers in chemical and biological systems. Hevesy was also the discoverer of hafnium. He was awarded the Nobel Prize for chemistry in 1943.

hexose A MONOSACCHARIDE that has six carbon atoms in its molecules, such as glucose. *See also* sugar.

hexose monophosphate shunt *See* pentose phosphate pathway.

Hill, (Sir) Archibald Vivian (1886–1977) British physiologist and biophysicist. He was awarded the Nobel Prize for physiology or medicine in 1922 for his work on the production of heat in muscle. The prize was shared with O. F. Meyerhof.

Hill, Robert (Robin) (1899–1991) British plant biochemist noted for his studies of the mechanism of photosynthesis.

Hill reaction The reaction, first demonstrated by Robert Hill in 1937, by which isolated illuminated chloroplasts bring about the reduction of certain substances with accompanying evolution of oxygen. For example, the blue dye dichlorophenol indophenol (DCPIP) may be reduced to a colorless substance. The reaction involves part of the normal LIGHT-DEPENDENT REACTIONS of photosynthesis. Electrons from water involved in noncyclic photophosphorylation are used to reduce the added substance. It provided support for the idea that a light-dependent reaction preceded reduction of carbon dioxide in PHOTOSYNTHESIS.

histamine An amine formed from the amino acid histidine by decarboxylation and produced mainly by the MAST CELLS in connective tissue as a response to injury or allergic reaction. It causes contraction of smooth muscle, stimulates gastric secretion of hydrochloric acid and pepsin, and dilates blood vessels, which lowers blood pressure and produces inflammation, itching, or allergic symptoms (such as sneezing).

histidine *See* amino acids.

histiocyte A wandering ameboid cell, capable of ingesting foreign particles, found in the matrix of connective tissue. It is a type of tissue MACROPHAGE.

histochemistry (cytochemistry) The study of the distribution of particular chemical compounds within cells and tissues by the use of specific staining techniques (e.g. phloroglucinol to stain lignin), radiolabeling of substrates taken up by cells, or other techniques.

histocompatibility The extent to which an organism's immune system will tolerate tissue grafts from another organism. Normally, a graft from an unrelated individual is recognized as foreign by the recipient's white blood cells because the marker molecules ('self' marker antigens) on the surface of the foreign cells differ from the recipient's marker molecules. Hence, the white cells are stimulated to mount an immune response against the foreign tissue. Only rarely (e.g. sometimes among close relatives) do two individuals share the same marker antigens, and are able to tolerate grafts of each other's tissues. The most important of these marker antigens are proteins encoded by a complex cluster of genes called the MAJOR HISTOCOMPATIBILITY COMPLEX (MHC). Two classes of these proteins are important in histocompatibility and immune responses: class I MHC proteins, which occur on most body cells and 'present' viral antigens to cytotoxic T-CELLS; and class II MHC proteins, which occur only on certain immune system cells, such as macrophages and B-CELLS, and are essential for activating T-helper cells and consequently antibody secretion by B-cells.

histogenesis The development of the tissues of an organism.

histology The study of tissues, especially at the microscopic level.

histone One of a group of relatively small proteins found in chromosomes, where they organize and package the DNA into chromatin. They contain a large proportion of positively charged basic amino acids, which interact with the negatively charged phosphate groups in DNA. Histones fall into five main groups, designated H1, H2A, H2B, H3, and H4. Most are very highly conserved – they show a high degree of similarity of amino acid composition across species as distantly related as garden peas and cattle. In the packaging of chromosomal DNA, histones form 'spools' (cores) around which the DNA molecule winds to form NUCLEOSOMES, which resemble beads on a string. H1 histones then take part in coiling this 'string' into a more condensed solenoid arrangement (*see* chromatin). It is thought that acetylation of lysine residues in core histones, by specific histone acetylase enzymes, controls the degree of condensation of the chromosomal DNA, and hence plays a role in regulating the level of gene transcription. Genes in fully condensed chromosomal regions are inaccessible to the enzymes and other factors required for transcription.

Hitchings, George Herbert (1905–98) American biochemist and pharmacologist noted for his introduction of a range of widely used synthetic drugs designed as antimetabolites, including the antifolate bactericidal agent co-trimoxazole, the immunosuppressants mercaptopurine and azathioprine, and the xanthine-oxidase inhibitor allopurinol. He was awarded the Nobel Prize for physiology or medicine in 1988 jointly with J. W. Black and G. B. Elion.

HIV (human immunodeficiency virus) A RETROVIRUS that causes AIDS in humans by infecting and ultimately destroying certain cells – T-helper cells and macrophages – that are vital for immunity against infections. The viral particles are spherical and contain genes in the form of RNA. The virus recognizes a suitable host cell by its surface markers, called CD4 proteins, and fuses with the cell membrane. Infected T-CELLS can bud off new virus particles from their surface. However, they can also fuse with uninfected cells carrying the same surface markers, forming multinucleate cells that are ineffective as immune cells. As increasing numbers of T-helper cells become infected, the effectiveness of the immune system diminishes and the patient becomes more and more prone to infections. There are two types of HIV: HIV-1, which occurs worldwide and is responsible for most cases of AIDS; and HIV-2, which is found mainly in Africa. *See also* AIDS.

HLA system (human leukocyte-associated antigen system) The system for designating the main HISTOCOMPATIBILITY genes and their antigens in humans. HLA antigens are found on the surface of human body cells, and are encoded by the group of linked genetic loci that make up the major histocompatibility complex in humans, located on chromosome 6. Hence, HLA antigens are crucial in determining the compatibility of tissue grafts between different persons, for example. There are six principal genetic loci in the HLA system: *A, B, C, DR, DQ,* and *DP*. The *A, B,* and *C* loci encode class I MHC proteins, whereas the *D* loci encode class II proteins (*see* major histocompatibility complex). Most of these loci are highly polymorphic, with large numbers of different alleles occurring among the members of human populations. Hence, there is enormous diversity in the genotypes at these loci, and in the antigens they encode. Consequently, there is often great difficulty in finding donors whose HLA antigens match those of a potential transplant recipient.

hnRNA *See* heterogeneous nuclear RNA.

hnRNP (heterogeneous nuclear ribonucleoprotein) *See* heterogeneous nuclear RNA.

Hodgkin, (Sir) Alan Lloyd (1914–98) British physiologist noted for his work on nerve-impulses. He was awarded the Nobel Prize for physiology or medicine in 1963 jointly with J. C. Eccles and A. F. Huxley.

Hodgkin, Dorothy (Mary) (1910–94) British x-ray crystallographer noted for her determination of the molecular structures of benzylpenicillin and cyanocobalamin and, in particular, of the structure of crystalline zinc insulin. She was awarded the Nobel Prize for chemistry in 1964.

Holley, Robert William (1922–93) American biochemist and molecular biologist noted for his determination of the structure of alanine transfer RNA (the first determination of the sequence of a nucleic acid). He was awarded the Nobel Prize for physiology or medicine jointly with H. G. Khorana and M. W. Nirenberg (1968) for their interpretation of the genetic code and its function in protein synthesis.

Holliday structure *See* crossing over.

holoblastic Describing the type of CLEAVAGE in which the entire zygote is converted into a multicellular blastula. It is typical of lightly and moderately yolked

eggs, such as those of echinoderms, amphibians, most annelids and mollusks, and placental mammals. *Compare* meroblastic.

holoenzyme A catalytically active complex made up of an APOENZYME and a prosthetic group. The former is the protein part of the enzyme that is responsible for the catalytic specificity but requires a cofactor or coenzyme to be functionally active.

HOM-C complex *See* homeotic gene.

homeobox A segment of DNA found in many genes concerned with controlling the development of organisms, including homeotic genes. It consists of 180 base pairs, and the sequence of bases is remarkably similar across a wide range of species, from yeasts to human beings. This suggests it arose early in evolutionary time and has been little changed since. The homeobox encodes a 60-amino acid peptide sequence, called a *homeodomain*, that enables the parent protein to bind to DNA. This is consistent with the role of homeotic proteins as transcription factors, binding to control regions of genes to regulate their expression.

homeodomain *See* homeobox.

homeotic gene Any of a class of genes that are crucial in determining the differentiation of tissues in different parts of the body during development. Mutations of such genes result in *homeosis* – the replacement of one body part with another. Homeotic genes encode proteins (*transcription factors*) that regulate the expression of other genes by binding to DNA. This binding capability can be pinpointed to a characteristic base sequence known as a HOMEOBOX. Homeotic genes have been intensively studied in the fruit fly *Drosophila*, in which they are called SELECTOR GENES. They come into play when the basic plan of body segments has been established in the fly embryo, and direct the development of particular groups of cells in each segment to form the organs, limbs, and other structures appropriate for that particular segment. In *Drosophila* there are eight linked homeotic genes on chromosome 3 – together called the *HOM-C complex*. This is split into two groups: five genes comprise the *antennapedia* complex, which controls development of the head and front thoracic segments, and three genes make up the *bithorax* complex, which governs the fate of cells in more posterior segments. Mice, humans, and other mammals have four clusters of homeotic genes, or *Hox genes* (from homeobox), located on separate chromosomes. Some of the Hox genes are homologous with *Drosophila* homeotic genes, and are arranged in the same order along the chromosome as their fruit-fly counterparts. In both *Drosophila* and the mouse, for example, the physical order of the genes in these complexes corresponds to the order in which they are expressed along the head-to-tail axis in the developing embryo. For example, the genes at the 'front' (3' end) of these clusters are called *lab* in *Drosophila* and *Hox1* in the mouse. Both *lab* and *Hox1* are expressed first in the body segment nearest to the head, and then sequentially in all segments to the tail region. The second gene is expressed initially in slightly more posterior segments, and so on along the gene cluster, until the last gene is expressed only in the tail region. *See also* differentiation; segmentation.

homoglycan   *See* glycan.

homograft (**allograft**)   A type of GRAFT between individuals of the same species.

homologous 1. (*Biology*) Describing organs or tissues that have similar structure and function in different species as a result of a common evolutionary origin.
2. (*Biochemistry*) Describing large molecules (e.g. proteins or nucleic acids) that have similar sequences of chemical groups (e.g. amino acids or bases) at corresponding positions in the molecule, reflecting a common evolutionary origin.
3. (*Genetics*) *See* homologous chromosomes.

homologous chromosomes   Chromosomes that pair at MEIOSIS. Each carries the same genes as the other member of the pair but not necessarily the same alleles for a given gene. During the formation of the germ cells only one member of each pair of homologs is passed on to the gametes. At fertilization each parent contributes one homolog of each pair, thus restoring the diploid chromosome number in the zygote. With the exception of the sex chromosomes – for example, in mammals the Y chromosome is much smaller than the X chromosome – the members of each homologous pair are similar to one another in size and shape.

homozygous   Describing an organism that has identical alleles for any specified gene or genes. *Compare* heterozygous.

Hopkins, (Sir) Frederick Gowland (1861–1947)   British biochemist noted for his discoveries of essential amino acids and vitamins. He also isolated glutathione and tryptophan. He was awarded the Nobel Prize for physiology or medicine in 1929. The prize was shared with Christiaan Eijkman.

Houssay, Bernardo Alberto (1887–1971)   Argentinian physiologist and endocrinologist noted for his researches on diabetes and on the adrenal, pituitary, and thyroid glands. He was awarded the Nobel Prize for physiology or medicine in 1947 for his discovery of the part played by the hormone of the anterior pituitary lobe in the metabolism of sugar. The prize was shared with C. F. and G. T. Cori.

Hox gene   *See* homeotic gene.

Hox genes   *See* homeotic gene.

Huber, Robert (1937–   )   German biochemist who worked on the structure of proteins. He was awarded the Nobel Prize for chemistry in 1988 with J. Deisenhofer and H. Michel.

Human Genome Project   An international project launched in 1989 with the aim of mapping and sequencing the entire human genome. The results are published in two stages: as *draft sequences*, mostly in the form of 10 000 bp fragments whose approximate chromosomal locations are known; and finished sequences, which are the full sequences of the gene-rich regions of each chromosome. Chromsome 23 was the first to be fully sequenced, completed in December 1999, and the goal is to provide finished sequences for all chromosomes by 2003.

The sequence information provided will help in the diagnosis and possibly the treatment of a wide range of diseases. These include not only hereditary disorders, such as Huntington's disease, but also many common ailments with a genetic component, such as heart disease and breast cancer. Sequencing the genes responsible for the development of these diseases enables the design of DNA probes that will identify susceptible individuals, so allowing preventive measures or routine check-ups. However, such knowledge has profound ethical, social, and legal implications, besides providing a host of commercial opportunities. For example, insurance companies may insist on a full genome check before agreeing terms for life insurance, even in healthy applicants. *See* chromosome map.

human immunodeficiency virus   *See* HIV.

humoral immunity   *See* immunity.

Hunt, R. Timothy (1943–) British cell biologist who was awarded the 2001 Nobel Prize for physiology or medicine jointly with L. H. Hartwell and P. M. Nurse for discoveries of the key regulators of the cell cycle.

Huxley, (Sir) Andrew Fielding (1917–) British physiologist noted for his work on the contraction of muscle and the conduction of nerve impulses. He was awarded the Nobel Prize for physiology or medicine in 1963 jointly with J. C. Eccles and A. L. Hodgkin.

Huxley, Hugh Esmor (1924–) British biophysicist noted for his study of the rela-

tion of ATP hydrolysis to the mechanics of muscle contraction.

hyaline cartilage   *See* cartilage.

hybridization   **1.** (*Biochemistry*) The association between two strands of nucleotides due to the formation of hydrogen bonds between complementary bases (i.e. base pairing). This yields a 'hybrid' molecule, which can consist of two DNA strands, one DNA and one RNA strand, or two RNA strands. *See also* in situ hybridization.
2. (*Genetics*) The creation of a hybrid, for example by cross-breeding two strains of a species.

hybridoma   A hybrid cell produced by fusing a normal antibody-producing lymphocyte (B-cell), derived from mouse or rat spleen, with a cancerous lymphocyte that is immortal. Such cells are cloned and used to produce potentially unlimited quantities of specific antibodies (*see* monoclonal antibody).

hydrocortisone   (**cortisol**)   A steroid hormone produced by the adrenal cortex having GLUCOCORTICOID activity.

hydrogen   An essential element in living tissues. It enters plants, with oxygen, as water and is used in building up complex reduced compounds such as carbohydrates and fats. Water itself is an important medium, making up 70–80% of the weight of organisms, in which chemical reactions of the cell can take place. Hydrogenated compounds, particularly fats, are rich in energy and on breakdown release energy for driving living processes. *See also* proton pump; proton-motive force.

hydrogen bond   A type of bond occurring between molecules. Hydrogen bonding takes place between oxygen, nitrogen, or fluorine atoms on one molecule, and hydrogen atoms joined to oxygen, nitrogen, or fluorine on the other molecule. The attraction is due to electrostatic forces. Hydrogen bonding is responsible for the properties of water, such as its high melting and boiling points. The formation of hydrogen bonds between water molecules and hydroxyl (–OH) and amino (–NH2) groups on other compounds makes such compounds soluble in water. Moreover, hydrogen bonds between component chemical groups within molecules are important in stabilizing the structure of large biomolecules such as proteins and nucleic acids.

hydrolase An enzyme that catalyzes a hydrolysis reaction. Digestive enzymes are an example. Hydrolases play an important part in rendering insoluble food material into a soluble form, which can then be transported in solution.

hydrolysis In general, a reaction between a compound and water, particularly one involving addition of the elements of water (H^+ or OH^- ions).

hydronium ion   *See* acid.

hydrophilic   *See* lyophilic.

hydrophobic   *See* lyophobic.

hydroxonium ion   *See* acid.

hyperplasia Enlargement of a tissue due to an increase in the number of its cells. For example, if part of the liver is removed, the remaining part may undergo hyperplasia in order to regenerate. *Compare* hypertrophy.

hyperpolarization An increase in the polarity of the potential difference across a membrane of a nerve or muscle cell, i.e. an increase in the resting potential. It is caused by the pumping of ions across the membrane so that differential concentrations are created on either side – the inside becoming more negative. As a result, a stronger stimulus is needed to evoke a response.

hyperthermophilic   *See* thermophilic.

hypertonic Describing a solution whose osmotic pressure is greater than that of another specified solution, the latter being HYPOTONIC. When separated by a semiper-

meable membrane (e.g. a cell membrane) water moves by osmosis into the hypertonic solution from the hypotonic solution. *Compare* isotonic.

hypertrophy Enlargement of a tissue or organ due to an increase in the size of its cells or fibers. An example is the enlargement of muscles as a result of exercise. *Compare* hyperplasia.

hypha 1. A fine tubular filament of a fungus that spreads to form a loose network termed a *mycelium*, or aggregates into fruiting bodies (e.g. toadstools). Hyphae may be branched or unbranched and may or may not have cross walls (*septa*) dividing them into cells. They are parasitic or saprophytic and the tips secrete enzymes to digest and penetrate the food supply.
2. A long thin extension of a bacterial cell, formed especially by budding bacteria such as *Hyphomicrobium*. In these a bud forms at the tip of the hypha, enlarges, and finally separates as a daughter cell.

hypodermis One or more layers of cells that may be found immediately below the epidermis of plants. It may be composed of thin-walled colorless cells and functions as water-storing tissue as in certain succulent leaves and the aerial roots of epiphytes. Alternatively in some species, the hypodermal cells possess heavily thickened walls and assist in mechanical protection of internal tissues as in pine leaves.

hypostasis The situation in which the expression of one gene (the *hypostatic gene*), is prevented in the presence of another, nonallelic, gene (the *epistatic gene*). *See* epistasis.

hypostatic gene   *See* hypostasis.

hypotonic Describing a solution whose osmotic pressure is less than that of another specified solution, the latter being HYPERTONIC. When separated by a semipermeable membrane (e.g. a cell membrane) water is lost by osmosis from the hypotonic to the hypertonic solution. *Compare* isotonic.

I

IAA (indole acetic acid) *See* auxin.

I band *See* sarcomere.

ICSH (interstitial-cell-stimulating hormone) *See* luteinizing hormone.

identical twins (monozygotic twins) Two offspring, produced during one birth, resulting from the division of a single fertilized egg. They are of the same sex and otherwise genetically identical, but may differ because of differences in nutrition, injuries, etc., either before or after birth. Identical triplets, quads, and quins are known, but multiple births in humans usually result from the simultaneous fertilization of several eggs.

idiogram *See* karyogram.

IFN *See* interferon.

Ig *See* immunoglobulin.

IGF *See* insulin-like growth factor.

Ignarro, Louis J. (1941–) American biochemist who was awarded the 1998 Nobel Prize for physiology or medicine jointly with R. F. Furchgott and F. Murad for work on the action of nitric oxide as a signaling molecule in the cardiovascular system.

IL *See* interleukin.

imbibition The phenomenon in which a substance absorbs a liquid and swells, but does not necessarily dissolve in the liquid. The process is reversible, the substance contracting on drying. Water is imbibed by many biological substances, including cellulose, hemicelluloses, pectic substances, lignin (all plant cell wall constituents), starch, and certain proteins. Dry seeds absorb water by imbibition, creating an *imbibition pressure*, which ruptures the seed coat (testa), and allows the embryo to emerge. Imbibition combined with osmosis is responsible for water uptake in growing plant cells, and water retention may be aided by swelling of mucilaginous materials, for example in succulent plants.

imbibition pressure *See* imbibition.

immune clearance (immune elimination) The rapid removal of antigen introduced into the body of an immune individual, as a result of its complexing with antibody.

immunity The ability of plants and animals to withstand harmful infective agents and toxins. It may be due partly to a number of nonspecific mechanisms, such as inflammation and phagocytosis or an impervious skin (*nonspecific immunity*). In vertebrates it is largely the result of specific immune responses, whereby certain proteins (ANTIBODIES) or immune cells (LYMPHOCYTES) present in the body combine with an introduced foreign substance (antigen) – *specific acquired immunity*. Specific acquired immunity includes *passive immunity*, where the antibody has been derived from another individual (e.g. from the mother to offspring), and *active immunity*, where the antibody is produced following stimulation with ANTIGEN (e.g. by vaccination or by exposure to infection). *Cell-mediated immunity* is achieved by T-CELLS, which perform various roles, including detecting and destroying virus-infected body cells. T-cells can also stimulate B-CELLS to produce antibodies, which circulate in the blood and lymph and provide *humoral immunity.*

Plants can show a localized *hypersensitive response* following infection with bacteria, fungi, and other pathogens. This involves changes in the composition of the plant's cell walls and synthesis of proteins and other metabolites that counter the effects of the invading pathogen. They can also develop a more general capability to resist pathogens, called *systemic acquired resistance.*

immunoassay Any of various techniques for measuring biological substances that depend on the substance acting antigenically and binding with a specific antibody. Most also require the addition of labeled antigen or antibody, bearing a radioisotope, fluorescent molecule, or enzyme, for example, to reveal the extent of the antigen–antibody binding, and enable the amount or concentration of the test sample to be established. *See* ELISA; radioimmunoassay.

immunoelectrophoresis A technique in which the components of a mixture are firstly separated by ELECTROPHORESIS on a gel. Then a solution of antibodies is added to a trough in the gel. The antibodies diffuse through the gel, and where they encounter their specific antigen they form a precipitate. This allows specific antigenic components to be identified.

immunoglobulin (Ig) Any of a group of proteins that take part in the immune responses of higher animals. They are produced by B-CELLS (B-lymphocytes), and act mainly as antibodies by binding to foreign antigens, such as bacteria, viruses, or toxins. Each consists of four polypeptide chains, two heavy chains, and two light chains, arranged to form a Y-shaped structure. The two arms of the 'Y' each bear an antigen-binding site. There are five classes, distinguished by the structure of their heavy chains.

Immunoglobulin A (IgA) is found mainly in secretions (e.g. saliva, tears, mucus) and its main role is to neutralize viruses and bacteria on the mucous surfaces of the body, such as those of the respiratory, gastrointestinal and urogenital tracts. *Immunoglobulin D* (IgD) occurs on the surface of lymphocytes and is thought to control the activation and suppression of B-cell activity. *Immunoglobulin E* (IgE) is normally present in very low concentrations in blood and connective tissue underlying epithelia, but its level is raised in allergies. It binds to mast cells, and in the presence of antigen causes histamine release from mast cells, with consequent inflammation and the other common symptoms of allergies.

Immunoglobulin G (IgG) is the main immunoglobulin of blood and tissue fluid. It binds to microorganisms, enhancing their engulfment by phagocytic cells, and neutralizes viruses and bacterial toxins. It also activates COMPLEMENT, leading to lysis of target cells, and can cross the placenta, affording protection to the fetus. *Immunoglobulin M* (IgM) is a star-shaped molecule comprising five of the basic Y-shaped units. It occurs within blood vessels and is most prominent early in the immune response, mopping up microorganisms or other antigens with its array of binding sites. It can also activate complement. *See also* immunity.

implantation (nidation) The attachment of the developing mammalian embryo to the wall of the uterus. The human embryo enters the uterus from the Fallopian tube, where it has been fertilized four days earlier. It is at the early BLASTOCYST stage of development. Cells on its outer surface (*see* trophoblast) break down the cells of the uterine wall and invade the mother's tissues, anchoring the growing embryo and making way for the development of the placenta.

incipient plasmolysis *See* plasmolysis.

incompatibility 1. The rejection of grafts, transfusions, or transplants between animals or plants of different genetic composition. *See* histocompatibility.
2. A mechanism in flowering plants that prevents fertilization and development of an embryo following pollination by the

same or a genetically identical individual. This self-incompatibility is due to interaction between alleles at one or more incompatibility genes. If an allele expressed in the stigma is also expressed by the pollen grain, then the pollen is either unable to germinate or its growth through the stigma is inhibited. This results in self-sterility, thus preventing inbreeding.

3. A genetically determined mechanism in some fungi that prevents sexual fusion between individuals of the same race or strain.

indole acetic acid (IAA) *See* auxin.

inducer 1. (*Biochemistry*) A substance that causes a cell to switch on or increase production of enzymes involved in its own uptake or metabolism. For example, in the bacterium *E. coli*, derivatives of lactose act as inducers of the *lac* operon. Their presence in the cell inactivates repression of the OPERON, thereby allowing expression of the genes encoding enzymes responsible for the uptake of lactose and its cleavage into glucose and galactose.
2. (*Embryology*) A chemical signal that influences the development of cells. *See* induction (def. 2).

inducible enzyme An enzyme that is produced by a cell only in the presence of its substrate. *Compare* constitutive enzyme.

induction 1. (*Biochemistry*) The production of enzymes by cells in response to the presence of an INDUCER (def. 1).
2. (*Embryology*) The process by which cells influence the development of other cells. Induction is crucial in laying down the basic body plan of an early embryo, and in more localized patterning of tissues and organs during subsequent development. Chemical signals (inducers) produced by one cell, or cell population (sometimes called an ORGANIZER), affect the developmental pathway taken by recipient cells. Some signals, for example Hedgehog proteins, remain tethered to the outside of the source cell and influence only neighboring cells, whereas others (e.g.

the TRANSFORMING GROWTH FACTORS) are freely diffusible in the extracellular matrix. Some can have an all-or-nothing effect, depending on whether a cell receives the signal or not, whereas others induce different fates in recipient cells depending on the signal concentration. This in turn depends on the distance of the recipient cell from the source cell. Moreover, the signaling can be unidirectional, or reciprocal. For example, reciprocal induction is important in the development of many internal organs, including lung, kidney, and pancreas. The responsiveness of cells to inductive signals varies both in space and time; cells that are responsive to such signals are described as competent. *See also* morphogen.

inflammation A defensive reaction of animal tissues to injury, infection, or irritation, characterized by redness, swelling, heat, and pain. An *acute inflammatory response* typically follows bacterial infection, for example, and is triggered by complement proteins. These bind to the microorganisms and also act as chemotactic agents, attracting phagocytic cells (neutrophils and macrophages) to the infection site: complement components increase the permeability of blood capillaries in the vicinity, so allowing an influx of neutrophils from the blood. The same components trigger tissue mast cells to release further chemical mediators of inflammation, including histamine, leukotrienes, and tumor necrosis factor (TNF). Tissue macrophages also release mediators, including interleukins and TNF, which sustain the inflammation. The process is regulated and resolved by complement regulatory proteins, and by cytokines such as prostaglandin E_2 and transforming growth factor b. Fibrin forms a clot to seal off the site, and phagocytic cells engulf microorganisms, thus destroying them.

inhibition The reduction or prevention of activation of an effector by means of inhibitory nerve impulses. Inhibitory synapses release a neurotransmitter that opens potassium (K^+) or chloride (Cl^-) channels in the postsynaptic membrane, causing either an outflow of K^+ ions from the postsynaptic cell or an inflow of Cl⁻ ions. This increases the polarization of the membrane and makes depolarization, and therefore formation of an action potential, less likely. Inhibitory synapses play an important part in central-nervous-system control of motor activity.

inhibitory postsynaptic potential (IPSP) A localized hyperpolarization of the postsynaptic membrane at an inhibitory synapse or a neuromuscular junction that tends to inhibit production of an action potential in the postsynaptic nerve fiber. It is due to an inhibitory neurotransmitter, released from the presynaptic membrane on stimulation by an impulse, causing an increase in membrane permeability to certain ions (*see* inhibition). The size of the IPSP depends on the amount of neurotransmitter released. *Compare* excitatory postsynaptic potential.

initial An undifferentiated plant cell that resides in a MERISTEM and divides to perpetuate itself while adding new cells to the plant body. Two basic groups exist: apical initials at root and shoot apices, and lateral meristem initials whose position depends on the location of the meristem. Vascular cambium, which lies between xylem and phloem, possesses two distinct types of initials: *ray initials* forming medullary rays, and *fusiform initials* producing xylem and phloem elements.

initiation *See* transcription; translation.

initiation complex 1. A complex assembly of proteins and RNAs that is formed at the start of translation during protein synthesis at ribosomes. The small ribosomal subunit associates with initiation factors, GTP, and the initiator transfer RNA to form a *preinitiation complex*. This then becomes positioned at the start site on a messenger RNA (mRNA) molecule, and the large ribosomal subunit binds to the small subunit, forming the complete initiation complex and enabling translation to proceed.
2. A large enzyme complex (holoenzyme) formed at the start of transcription of eukaryotic genes. It comprises RNA polymerase II and as many as 60–70 polypeptides, including transcription factors, and binds to the promoter site.

initiation factor Any of various proteins that enable the formation and correct positioning of the INITIATION COMPLEX that precedes translation of messenger RNA (mRNA) during protein synthesis at ribosomes. These factors are designated IF1, IF2, etc. in prokaryotes, and eIF1, eIF2, etc. in eukaryotes.

inositol An optically active cyclic sugar alcohol. It is a component of the vitamin B complex and is required for growth in certain animals and microorganisms. The stereoisomer *myo*-inositol is a component of the phospholipid PHOSPHATIDYLINOSITOL, an important constituent of cell membranes, and also a precursor of several second messengers, including INOSITOL 1,4,5-TRISPHOSPHATE (IP₃).

inositol 1,4,5-trisphosphate (IP₃) An intracellular signaling molecule, derived from phosphatidylinositol, that acts as a second messenger in various cell signaling pathways. It is water soluble and can diffuse through the cell cytoplasm to its target. For example, it can bind to IP₃-sensitive calcium channels in the endoplasmic reticulum (ER). Opening of the latter releases calcium ions from the ER, which mediates various cellular processes. Hormone-induced cell responses often involve such an IP₃-triggered rise in intracellular calcium ions. IP₃ is hydrolyzed, and thus inactivated, within a second of its formation.

insertion sequence *See* transposon.

in situ hybridization A technique in which a DNA PROBE is applied to a preparation of chromosomes to identify regions of complementary base sequence. It is used to find the chromosomal location of particular genes or other loci. The DNA of the chromosomes is partially denatured, and bathed in a solution containing the probe, which carries a radioactive or fluorescent label. The probe hybridizes to any complementary region of the chromosome, and the label reveals its location when excess probe is washed away. For example, *fluorescence in situ hybridization* (FISH) reveals the bound probe as spots of fluorescent light. In situ hybridization can also be adapted to localize particular messenger RNAs within, say, a developing embryo, and so reveal patterns of gene expression.

insulin A polypeptide hormone, produced by the B-cells of the islets of Langerhans in the pancreas, that controls the metabolism of glucose. Lack of insulin results in diabetes, but excess insulin leads to coma. Its secretion is stimulated by high blood levels of glucose and amino acids after a meal. Insulin promotes the uptake of blood glucose by muscle cells and fat cells (adipocytes), and activates the synthesis of enzymes required to manufacture glycogen (in liver cells) and fats (in adipocytes). Many cells also require insulin as a growth factor. Insulin is used therapeutically in the treatment of diabetes mellitus.

insulin-like growth factor (IGF) Any of a group of polypeptides that are structurally related to insulin and act as growth-promoting agents. They stimulate cell division and protein synthesis, and also promote glycogen formation in muscle and fat deposition. Their production is itself stimulated by growth hormone (somatotropin), and they include IGF-I (somatomedin C) and IGF-II (somatomedin A).

integrin *See* cell adhesion molecule.

intercalary meristem *See* meristem.

intercellular Situated or occurring between cells. *Compare* intracellular.

interferon (IFN) Any of a group of proteins produced by animal cells in response to infection by viruses or other agents. They act to inhibit viral replication within the cell; some also have antibacterial and anticancer properties. α-Interferons (IFNα) are produced by leukocytes, whereas IFNβ is produced by fibroblasts and other cell types. IFNγ is produced by T-cells, and stimulates macrophages to destroy ingested microorganisms. Interferons are also produced commercially for various therapeutic applications, using recombinant DNA technology.

interleukin (IL) Any of a class of CYTOKINES that act as chemical messengers between various types of white blood cells (leukocytes). All are soluble proteins or glycoproteins. For example, interleukin-1 (IL-1) is secreted by antigen-presenting macrophages, and stimulates T-cells. These in turn secrete IL-2, which promotes proliferation of T-cells (i.e. self-activation) and differentiation of B-cells to antibody-secreting plasma cells. Around 20 other interleukins have been described. Interleukin 11 acts as a growth factor for blood cell precursors, and is manufactured for therapeutic use by recombinant DNA techniques. *Compare* lymphokine.

intermediate filament (IF) Any of a group of fibrous proteins that provide support for cells as components of the CYTOSKELETON. They typically have a diameter of about 10 nm, intermediate between the narrower microfilaments and the wider microtubules. Various types are found in different cell types. For example, IFs made of keratin occur in epithelial cells. When such cells die, the keratin IFs form hair, nails, and wool. Other types include vimentin IFs in leukocytes and fibroblasts, and desmin IFs in muscle cells. In general, each IF consists of helically coiled polypeptides, forming rodlike arrays that are joined end to end in long protofibrils. Each IF consists of four protofibrils, arranged side by side. IFs stabilize cell membranes, and help in forming an internal framework for the cell. For example, the inner surface of the nuclear envelope is supported by a network of IFs composed of the protein lamin; other IFs extend through the cytoplasm and run underneath the plasma membrane. They also participate in certain cell junctions (*see* desmosomes).

internal environment The medium surrounding the body cells of multicellular animals, i.e. the intercellular fluid. In vertebrates its composition is kept relatively constant by the mechanisms of homeostasis.

interneuron A neuron that connects sensory neurons and motor neurons. Interneurons are generally located in the central nervous system.

internode 1. (*Zoology*) The region of a medullated nerve axon between two nodes of Ranvier. It is covered with a myelin sheath.
2. (*Botany*) The region of the stem between two nodes.

interphase The stage in the CELL CYCLE when the nucleus is not in a state of division. Interphase is divisible into three phases – G_1, S, and G_2 – each characterized by different metabolic events.

interstitial-cell-stimulating hormone *See* luteinizing hormone.

intine (**endosporium**) The inner layer of the cell wall surrounding the pollen grains of angiosperms and gymnosperms. By contrast with the outer cuticularized layer (exine), the intine is thin and composed of cellulose. The pollen tube, which emerges during germination of the pollen grain, is an outgrowth of the intine.

intracellular Occurring or situated inside a cell, within the plasma membrane. *Compare* intercellular.

intrinsic factor A glycoprotein secreted by the stomach lining that is required for the absorption of vitamin B_{12} (cobalamin). It binds the vitamin in the small intestine and is then itself bound to a specific receptor on the intestinal mucosa, where the vitamin is released and transported across the mucosal cell into the bloodstream. Deficiency of intrinsic factor leads to malabsorption of vitamin B_{12} and hence pernicious anemia.

intron A noncoding DNA sequence that occurs between coding sequences (exons) of a gene. Introns are found in most eukaryotic nuclear genes, in some cytoplasmic genes, in some genes of prokaryotes, particularly in archaea and cyanobacteria, and even in certain viral genes. Mature functional messenger RNA (mRNA) does not contain introns, these being removed from the initial transcript (*pre-mRNA*) during processing. In each case the pre-mRNA is cleaved, the intron is removed as a loop, and the adjoining exons are spliced together. In eukaryotes the process is generally performed in the nucleus by a complex of proteins called a *spliceosome*. However, in prokaryotes, in some eukaryotic protists (e.g. *Tetrahymena*), and in the cytoplasmic genes of mitochondria and chloroplasts, the pre-mRNA can act as its own enzyme (*see* ribozyme) to catalyze splicing of its introns (i.e. self-splicing), without the involvement of any proteins.

Introns were discovered in 1977, and were originally regarded as 'junk DNA' by many. However, they are thought to have evolutionary significance. By acting as 'spacers' for exons in genes they enable *exon shuffling* – recombination or rearrangement of exons – and hence the rapid evolution of proteins with different combinations of structural domains, and the corresponding functional groups. Certain types of intron can also behave as mobile DNA elements, inserting themselves at different sites within the DNA molecule. Their transposition within the genome might alter the expression of neighboring genes. Moreover, the introns of certain large eukaryotic genes contain embedded genes – in effect 'genes within genes'.

intussusception The incorporation of cellulose molecules into the existing cell wall of a plant cell, giving an increase in wall area. *Compare* apposition.

inulin A polysaccharide food reserve of some higher plants, found particularly in Jerusalem artichoke and other members of the Compositae. It is a polymer of fructose.

invagination 1. The formation of a pocket of tissue during embryonic development, especially the intucking of the early gut (archenteron).
2. Any infolding, for example, of a cell membrane or epithelium.

inversion *See* chromosome mutation.

inverted repeat A nucleotide sequence that is repeated in reverse order. It can involve a single RNA or DNA strand, opposite strands of a double-stranded DNA molecule, or both DNA strands in different regions of the molecule. An example on a single strand would be:
    TGTTGTAGCCAAGTGTTGT.

in vitro Literally 'in glass'; describing experiments or techniques performed in laboratory apparatus rather than in the living organism. Cell tissue cultures and *in vitro* fertilization (to produce test-tube babies) are examples. *Compare in vivo.*

in vivo Literally 'in life'; describing processes that occur within the living organism. *Compare in vitro.*

involuntary muscle *See* smooth muscle.

iodine A trace element essential in animal diets mainly as a constituent of the thyroid hormones. Iodine is not essential to plant growth although it is accumulated in large amounts by the brown algae. *See also* staining.

ion channel A protein that permits the passage of ions through a cell membrane, by crossing the membrane to form a pore. Various types occur widely within the plasma membrane and also in certain internal cell membranes, such as the tonoplast surrounding plant cell vacuoles, and the sarcoplasmic reticulum of skeletal muscle cells. Ion channels determine the electrical and chemical environment inside cells, including the resting potential, and are crucial for the excitability of cells such as nerve and muscle cells. They show varying degrees of selectivity; some allow passage only to one particular ion, whereas others may admit two or more ions of similar size and electrical charge (e.g. K^+ and Na^+). *Ungated ion channels* are always open, and the passage of ions depends mainly on the electrochemical gradients across the membrane. *Gated ion channels* can open and close; the main types are LIGAND-GATED ION CHANNELS, which open and close in response to binding of a signal molecule to the channel protein, and VOLTAGE-GATED ION CHANNELS, which open and close in response to changes in membrane potential. Yet others open and close due to binding of intracellular signals, such as G protein.

IPSP *See* inhibitory postsynaptic potential.

iron An essential nutrient for animal and plant growth. It is contained in the protein HEMOGLOBIN, which gives the color to erythrocytes (red blood cells) and is responsible for oxygen transport from the lungs. Iron deficiency leads to anemia. Iron is also found in other porphyrins, in cytochromes (which are important components of the electron-transport chain), and as a cofactor for certain enzymes. The iron-containing enzyme complex dinitrogenase is essential for the biological fixation of nitrogen by bacteria.

islets of Langerhans *See* pancreatic islets.

isoantigen A type of antigen that induces antibody production in genetically different individuals of the same species but not in the individual itself.

isoenzyme (isozyme) One of a group of enzymes that have similar catalytic properties but different chemical compositions, for example in the make-up of their polypeptide subunits. Hence, isoenzymes show differences in physical properties such as isoelectric point and electrophoretic mobility, and can be separated using appropriate techniques. Many enzymes are known to have isomeric forms; for example, lactate dehydrogenase has five forms. Variations in the isoenzyme constitution of individuals can be distinguished by electrophoresis.

isogamete *See* gamete.

isogamy The sexual fusion of gametes of similar size and form. It occurs in fungi and some protists. *Compare* anisogamy.

isograft (**syngraft**) A GRAFT between individuals that are genetically identical.

isoleucine *See* amino acids.

isotonic Designating a solution with an osmotic pressure or concentration equal to that of a specified other solution, usually taken to be within a cell. It therefore neither gains nor loses water by osmosis. *Compare* hypertonic; hypotonic.

isotope One of two or more atoms of the same element that differ in atomic mass, having different numbers of neutrons. For example ^{16}O and ^{18}O are isotopes of oxygen, both with eight protons, but ^{16}O has eight neutrons and ^{18}O has ten neutrons. A natural sample of most elements consists of a mixture of isotopes. Many isotopes are radioactive and can be used for labeling purposes. The isotopes of an element differ in their physical properties and can therefore be separated by techniques such as fractional distillation, diffusion, and electrolysis.

isozyme *See* isoenzyme.

Jacob, François (1920–) French molecular biologist who postulated the existence of messenger RNA. He also investigated genetic regulatory mechanisms in protein synthesis. He was awarded the Nobel Prize for physiology or medicine in 1965 jointly with J. L. Monod and A. M. Lwoff.

Jerne, Niels Kaj (1911–94) Danish microbiologist and immunologist who studied antibody generation. He was awarded the Nobel Prize for physiology or medicine in 1984 jointly with G. J. F. Köhler and C. Milstein.

jumping gene *See* transposon.

junk DNA Repetitive sequences of eukaryotic DNA that apparently have no useful function. *See* selfish DNA.

juvenile hormone (neotonin) A hormone secreted by endocrine glands associated with the brain in insects that prevents metamorphosis into the adult form and maintains the presence of larval characteristics. The exact mechanism of juvenile hormone action is not clear, but it appears to modify the effect of the molting hormone, ecdysterone.

K

Kandel, Eric R. (1929–) American biochemist noted for his work on signal transduction in the nervous system. He shared the 2000 Nobel Prize for physiology or medicine with P. Greengard and A. Carlsson.

Karrer, Paul (1889–1971) Swiss organic chemist noted for the synthesis of vitamins A, B$_2$, and E. He was awarded the Nobel Prize for chemistry in 1937. The prize was shared with W. N. Haworth.

karyogamy The fusion of two nuclei that exist within a common cytoplasm, as occurs in the formation of the zygote from two gametes (*see* fertilization). The process also occurs within the multinucleate plasmodium of slime molds belonging to the Myxomycota. *Compare* plasmogamy.

karyogram (**idiogram**) The formalized layout of the KARYOTYPE of a species, often with the chromosomes arranged in a certain numerical sequence.

karyokinesis *See* mitosis.

karyotype The physical appearance of the chromosome complement of a given species. A species can be characterized by its karyotype since the number, size, and shape of chromosomes vary greatly between species but are fairly constant within species.

Katz, (Sir) Bernard (1911–) German-born British biophysicist and neurophysiologist who worked on the physiology of muscle and nerve, especially the mechanism of release of acetylcholine by neural impulses. He was awarded the Nobel Prize for physiology or medicine in 1970 jointly with J. Axelrod and U. S. von Euler.

Kendall, Edward Calvin (1886–1972) American chemist who first isolated the thyroid hormone thyroxine (1915) and also isolated a number of hormones of the adrenal cortex and studied their action. He was awarded the Nobel Prize for physiology or medicine in 1950 jointly with P. S. Hench and T. Reichstein.

Kendrew, (Sir) John Cowdery (1917–97) British biophysicist and molecular biologist who determined the three-dimensional structure of the myoglobin molecule by x-ray diffraction. He was awarded the Nobel Prize for chemistry in 1962; the prize was shared with M. F. Perutz for their studies of the structures of globular proteins.

keratin One of a group of fibrous insoluble sulfur-containing proteins (scleroproteins) found in ectodermal cells of animals, as in hair, horns, and nails. Leather is almost pure keratin. There are two types: α-keratins and β-keratins. The former have a coiled structure, whereas the latter have a beta-pleated sheet structure. *Cytokeratins* are among the proteins that make up INTERMEDIATE FILAMENTS, key components of the cytoskeleton.

keratinization (**cornification**) A process occurring in vertebrate epidermis and epidermal structures in which KERATIN replaces the cytoplasm of a cell. For example, the cornified outer layer of the epidermis of the skin consists of dead horny cells. Hairs and nails also consist of keratinized cells.

ketohexose A ketose SUGAR with six carbon atoms, for example, fructose.

ketone body One of a group of organic substances formed in fat metabolism, mainly in the liver. Examples are acetoacetic acid and acetone. If the body has little or no carbohydrate as a respiratory substrate, *ketosis* occurs, in which more ketone bodies are produced than the body can use.

ketopentose A ketose SUGAR with five carbon atoms, for example, ribulose.

ketose A SUGAR containing a ketone (=CO) or potential ketone group.

ketosis *See* ketone body.

Khorana, Har Gobind (1922–) Indian organic chemist and biochemist. He was awarded the Nobel Prize for physiology or medicine in 1968 jointly with R. W. Holley and M. W. Nirenberg for their interpretation of the genetic code and its function in protein synthesis.

killer cell *See* natural killer cell; T-cell.

kilobase Symbol: kb A unit of length of a polynucleotide (i.e. DNA or RNA) equal to 1000 bases of a single strand, or 1000 base pairs (bp) of a double strand. *Compare* bp.

kinase 1. Any enzyme that transfers a phosphate group, usually from ATP. Kinases are important components of many intracellular signaling pathways. Addition of phosphate groups generally changes the structural conformation of a protein so that its enzymic activity is 'switched on'. *See* protein kinase.
2. An enzyme that activates the inactive form of other enzymes. For instance, when trypsinogen, the inactive form of trypsin, comes in contact with enterokinase, active trypsin is released.

kinesin A motor protein that transports cargoes along microtubules inside cells. The kinesin molecule has a pair of globular 'head' domains, which alternately bind to the microtubule in a walking action, using energy supplied by the hydrolysis of ATP. At the other end of the molecule, a 'tail' region binds to a cargo, such as a membrane-bound vesicle or cell organelle. Such a mechanism is responsible for transporting synaptic vesicles along the length of nerve cell axons from the cell body to the synapses, for example. Kinesins are also involved in the assembly of the mitotic spindle, and segregation of chromosomes.

kinetin (6-furfurylaminopurine) An artificial CYTOKININ found in extracts of denatured DNA; the first of the cytokinins to be isolated.

kinetochore A platelike structure within the CENTROMERE of a chromosome by which sister chromatids attach to spindle microtubules during mitosis. It becomes apparent during late prophase, when the chromosomes have condensed in readiness for mitosis. Proteins making up the kinetochore interact with the DNA of the centromere region (centromeric DNA) and with the microtubules. Hence, microtubules from one pole attach to one sister chromatid, while those from the opposite pole attach to the other sister chromatid. During anaphase, the sister chromatids are pulled to opposite poles of the cell.

kinetosome *See* basal body.

kinin 1. Any of a class of peptides that are formed from precursors in blood or other tissues and function as local hormones. They include the angiotensins, which constrict blood vessels and raise blood pressure; bradykinin, which dilates blood vessels and lowers blood pressure; and tachykinins, which have a similar effect and also stimulate salivation and tear production.
2. A former name for cytokinin.

Klug, (Sir) Aaron (1926–) Lithuanian-born British molecular biologist distinguished for his determination of the structure of transfer RNA and for his work on three-dimensional structure of complexes of proteins and nucleic acids. He was awarded the Nobel Prize for chemistry in 1982 for his development of crystallographic electron microscopy and his work on the structures of biologically important nucleic acid–protein complexes.

knockout *See* gene knockout.

Köhler, Georges Jean Franz (1946–95) German immunologist famous for his discovery of the use of cell fusion to produce a clonal cell population capable of producing unlimited quantities of a pure antibody. He was awarded the Nobel Prize for physiology or medicine in 1984 jointly with N. K. Jerne and C. Milstein.

Kornberg, Arthur Kornberg (1918–) American biochemist who discovered the enzyme DNA polymerase and used it to synthesize segments of DNA molecules and viral DNA. He was awarded the Nobel Prize for physiology or medicine in 1959 jointly with S. Ochoa for their discovery of

the mechanisms in the biological synthesis of ribonucleic acid and deoxyribonucleic acid.

Kossel, Karl Martin Leonhard Albrecht (1853–1927) German chemist noted for his discovery of adenine, cytosine, thymine, and uracil as breakdown products of nucleic acids. He also discovered histone and histidine. He was awarded the Nobel Prize for physiology or medicine in 1910 in recognition of his work on cell chemistry.

Krebs, Edwin Gerhard (1918–) American biochemist. He was awarded the Nobel Prize for physiology or medicine in 1992 jointly with Edmond H. Fischer for their discoveries concerning reversible protein phosphorylation as a biological regulatory mechanism.

Krebs, (Sir) Hans Adolf (1900–81) German-born British biochemist renowned for his work on metabolism and, in partic-

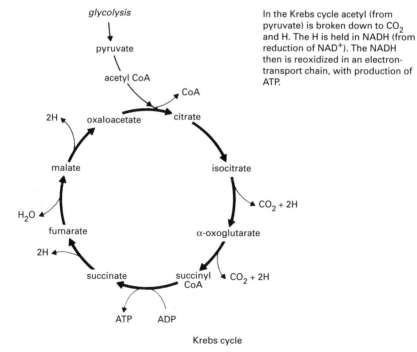

In the Krebs cycle acetyl (from pyruvate) is broken down to CO_2 and H. The H is held in NADH (from reduction of NAD^+). The NADH then is reoxidized in an electron-transport chain, with production of ATP.

Krebs cycle

ular, for the discovery of the cyclic metabolic pathways named after him. He was awarded the Nobel Prize for physiology or medicine in 1953 for his discovery of the TCA cycle. The prize was shared with F. A. Lipmann.

Krebs cycle (citric acid cycle; tricarboxylic acid cycle) A cycle of reactions that is universally important in living cells for respiration and biosynthesis. It forms the second stage of AEROBIC RESPIRATION, in which pyruvate, produced by glycolysis, is oxidized to carbon dioxide and water, with the production of large amounts of energy. In eukaryotic cells it takes place in the matrix of mitochondria, whereas in prokaryotic cells the Krebs cycle enzymes are associated with the plasma membrane.

In eukaryotes, pyruvate is transported from the cytosol into the mitochondria, where it combines with coenzyme A to form acetyl CoA. This reaction, catalyzed by pyruvate dehydrogenase, also produces carbon dioxide and the reduced coenzyme NADH. The two-carbon acetyl group reacts with the four-carbon oxaloacetate to form the six-carbon compound citrate, which is then decarboxylated in a series of reactions to reconstitute oxaloacetate. Most ATP production is coupled to the ELECTRON-TRANSPORT CHAIN. The operation of this depends on the generation of reduced coenzymes, NADH and $FADH_2$, by the Krebs cycle. The Krebs cycle itself produces only one GTP (similar to ATP) by substrate-level phosphorylation, per pyruvate entering the cycle. However, the reduced coenzymes it generates enter the electron-transport chain and yield another 11 ATP. Since two pyruvate molecules per glucose enter the cycle from glycolysis, 24 ATP (11 + 11 + 2) are produced in all. If the NADH generated by the pyruvate→acetyl CoA reaction is included, this adds a further 6 ATP, making a total of 30 ATP per glucose.

The Krebs cycle also plays a central role in providing building blocks for a range of biosynthetic pathways, including synthesis of various amino acids, cytochromes, and fatty acids. Oxaloacetate can also be converted to glucose, via phosphoenolpyruvate (*see* gluconeogenesis).

Kuhn, Richard Johann (1900–67) Austrian-born German chemist and biochemist. He is noted for his work on carotenoids and vitamins, including the synthesis of vitamins A and B_2 and the synthesis of the coenzyme flavin mononucleotide. He was awarded the Nobel Prize for chemistry in 1938.

labeling   The technique of using isotopes (usually RADIOACTIVE ISOTOPES) or other recognizable chemical groups to investigate biochemical reactions, or to identify the location of labeled compounds within cells. For instance, a compound can be synthesized with one of the atoms replaced by a radioactive isotope of the element. The radioactivity can then be used to follow the course of reactions involving this compound. Other commonly used labels include enzymes, which can be detected by their catalytic activity, and fluorescent reagents, such as fluorescein. Electron-dense labels (e.g. ferritin) can be attached specifically to certain structures in order reveal their location under the electron microscope.

lactic acid   A syrupy liquid occurring in sour milk as a result of FERMENTATION by lactobacillii. It is produced (L-form only) during ANAEROBIC RESPIRATION in animals as the end product of glycolysis.

lactic acid bacteria   A group of bacteria that ferment carbohydrates in the presence or absence of oxygen, with lactic acid always a major end product (*see* fermentation). They have a high tolerance of acid conditions. Lactic acid bacteria are involved in the formation of yoghurt, cheese, sauerkraut, and silage. They can occur as spoilage organisms and some are pathogenic, causing infections of the nasopharynx.

lagging strand   *See* DNA replication.

lambda phage (λ **phage**)   A bacteriophage that infects the bacterium *E. coli*. It is one of the best studied of all viruses, and is widely used as a cloning vector. The virus particle consists of an iscosahedral head, which contains the viral DNA genome, and a long tail, used by the virus to infect its bacterial host. When inside its host, lambda phage can undergo either LYSOGENY, in which the viral DNA is integrated into the bacterial chromosome, or lytic growth. In the latter, the viral DNA is replicated by the host and about 100 new virus particles are assembled, killing the host cell and releasing the progeny viruses.

In constructing a lambda cloning vector, only the viral genes needed for infection and lytic growth are retained. Other segments of the genome can be replaced with up to roughly 25 kb of foreign DNA for cloning. Lambda virions carrying foreign DNA are assembled *in vitro*, and allowed to infect *E. coli* cells plated in petri dishes. Inside each infected host cell, the vector DNA, including the recombinant fragment, is replicated numerous times. The progeny virions are released to form clear areas (plaques) on the lawn of bacterial cells. Such vectors are particularly well suited to preparing genomic libraries (*see* DNA library). For example, the entire human genome can be represented by 20–30 petri dishes, each containing some 5×10^4 lambda plaques.

lamella   1. A layer of photosynthetic membranes (thylakoids) in CHLOROPLASTS or in the cytoplasm of certain photosynthetic bacteria.
2. Any thin platelike structure, for example the layers of calcified matrix in bone.

lamellipodium   (*pl.* **lamellipodia**)   A broad flaplike protrusion of a cell, seen typically at the leading edge of moving cells such as fibroblasts and keratinocytes.

laminin A large cross-shaped protein found in the extracellular matrix, especially where this forms a thin fibrous network underlying epithelial and endothelial cells (*see* basal lamina). It is a multiadhesive matrix protein, binding to cells via cell adhesion molecules on their surface and linking them to collagen fibers and other matrix components.

lampbrush chromosome An extended chromosome structure found in the oocytes of certain animals, notably amphibians, during the prophase of meiosis. In those species that show a great increase in nuclear and cytoplasmic volume during prophase, the lampbrush chromosomes may measure up to 1 mm in length and 0.02 mm in width. Such chromosomes consist of two central strands along which fine loops extend laterally. The loops are associated with an RNA matrix and are sites of active transcription.

Landsteiner, Karl (1868–1943) Austrian-born American pathologist and immunologist distinguished for his discoveries of the human blood groups. He designated factors A, B, and O (1901), factors M and N (1927), and factor Rh (1940). He was awarded the Nobel Prize for physiology or medicine in 1930.

Langerhans, Paul (1847–88) German physician and anatomist who first described the *islets of Langerhans*.

larva (*pl.* **larvae**) The young immature stage into which many animals hatch after development of the fertilized egg. Larvae are independent and self-sustaining but differ appreciably from the adult in structure and mode of life and are usually incapable of sexual reproduction. Development into the adult is by metamorphosis. Most invertebrates have a larval stage (examples include the caterpillar of butterflies and moths and the ciliated planktonic larvae of many marine species); the tadpole of frogs is an example of a vertebrate larva.

latent virus A virus that can remain inactive in its host cell for a considerable period after initial infection. The viral nucleic acid becomes integrated in the host chromosome and multiplies with it. Eventual replication inside the host cell may be triggered by such factors as radiation and chemicals. An example of a latent virus is herpes simplex. *See also* provirus.

latex A liquid found in some flowering plants contained in special cells or vessels called LATICIFERS. It is a complex variable substance that may contain terpenes (e.g. rubber), resins, tannins, waxes, alkaloids, sugar, starch, enzymes, crystals, etc., and discourages browsing of the plant by herbivores. Commercial rubber comes from the latex of rubber plants; opium comes from alkaloids found in the latex of the opium poppy.

laticifer A latex-containing structure found in certain plants, for example, rubber, poppy, and euphorbia. Laticifers may be formed by fusion of cells to give vessels, or by the elongation and branching of a single coenocytic cell. *See* latex.

L-DOPA L-3,4-dihydroxyphenylalanine; an intermediate in the synthesis of dopamine, norepinephrine, and epinephrine and in the conversion of tyrosine to melanin pigments. L-DOPA is used to treat Parkinson's disease, a primary cause of which is a deficiency of dopamine in the brain cells.

leading strand *See* DNA replication.

lecithin *See* phosphatidylcholine.

Lederberg, Joshua (1925–) American geneticist. He was awarded the Nobel Prize for physiology or medicine in 1958 for his discoveries concerning genetic recombination and the organization of the genetic material of bacteria. The prize was shared with G. W. Beadle and E. L. Tatum.

Leloir, Luis Federico (1906–87) Argentinian biochemist who first isolated uridine diphosphate glucose and discovered the role of sugar nucleotides in interconversion of sugars and in polysaccharide formation. He was awarded the Nobel Prize for chemistry in 1970.

leptotene In MEIOSIS, the stage in early prophase I when the chromosomes, already replicated, start to condense and appear as fine threads, although sister chromatids are not yet distinct. The spindle starts to form around the intact nucleus.

leucine *See* amino acids.

leukocyte (**white blood cell**) A type of blood cell that has a nucleus but no pigment. White cells are larger and less numerous than red cells (about 6000–8000 per cubic millimeter of blood). They are important in defending the body against disease because they engulf bacteria and produce antibodies. They are all capable of ameboid movement. There are several types of leukocytes, and they can be divided into two groups according to the presence or absence of granules in the cytoplasm: *granulocytes*, which comprise NEUTROPHILS, EOSINOPHILS and BASOPHILS; and *agranulocytes*, which include LYMPHOCYTES and MONOCYTES. The most numerous are the neutrophils (70%) and lymphocytes (25%). Leukocytes have a very short lifespan and are continuously produced in the MYELOID TISSUE of the red marrow.

leukoplast A colorless PLASTID, i.e. one not containing chlorophyll or any other pigment.

leukotrienes A group of substances that are produced by certain cell types to serve as local hormones. Structurally related to the prostaglandins, they are derived from polyunsaturated fatty acids, and all contain at least three double bonds. For example, leukotriene B_4 is secreted by activated mast cells and promotes increased permeability of blood vessels and infiltration of certain types of white blood cells (basophils, neutrophils, and eosinophils), while leukotriene C_4 causes constriction of blood capillaries in various tissues, including the heart and brain.

Levi-Montalcini, Rita (1909–) Italian cell biologist. She was awarded the Nobel Prize for physiology or medicine in 1986 jointly with S. Cohen for their discoveries of growth factors.

levorotatory Describing compounds that rotate the plane of polarized light to the left (anticlockwise as viewed facing the oncoming light). *Compare* dextrorotatory.

Lewis, Edward B. (1918–) American biochemist who shared the 1995 Nobel Prize for physiology or medicine with C. Nüsslein-Volhard and E. F. Wieschaus for work on the genetic control of early embryonic development.

LH *See* luteinizing hormone.

Libby, Willard Frank (1908–80) American chemist. He was awarded the Nobel Prize for chemistry in 1960 for the invention of radiocarbon dating.

ligand 1. (*Chemistry*) An ion, atom, or chemical group that is attached to a central metal atom and supplies a pair of electrons to it, forming a type of covalent bond called a coordinate bond.
2. (*Cell Biology*) Any molecule that binds specifically to a site on a cell receptor molecule, causing a change in the shape of the receptor and triggering some response in the cell.

ligand-gated ion channel An ion channel whose opening is activated by binding of a signal molecule to the channel protein. On arriving at the cell, the signal molecule acts as a ligand, binding specifically and tightly to the channel protein's receptor site, which is situated on the exterior region of the channel protein. Binding causes a change in the conformation of the channel protein, which opens the channel. Many neurotransmitters act as ligands for ion-channel receptors. For example, at neuromuscular junctions, the binding of acetylcholine to cation channels causes the influx of sodium ions and consequent depolarization of the muscle cell membrane, which leads to muscle contraction. Other types of ligand-gated ion channels occur at synapses and in the brain.

ligase An enzyme that catalyzes the bond formation between two substrates at the expense of the breakdown of ATP or some other nucleotide triphosphate. *See* DNA ligase.

light-dependent reactions The reactions of photosynthesis that convert light energy into the chemical energy of NADPH and ATP. *See* photosynthesis.

light green *See* staining.

light-harvesting complex *See* photosystem.

light microscope A microscope that uses visible light as the source of radiation to view an image. Most modern light microscopes are compound microscopes, which incorporate two or more sets of lenses (*see* microscope). The limit of resolution is 200 nm; i.e. a light microscope cannot distinguish two objects if they are less than 200 nm apart, regardless of the magnification. This is because of constraints imposed by, among other factors, the wavelength of visible light. Preparations of cells and tissues must generally be fixed, sectioned, and stained to enable particular structures to be distinguished through conventional light microscopes. One powerful technique is FLUORESCENCE MICROSCOPY. This employs fluorescent dyes to highlight particular molecules, and can be used in living as well as fixed specimens. PHASE-CONTRAST MICROSCOPY exploits small differences in refractive index between different components of a specimen, and gives good definition of unstained living specimens.

limiting layer *See* meristoderm.

LINES (long interspersed elements) A class of mobile elements that are abundant in mammals and act as nonviral RETRO-TRANSPOSONS, inserting copies of themselves at new sites within the genome. In humans, each sequence is typically about 6–7 kb long, and there are numerous copies of each. For example, the most common are the L1 LINES, of which there are around 600 000 copies in the human genome, representing some 15% of the total DNA. The exact mechanism by which LINES copy and insert themselves is unknown. The LINE sequence encodes a reverse transcriptase enzyme, which is believed to synthesize a complementary DNA (cDNA) copy of the RNA transcript. The cDNA then probably serves as a template for the synthesis of a double-stranded DNA copy, which is inserted into the cell's DNA.

linkage The occurrence of genes together on the same chromosome so that they tend to be inherited together and not independently. Groups of linked genes are termed *linkage groups* and the number of linkage groups of a particular organism is equal to its haploid chromosome number. Linkage groups can be broken up by CROSSING OVER at meiosis to give new combinations of genes. The closer together on a chromosome two genes are, the more strongly linked they are, i.e. there is less chance of a cross over between them.

linkage group *See* linkage.

linkage map *See* chromosome map.

lipase Any of various enzymes that catalyze the hydrolysis of fats to fatty acids and glycerol. Lipases are present in the pancreatic juice of vertebrates.

lipid Any of a diverse group of substances that are characteristically soluble in organic solvents, such as ether and benzene, but insoluble in water. Lipids play various roles in living organisms. TRIGLYCERIDES, which are esters of glycerol and fatty acids, make up the fats and oils that serve as energy reserves in plants and animals. Waxes are esters of long chain monohydric alcohols and fatty acids; they provide a waterproof covering for the external surfaces of many organisms (*see* wax). PHOSPHOLIPIDS form another biologically important class of lipids. These are diglycerides, in which only two of the hydroxyl groups of glycerol are esterified with a fatty acid, the third being linked to a phosphate group. Phospholipids are major components of cell membranes. Membranes often contain GLYCOLIPIDS as well. These are similar to phospholipids but contain the sugar galactose instead of the phosphate group. STEROIDS, and their alcohol derivatives (sterols), are also considered as lipids. The most abundant example is cholesterol, which is found in animal cell membranes and is the precursor for the production of steroid hormones and vitamin D. *See also* lipoprotein.

lipidoplast *See* elaioplast.

Lipmann, Fritz Albert (1899–1986) German-born American biochemist. He was awarded the Nobel Prize for physiology or medicine in 1953 for his discovery of coenzyme A and its importance for intermediary metabolism. The prize was shared with H. A. Krebs.

lipoprotein Any conjugated protein formed by the combination of a protein with a lipid. In the blood of humans and other mammals, cholesterol, triglycerides, and phospholipids associate with various plasma proteins to form lipoproteins. These are spherical particles, 7.5–70 nm in diameter. Each particle consists typically of an outer single layer (monolayer) of phospholipid, surrounding a core of cholesterol esters (mainly cholesterol and a long-chain fatty acid). Within the outer surface are free cholesterol molecules and proteins. Such lipoproteins are placed in several classes, according to their density. The largest lipoproteins in this size range are the *very low-density lipoproteins* (VLDLs), which are formed in the liver and contain up to about 20% cholesterol. *Low-density lipoproteins* (LDLs) are formed in plasma from VLDLs and contain over 50% cholesterol. LDLs transport cholesterol from the liver to peripheral tissues, and high blood levels of these lipoproteins are thought to be a factor in the development of fatty arterial deposits and cardiovascular disease. *High-density lipoproteins* (HDLs), with about 20% cholesterol, are the smallest of the plasma lipoproteins, and transport cholesterol from tissues to the liver for excretion in bile.

The largest of all plasma lipoproteins are the *chylomicrons*, which act as vehicles for the absorption of fat from the intestine. Measuring up to 100 nm in diameter, they are formed in the intestinal mucosa and contain mostly triglycerides, with relatively small amounts of cholesterol and protein. They enter the lacteals of the intestinal villi and are conveyed via the lymphatic system to the bloodstream. *See* low-density lipoprotein.

liposome A microscopic spherical sac consisting of a lipid envelope enclosing fluid. They are made in the laboratory by adding an aqueous solution to a gel of complex lipids, and have a wall consisting of a double layer of lipids similar to that of cell membranes. Liposomes are used experimentally as models of cells or cell organelles, and also in medicine to deliver toxic drugs to target tissues in the body.

locus (*pl.* loci) The position of a gene on a chromosome. Alleles of the same gene occupy the equivalent locus on homologous chromosomes.

Loewi, Otto (1873–1961) German-born American physiologist. He shared the Nobel Prize for physiology or medicine in 1936 with H. H. Dale for their work on the chemical transmission of nerve impulses.

lomasome An infolding of the plasma membrane found particularly in fungal hyphae and spores, and also in some algae and higher plants. Lomasomes may be concerned with the formation or breakdown of the cell wall, or with secretion or endocytosis.

long interspersed elements *See* LINES.

lophotrichous Describing bacteria that possess a tuft of flagella.

low-density lipoprotein (LDL) A spherical particle, typically about 20–25 nm in diameter, that is found in blood plasma and transports cholesterol to tissue cells. It is bounded by a single layer of phospholipid and free cholesterol, which encloses a core of cholesterol esterified to a long-chain fatty acid. Embedded in the surface layer is a single large protein, called apo-B, which assists in binding of the LDL to cell-surface receptors. LDLs are taken into cells by receptor-mediated ENDOCYTOSIS. The cholesterol is incorporated into cell membranes or stored as lipid droplets. High concentrations of LDLs in the blood have been associated with an increased risk of atherosclerosis ('hardening of the arteries').

LTH *See* luteotropic hormone.

luciferase An enzyme that catalyzes the oxidation of luciferin to produce the light of bioluminescent reactions, such as occur in fireflies and some bacteria (*see* bioluminescence). It is also used as a reporter molecule in certain antibody assays of proteins or other molecules. The luciferase is linked to a specific antibody, which binds to a target molecule. When ATP and luciferin are added, the appearance and amount of light emitted corresponds to the presence and amount of the target molecule.

luciferin Any substance that acts as substrate for an enzyme (*luciferase*) that catalyzes a light-emitting reaction in living organisms (bioluminescence). Luciferens vary widely in different organisms.

lumen 1. The central space that remains, surrounded by cell walls, in a cell that has lost its living contents (e.g. in xylem elements).
2. The central cavity or canal within a tube, duct, organelle, or similar structure.

Luria, Salvador Edward (1912–91) Italian-born American physician and biologist. He was awarded the Nobel Prize for physiology or medicine in 1969, jointly with M. Delbrück and A. D. Hershey, for discoveries concerning the replication mechanism and the genetic structure of viruses.

luteinizing hormone (LH; interstitial-cell-stimulating hormone; ICSH) A glycoprotein hormone secreted by the anterior pituitary lobe under regulation of the hypothalamus. In female mammals it stimulates secretion of estrogen, ovulation, and formation of corpora lutea. In male mammals it stimulates interstitial cells in the testes to secrete androgens. *See also* gonadotropin.

luteotrophic hormone *See* luteotropic hormone.

luteotropic hormone (luteotrophic hormone; LTH) *See* prolactin.

Lwoff, André Michel (1902–94) French microbiologist. He was awarded the Nobel Prize for physiology or medicine in 1965, jointly with F. Jacob and J. L. Monod, for work on the genetic control of enzyme and virus synthesis.

lyase An enzyme that catalyzes the separation of two parts of a molecule with the formation of a double bond in one of them. For example, fumarase catalyzes the interconversion of malic acid and fumaric acid.

lymph The fluid contained within the vessels of the lymphatic system. It is derived from tissue fluid that is drained from intercellular spaces, and is similar to plasma but with a lower protein concentration and contains cells (mainly lymphocytes), bacteria, etc. It is colorless except in the region of the small intestine where absorbed fat gives the lymph a milky appearance.

lymphatic tissue *See* lymphoid tissue.

lymph node (lymph gland) One of numerous flat oval structures distributed along the lymphatic vessels and clustered in certain regions, such as the neck, armpits, and groin. Lymph nodes are composed of lymphatic tissue and contain white blood cells of the immune system (*see* lymphocytes). They act as defense posts against the spread of infection, their lymphocytes engulfing bacteria and other foreign materials from the lymph; the nodes may become inflamed and enlarged as a result.

lymphoblast *See* lymphocyte.

lymphocyte A relatively small white blood cell (leukocyte), characterized by a very large nucleus and a small amount of clear (agranular) cytoplasm. Lymphocytes comprise 25% of all leukocytes and produce antibodies, important in defense against disease (*see* immunity). Lymphocytes are derived from the division of precursor cells (*lymphoblasts*), which themselves derive from hematopoietic stem cells in lymphoid tissue (i.e. lymph nodes, thymus, tonsils, and spleen). During infection, antigens stimulate B-lymphocytes, or B-CELLS, in the lymphoid tissue to multiply rapidly, and the resulting PLASMA CELLS are released into the bloodstream to produce the appropriate antibody. This antibody response is regulated by the other main class of lymphocytes, the T-CELLS. These are also responsible for direct killing of infected cells, and other aspects of immunity.

lymphoid tissue (lymphatic tissue) Tissue in which lymphocytes are produced, found in the lymph nodes, tonsils, spleen, and thymus. It consists of a delicate network of cells through which lymph flows continuously. Lymphocytes have a life span of only a few days and must be constantly replaced. When an antigen enters lymphoid tissue, it is 'recognized' by one particular type of lymphocyte, which then multiplies rapidly; the resulting PLASMA CELLS circulate in the blood, producing the necessary ANTIBODY for that antigen. Lymphoid tissue also contains numerous MACROPHAGES, which ingest foreign particles, especially bacteria, hence the lymph nodes act as filters to remove bacteria from the lymph.

lymphokine Any of various CYTOKINES that are released by lymphocytes when activated by encountering their specific antigen. They serve as chemical signals to other cell types involved in the immune response, such as macrophages, neutrophils, and ba-

sophils. Examples include *macrophage migration inhibition factor*, which inhibits the movement of macrophages, and *tumor necrosis factor*, which stimulates a range of responses in body cells. *Compare* interleukin.

Lynen, Feodor (1911–79) German biochemist. He was awarded the Nobel Prize for physiology or medicine in 1964, jointly with K. E. Bloch, for discoveries concerning the mechanism and regulation of the cholesterol and fatty acid metabolism.

lyophilic Solvent attracting. When the solvent is water, the word *hydrophilic* is often used. In aqueous or other polar solutions, ions or polar groups are lyophilic. For example, the phosphate (PO_3^{2-}) group on a phospholipid is the lyophilic (hydrophilic) part of the molecule. In *lyophilic colloids* the dispersed particles have an affinity for the solvent, and the colloids are generally stable. *Compare* lyophobic.

lyophobic Solvent repelling. When the solvent is water, the word *hydrophobic* is used. In aqueous or other polar solvents, the lyophobic group will be nonpolar. For example, the long hydrocarbon 'tail' of a phospholipid is the lyophobic (hydrophobic) part. In *lyophobic colloids* the dispersed particles are not solvated and the colloid is easily solvated. Gold and sulfur sols are examples. *Compare* lyophilic.

lysine *See* amino acids.

lysis The bursting of a cell or cell organelle due to rupture of the enclosing membrane(s). If animals cells are placed in a very dilute (hypotonic) aqueous solution, they take in water by osmosis, so that they swell and burst. This osmotic lysis is induced, for example, by complement proteins when they bind to invading foreign cells, such as bacteria, as part of the immune response. Lysis also occurs in virus-infected cells, following viral replication (*see* lytic cycle). Under normal conditions, body cells do not undergo lysis following when they reach the end of their useful life, but are disposed of in a manner that does not release the cell contents (*see* apoptosis).

lysogeny 1. The formation of an intercellular space in plants by dissolution of cells. *Compare* schizogeny.
2. A phage–bacterium relationship in which lysing of the bacterium does not occur. The PHAGE (known as a *temperate phage*) penetrates the host cell and its nucleic acid becomes integrated into the bacterial DNA. In this state the phage is termed a *prophage,* and the cell is described as *lysogenic.* Most of the viral genes are repressed and both bacteria and phage reproduce together, producing infected daughter cells. In a process called *induction,* certain environmental factors can cause the phage to leave the host DNA and resume the lytic cycle.

lysosome An organelle, found in animal cells and certain protists, that contains various hydrolytic enzymes responsible for degrading aged or defective cellular components, or materials taken in by the cell from outside. Lysosomes are bounded by a single membrane, and can fuse with other membranous vesicles. Hence, they can contribute enzymes to food vacuoles, as in *Amoeba,* or to similar vacuoles formed in white blood cells during phagocytosis. The interior of a lysosome is maintained at a pH of about 4.8, providing the acidic conditions under which the lysosomal enzymes function best. If these enzymes escape, they are virtually inactive at the neutral pH of the cytosol, thereby safeguarding the cell's contents from destruction. The lysosomal membrane contains proton pumps, which pump hydrogen ions into the interior of the lysosome, and also chloride channels, which allow the entry of chloride ions. Consequently, the net result is the influx of hydrochloric acid (HCl), which gives the acidic conditions.

Lysosomal enzymes are synthesized in the rough endoplasmic reticulum, from whence they enter the Golgi apparatus. Here they are sorted and labeled for despatch to lysosomes, via transport vesicles. *Primary lysosomes* are roughly spherical and do not contain debris. They fuse with other organelles that contain material destined for digestion, forming a *secondary lysosome,* in which debris is apparent and enzymic digestion takes place. For example, a bacterial cell taken in by phagocytosis is first incorporated in a phagosome, which then fuses with a primary lysosome. Food material internalized by endocytosis enters ENDOSOMES, which subsequently undergo fusion with lysosomes. In plant cells, the VACUOLE has an acid pH and contains an array of hydrolytic enzymes, enabling it to perform the role played by lysosomes in animal cells. *See also* autophagy.

lytic cycle *See* virus.

M

McClintock, Barbara (1902–92) American geneticist noted for her studies on chromosome breakage and reunion and her proposal of the existence of transposable elements. She was awarded the Nobel Prize for physiology or medicine in 1983.

Macleod, John James Richard (1876–1935) British physiologist and biochemist. He shared the Nobel Prize for physiology or medicine in 1923 with F. G. Banting for the discovery of insulin.

macromolecule   A very large molecule, usually a polymer, having a very high molecular mass. Proteins and nucleic acids are examples.

macronucleus (meganucleus)   The larger of the two nuclei found in certain protists, particularly ciliates (e.g. *Paramecium*). The smaller of the two is the *micronucleus*. The macronucleus contains multiple copies of the DNA needed for normal (nonreproductive) cell metabolism (i.e. is polyploid). It contains nucleoli, is variable in form, and divides amitotically. It degenerates during sexual reproduction and is reconstituted from the micronuclear chromosomes that produce the zygote. The multiple DNA copies are probably required to produce sufficient RNA for protein synthesis in the relatively large volume of cytoplasm.

The diploid *micronucleus* contains much less DNA and can undergo normal nuclear division (mitosis or meiosis). It is involved in sexual reproduction, when two individuals unite by cytoplasmic bridges (conjugation) and exchange micronuclei.

macrophage   A large ameboid cell that can engulf, ingest, and destroy bacteria, damaged cells, and worn-out red blood cells. This process, called PHAGOCYTOSIS, is an important part of the body's defense against disease. Macrophages are found free ('wandering') in the tissues, in the blood (as monocytes), in connective tissue (as histiocytes), in the lining of the blood sinusoids of the liver (as Kupffer cells), and in LYMPHOID TISSUE. Macrophages can also promote INFLAMMATION by releasing cytokines such as interleukin 1 and tumor necrosis factor. They make up the MONONUCLEAR PHAGOCYTE SYSTEM.

macrosclereid   An elongated rod-shaped cell (a type of SCLEREID), many of which form the outer protective layer in the seed coat (testa) and fruit wall of some plants.

magnesium   An element that is essential for the growth and maintenance of plants, animals, and other organisms. It is contained in the chlorophyll molecule and is thus essential for photosynthesis. In animals it is found in bones and teeth. As magnesium carbonate it is found in large quantities in the skeletons of certain marine organisms, and in smaller quantities in the muscles and nerves of higher animals. It is an essential cofactor for certain phosphate enzymes, for example phosphotransferases, including kinases. High concentrations of magnesium ions, Mg^{2+}, are needed to maintain ribosome structure.

major histocompatibility complex (MHC)   A large cluster of related genes found in mammals and encoding cell-surface proteins (MHC proteins) that play several vital roles in the immune system. They form the markers (i.e. 'self' antigens) of the MHC system of self-recognition, which prevents cells of the immune system (lymphocytes) attacking self tissue, and they serve to 'present' foreign antigens to lymphocytes. In humans, the MHC complex consists of numerous genes located on chromosome 6; the genes and the antigens they encode are called the *HLA system*.

MHC genes generally fall into three classes, with protein products having different roles. *Class I MHC proteins* are found on virtually all cells. Within the cell they attach themselves to processed foreign antigen, for example from an invading virus, and migrate to the cell surface where they are recognized by cytotoxic T-CELLS, which are stimulated to destroy the infected cell. *Class II MHC proteins* occur only on certain cells of the immune system, notably macrophages and B-CELLS. They combine with foreign antigen such as debris from bacteria phagocytosed by the macrophage, and are recognized by another set of T-cells called T-helper cells. The latter stimulate the macrophages to destroy the foreign antigens they contain, and also activate antibody secretion by B-cells. *Class III MHC proteins* are components of the complement system.

A large number of alleles exist for the MHC genes, which creates immense genetic variability in the MHC proteins. Hence, only close relatives are likely to have similar class I MHC proteins, i.e. show some degree of histocompatibility. As a consequence, a graft from such a relative is more likely to be tolerated by the immune system. Tissue grafts from unrelated individuals have different class I MHC proteins, and are recognized as foreign and killed by the recipient's cytotoxic T-cells, causing rejection of the graft.

malic acid A colorless crystalline carboxylic acid, which occurs in acid fruits such as grapes and gooseberries. In biological processes malate ion is an intermediate component of the KREBS CYCLE.

manganese *See* trace element.

mannitol A soluble sugar alcohol (carbohydrate) found widely in plants and forming a characteristic food reserve of the brown algae. It is a hexahydric alcohol, i.e. each of the six carbon atoms has an alco-

hol (hydroxyl) group attached. In medicine it is used as a diuretic to treat fluid retention.

mannose A simple sugar found in many polysaccharides. It is an aldohexose, isomeric with glucose.

marker gene A gene of known location and function which can therefore be used to establish the relative positions and functions of other genes. During gene transfer, a marker gene may be linked to the transferred gene to determine whether or not the transfer has been successful. *See* chromosome map; genetic marker.

Martin, Archer John Porter (1910–) British biochemist and physical chemist. He shared the Nobel Prize for chemistry in 1952 with R. L. M. Synge for their invention of partition chromatography.

mast cell A type of white blood cell (leukocyte) with granular cytoplasm found within connective tissue, for example beneath the skin and around blood vessels. Mast cells bind certain immunoglobulin antibodies (IgE) to their cell membrane, and when a specific antigen is encountered this triggers the mast cell to release chemical mediators, such as histamine, tumor necrosis factor, and leukotriene B_4, from its granules. These substances increase vascular permeability at the site, causing inflammation and attracting other types of immune cell. Mast cells are also activated by complement proteins, and are responsible for causing the symptoms of various allergies and other hypersensitivity reactions. *See also* basophil.

mastigoneme A small lateral projection occurring on certain types of undulipodia (flagella).

mastigote (flagellate) Any eukaryotic microorganism that moves by means of one or more undulipodia (flagella).

Maxam–Gilbert method *See* DNA sequencing.

mechanoreceptor A receptor that responds to a mechanical stimulus, for example touch, pressure, or sound.

Medawar, (Sir) Peter Brian (1915–87) British zoologist and immunologist. He shared the Nobel Prize for physiology or medicine in 1960 with F. M. Burnet for their discovery of acquired immunological tolerance.

medulla *See* pith.

megabase Symbol: Mb A unit of length used for measuring polynucleotides (i.e. DNA or RNA) and equal to 10^6 bases or base pairs. *Compare* kilobase.

megakaryocyte *See* platelet.

meganucleus *See* macronucleus.

meiosis A type of nuclear division leading to the production of daughter nuclei with half the genetic complement of the parent cell. Hence, one diploid parent cell undergoing meiosis gives rise to four haploid daughter cells, which can potentially act as gametes. Fusion of two gametes at fertilization restores the diploid chromosome complement.

Meiosis consists of two divisions during which the chromosomes replicate only once. Like MITOSIS the stages prophase, metaphase, and anaphase can be recognized. However, during prophase homologous chromosomes become paired up forming bivalents. At the end of prophase genetic material may be exchanged between the chromatids of homologous chromosomes, by a process called CROSSING OVER. Meiosis also differs from mitosis in that after anaphase, instead of nuclear membranes re-forming, there is a second division, which may be divided into metaphase II and anaphase II. The second division ends with the formation of four haploid nuclei, which develop into gametes.

Meiosis gives rise to genetic variation in two ways: (1) there is 'shuffling', or recombination, of maternal and paternal genes during crossing over; and (2) there is independent assortment of maternal and paternal chromosomes to the gametes, due to the random orientation of chromosome pairs on the spindle during metaphase I.

melanin One of a group of pigments found in animals and plants, derived from the amino acid tyrosine. The colors range from black through brown to yellow, orange, or red. In animals melanin occurs in melanophores (pigment cells) in the skin, usually below the epidermis. It gives color to the skin, hair, and eyes of animals and causes color in various seedlings and roots of plants. The absence of the enzyme tyrosinase in animals leads to a condition known as *albinism*, in which no pigment develops in the eyes, skin, or hair.

melanocyte-stimulating hormone (MSH) A peptide hormone produced by the anterior pituitary gland in vertebrates. It has a marked action on pigmentation in the skin of amphibians and reptiles, but its physiological role in mammals is unclear.

membrane A structure consisting mainly of lipid and protein (lipoprotein) surrounding all living cells as the PLASMA MEMBRANE, and also found surrounding organelles within cells. Membranes function as selectively permeable barriers, controlling passage of substances between the cell and its organelles, and the environment. They are typically 7.5–10 nm in thickness with two regular layers of lipid molecules (a *bilayer*) containing various types of protein molecules. Some proteins penetrate through the membrane, others are associated with the inner or outer face; some float freely over the surface while others remain stationary (*see* fluid mosaic model). Various types of transport proteins and protein-lined channels convey ions and small molecules across the membrane – either actively, with the expenditure of energy (usually in the form of ATP), or passively. Larger molecules or particles enter or leave cells by endocytosis or exocytosis, respectively. The outer face of plasma membranes also contains receptor proteins, which bind signal molecules from the cell's environment and activate signal transduc-

Prophase

Leptotene

Chromosomes appear as single uncoiled threads

Zygotene

Homologous chromosomes attract each other, coming together to form bivalents

Pachytene

Chromosomes shorten by coiling and individual chromatids become distinguishable, giving tetrads

Diplotene — Diakinesis

Homologous chromosomes repel each other at the centromeres, remaining attached only at chiasmata

Metaphase I

Nuclear membrane breaks down, spindle forms, and bivalents align themselves along the spindle equator

Anaphase I

Homologous chromosomes continue to repel each other, the homologues of each pair moving to opposite ends of the spindle

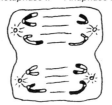 *(Metaphase II — Anaphase II)*

The haploid number of chromosomes gathers at either end of the spindle

Metaphase II — Anaphase II

Two spindles form at right angles to the first and chromatids separate

Telophase

A nuclear membrane forms around each group of daugher chromosomes to form four haploid nuclei and the cytoplasm divides forming four gametes

Meiosis

tion proteins on the inner face or in the cell's interior.

The lipids are mostly PHOSPHOLIPIDS. These are polar molecules, i.e. one end (the phosphate end) is *hydrophilic* (water-loving) and faces outwards, while the other end (two fatty acid tails) is *hydrophobic* (water-hating) and faces inwards. Short chains of sugars may be associated with the proteins or lipids forming glycoproteins and glycolipids. The particular types of carbohydrates, lipids, and proteins determine the characteristics of the membrane, affecting, for example cell–cell recognition (as in embryonic development and immune mechanisms), permeability, and hormone recognition. Membranes may contain efficient arrangements of molecules involved in certain metabolic processes, for example electron transport and phosphorylation (ATP production) in mitochondria and chloroplasts. *See* osmosis; freeze fracturing.

membrane hybridization assay A technique for detecting a particular cloned DNA fragment among a collection of different cloned fragments, commonly used in screening a DNA LIBRARY. The double-stranded DNA fragments are 'melted' by heating them in a dilute salt solution. This breaks the hydrogen bonds between the two strands of each fragment. The single strands are then irreversibly bound to a treated nylon membrane or nitrocellulose filter. The membrane is incubated with a radioactively labeled probe (DNA or RNA), which will undergo base pairing (hybridization) with any complementary DNA sequence attached to the membrane. After washing away excess probe, the remaining bound probe reveals the location of the sought-for fragment (detected by autoradiography).

membrane potential The difference in electrical potential (voltage) that exists across a cell membrane due to the unequal distribution of positively and negatively charged ions and molecules on either side of the membrane. Perhaps most significant to the life of a cell is the RESTING POTENTIAL across the plasma membrane when the cell is in an unexcited state. *See also* action potential.

memory cell *See* B-cell.

menaquinone *See* vitamin K.

Mendel, Gregor Johann (1822–84) Austrian monk and botanist who did early work on genetics.

Mendelism The theory of inheritance according to which characteristics are determined by particulate 'factors', or genes, that are transmitted by the germ cells. It is the basis of classical genetics, and is founded on the work of Mendel in the 1860s. *See* Mendel's laws.

Mendel's laws Two laws formulated by Mendel to explain the pattern of inheritance he observed in plant crosses. The first law, the *Law of Segregation*, states that any character exists as two factors, both of which are found in the somatic cells but only one of which is passed on to any one gamete. The second law, the *Law of Independent Assortment*, states that the distribution of such factors to the gametes is random; if a number of pairs of factors is considered each pair segregates independently.

Today Mendel's 'characters' are termed genes and their different forms (factors) are called alleles. It is known that a diploid cell contains two alleles of a given gene, each of which is located on one of a pair of homologous chromosomes. Only one homolog of each pair is passed on to a gamete. Thus the Law of Segregation still holds true. Mendel envisaged his factors as discrete particles but it is now known that they are grouped together on chromosomes. The Law of Independent Assortment therefore only applies to pairs of alleles found on different chromosomes.

Menten, Maude Leonora (1879–1960) Canadian physician and biochemist, noted for her collaboration with Leonor Michaelis in the formulation of Michaelis–Menten kinetics of enzyme action.

meristem A distinct region of actively dividing cells in the body of a plant that is primarily concerned with growth. Plants contain numerous meristems. In active meristems separation occurs between the cell that remains meristematic (the INITIAL) and the cell ultimately being differentiated. Two basic meristematic groups are the *primary apical meristems* at root and shoot apices, and the *secondary lateral meristems*, which include vascular and cork cambia, and are responsible for increasing the girth of stems and roots. *Intercalary meristems* occur at other locations, for example, at the leaf sheath bases and internodes of grasses and other monoocotyledons. Meristematic cells are typically small, with a large nucleus and dense protoplast. Apical meristems comprise longitudinal files of meristematic cells, whose roles are to form the primary root or shoot tissues, and to elongate the root or shoot. These cell files form three distinct longitudinal zones: *protoderm*, which differentiates into epidermal tissues; *procambium*, which gives rise to the primary phloem and xylem; and *ground meristem*, which differentiates into pith, cortex, and mesophyll.

meristoderm (limiting layer) The outermost cellular layer of the thallus of certain brown algae, including kelps. It consists of small densely packed rectangular cells containing brown pigmented plastids and covered by a mucilaginous layer to prevent desiccation. The meristoderm maintains its meristematic activity and assists the outer cortical layers in adding to the thickness of the thallus.

meroblastic Describing the type of incomplete CLEAVAGE that occurs in very yolky eggs (e.g. of birds and sharks), in which the egg cytoplasm but not the yolk divides. *Compare* holoblastic.

Merrifield, Robert Bruce (1921–) American peptide chemist and biochemist. He was awarded the Nobel Prize for chemistry in 1984 for his development of methods of chemical synthesis on a solid matrix.

mesenchyme A loose network of cells derived from either mesoderm or endoderm and usually underlying epithelial layers in animal embryos. It is sometimes divided into *primary mesenchyme*, the first cells to invade the blastocoel, and *secondary mesenchyme*, later contributions of diffuse cells from other sources. Mesenchymal cells interact with overlying epithelium in the development of glands and internal organs such as kidney, gut, and pancreas. This interaction typically involves reciprocal INDUCTION, such that each cell type influences the other as together they form the various parts of a particular organ.

mesoderm The GERM LAYER in an early animal embryo from which the muscles, reproductive organs, circulatory system, and, in vertebrates, the skeleton and excretory system usually develop. At gastrulation the mesoderm comes to lie between ectoderm on the outside and endoderm lining the gut. Either side of the notochord and neural tube, blocks of mesoderm called SOMITES develop, arranged in a series of paired segments running from front to rear. These spread laterally and ventrally, and give rise to epithelial sheets of cells and wandering cells known as mesenchyme. In most animals the coelom divides the mesoderm into an outer *somatopleure* under the skin and an inner *splanchnopleure* around the gut.

mesophyll Specialized tissue located between the epidermal layers of the leaf. Veins, supported by sclerenchyma and collenchyma, are embedded in the mesophyll. *Palisade mesophyll* consists of cylindrical cells, at right angles to the upper epidermis, with many chloroplasts and small intercellular spaces. It is the main photosynthesizing layer in the plant. *Spongy mesophyll*, adjacent to the lower epidermis, comprises interconnecting irregularly shaped cells with few chloroplasts and large intercellular spaces that communicate with the atmosphere through pores (stomata) allowing gas exchange between the cells and the atmosphere. The distribution of mesophyll tissue varies in different leaves depending on the environment in which the plant lives.

mesosome An extensive invagination of the plasma membrane of certain bacteria, associated with DNA synthesis and protein secretion. Its precise function is uncertain.

mesothelium The tissue, consisting of one or more layers of cells, that lines a coelomic cavity.

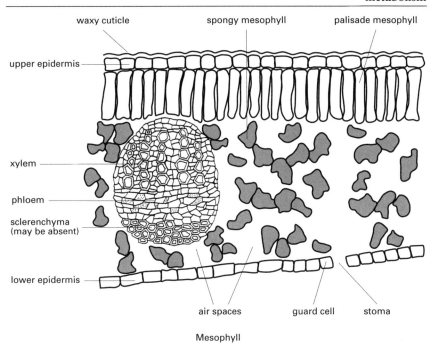

upper epidermis

waxy cuticle · spongy mesophyll · palisade mesophyll

xylem

phloem

sclerenchyma
(may be absent)

lower epidermis

air spaces · guard cell · stoma

Mesophyll

messenger RNA (mRNA) The form of RNA that transfers information from protein-coding genes in the DNA to the ribosomes for protein synthesis. It represents the mature functional transcript of the genetic information, and in eukaryotes is formed by processing of the newly transcribed RNA within the nucleus (*see* pre-mRNA; transcription). Processing is performed by SPLICEOSOMES, which remove the noncoding sequences (introns) and splice the coding sequences (exons) together. The information is carried by the sequence of bases in the constituent nucleotides of the linear mRNA molecule (*see* genetic code). In eukaryotes, the mRNA associates with various proteins to form *messenger ribonucleoproteins* (mRNPs). These associate also with nuclear export signals, which carry the mRNPs through pores in the nuclear membrane (*see* nuclear pore complex). The mRNAs then associate with ribosomes in the cytosol, while the RNPs dissociate and return to the nucleus. Each ribosome then 'reads' the message carried by the mRNA, and assembles the appro-

priate polypeptide, in the process called TRANSLATION.

In prokaryotes, the DNA has very few noncoding regions, and is transcribed directly into mRNA. This in turn is immediately translated into protein, even while the latter part of the same mRNA molecule is still being formed. Moreover, much of prokaryotic DNA is organized into OPERONS, comprising closely linked arrays of functionally related genes. Each operon is transcribed as a single mRNA, which thus encodes several different proteins, not a single protein or polypeptide as in eukaryotic mRNAs. *See* RNA.

metabolic pathway *See* metabolism.

metabolism The chemical reactions that take place in cells. The molecules taking part in these reactions are termed *metabolites*. Some metabolites are synthesized within the organism, while others have to be taken in as food. Metabolic reactions, particularly those that produce energy, keep cells alive, and characteristically occur in small steps, comprising a *meta-*

bolic pathway. Metabolic reactions involve the breaking down of molecules to provide energy (catabolism) and the building up of more complex molecules and structures from simpler molecules (anabolism).

metabolite A substance that takes part in a metabolic reaction, either as reactant or product. Metabolites are thus intermediates in metabolic pathways. Some are synthesized within the organism itself, whereas others have to be taken in as food. *See also* metabolism.

metachronal rhythm A pattern of movement shown by cilia and parapodia of certain polychaetes, etc., in which each beats one after the other in regular succession and gives the appearance of wave motion. A wave passing forwards may propel the surrounding medium backwards or the organism forwards.

metalloporphyrin *See* porphyrins.

metaphase The stage in mitosis and meiosis when the chromosomes become aligned along the equator of the nuclear spindle.

metaplasia The change from one tissue type to another as seen in response to certain diseases or abnormal conditions. For example the epithelium of the respiratory tract may show metaplasia in response to irritants, such as cigarette smoke.

methionine *See* amino acids.

methylene blue *See* staining.

Meyerhof, Otto Fritz (1884–1951) German physiologist and biochemist. He was awarded the Nobel Prize for physiology or medicine in 1922 for his discovery of the relationship between the consumption of oxygen and the metabolism of lactic acid in the muscle. The prize was shared with A. V. Hill.

MHC *See* major histocompatibility complex.

micelle A small spherical particle that forms spontaneously when a suspension of phospholipids is dispersed in aqueous solution. The hydrophilic heads are oriented to face the exterior aqueous phase, while the hydrophobic tail regions are directed toward the interior.

Michaelis, Leonor (1875–1949) German physician and biochemist; noted for his development of the theory of enzyme–substrate combination.

Michel, Hartmut (1948–) German biophysicist. He was awarded the Nobel Prize for chemistry in 1988 jointly with J. Deisenhofer and R. Huber.

microbiology The study of microscopic organisms (e.g. bacteria and viruses), including their interactions with other organisms and with the environment. Microbial biochemistry and genetics are important branches, due to the widespread use of microorganisms in biotechnology and genetic engineering.

microbody A common organelle of plant and animal cells, bounded by a single membrane, spherical, and usually about 0.2–1.5 μm in diameter. Microbodies originate from the endoplasmic reticulum and contain enzymes that oxidize various substrates using molecular oxygen. Unlike aerobic respiration in mitochondria, this oxidation is not linked directly to ATP production. *Peroxisomes* occur in most animal cells and are numerous in photosynthetic cells of plant leaves. Oxidation of organic compounds results in the formation of hydrogen peroxide (H_2O_2), which is highly reactive and toxic. Hence, the peroxisome contains high levels of the enzyme catalase, which breaks down hydrogen peroxide into water and oxygen. Peroxisomes are generally the main site of fatty acid oxidation in eukaryote cells, a process that yields acetyl groups, which serve as building blocks for biosynthetic reactions. In animals, toxic molecules absorbed from the bloodstream can also be degraded using peroxisomal enzymes, especially within cells of the liver and kidney. In plants, peroxisomes are concerned with glycolate metabolism in PHOTORESPIRATION and contain high levels of glycolate oxidase and other associated enzymes. Glycolate comes from chloroplasts and products such as glycine are passed to mitochondria. Hence these three organelles often appear close together.

Another type of microbody, called a *glyoxysome*, is common in seeds, especially in lipid-rich food reserves, for example in the endosperm of castor-oil seeds. Glyoxysomes contain enzymes of the GLYOXYLATE CYCLE, transaminases, and enzymes associated with β-oxidation of fatty acids, and play a major role in conversion of lipids to sucrose during germination and seedling growth.

microdissection The technique of dissecting under a microscope using fine mechanically manipulated instruments or laser beams. Such techniques are often used in microsurgery, or when dealing with living organisms at the cellular level. *See* micromanipulation.

microfilament A minute fiber, about 7–9 nm in diameter, that is one of the three main types of fiber comprising the CYTOSKELETON in eukaryotic cells, and has roles in cell motion and shape. Microfilaments are made of two helically twisted strands of globular subunits of the protein ACTIN, almost identical to the actin of muscle. They can undergo rapid extension or shortening by subunit assembly or disassembly, or form complex three-dimensional networks. Individual microfilaments are linked together by ACTIN-CROSS-LINKING PROTEINS, such as fascin and filamin, to form tight bundles or looser networks of fibers. A planar network of microfilaments underlies the plasma membrane, to which it is attached by binding proteins. A three-dimensional microfilament network permeates the cell interior to provide a structural framework for the cell's contents. Bundles of microfilaments are often associated with protrusions of the cell surface, such as microvilli, and with lamellipodia and pseudopodia formed by motile cells or phagocytes.

Muscle is a specialized tissue that contracts as a result of the interaction between thin filaments, which are modified actin microfilaments, and thick filaments, made of myosin, which serves as a 'motor' protein (*see* sliding filament model). However, actin filaments interact with myosin in various other ways in noncontractile cells. For example, in dividing cells actin and myosin filaments form a CONTRACTILE RING that encircles and pinches the cell so that a cleavage furrow forms and divides the daughter cells. A similar contractile ring occurs permanently as a tensioning girdle in epithelial cells, whereas other cell types often contain contractile bundles in regions where they contact other cells or a substrate. Myosin molecules can also move along actin filaments while carrying membrane-bound vesicles, thus transporting materials within the cell.

micrograph A photograph taken with the aid of a microscope. *Photomicrographs* and *electron micrographs* are produced using optical microscopes and electron microscopes respectively.

micromanipulation The manipulation of microscopic objects, such as cells and subcellular structures. Many procedures in experimental cell biology, medicine, and biotechnology, involve the use of techniques whereby cells or their contents are sampled or altered in a precise manner that does not impair viability of the cell. Such micromanipulation is necessarily performed with the aid of a high-power optical microscope, and requires specially designed instruments, which are usually held by mechanical manipulators guided by the operator. An established technique is to hold cells in place by means of micropipettes, made of very fine glass tubing, to which slight suction is applied. Microinjectors can be used to insert material, such as DNA, into the cell, or to extract material for sampling. More recently, lasers have been adapted for use in microdissection and micromanipulation. The laser is interfaced with the operator's microscope and focused onto the specimen to give a spot of the order of 1–0.5 μm diameter. Ultraviolet lasers are used for microablation or high-precision cutting, for example to create holes in cell membranes or to dissect undulipodia. Infrared lasers are used as optical tweezers: minute forces generated by diffraction of the focused beam cause microscopically small objects to become trapped in the beam – a phenomenom known as *optical trapping*. Hence laser tweezers can be used to isolate and move single cells, or to hold organelles within cells.

micrometer[1] In microscopy, a device for measuring the size of an object under the microscope. An eyepiece micrometer (*graticule*) of glass or transparent film, with a scale etched or printed on it, is placed in the eyepiece so that both the object to be measured and the scale are in focus. The scale of the graticule changes at different magnifications and it must therefore be calibrated against a stage micrometer, which is contained in a glass slide and placed on the microscope stage.

micrometer[2] Symbol: μm A unit of length equal to 10^{-6} meter (one millionth of a meter). It is often used in measurements of cell diameter, sizes of bacteria, etc. Formerly, it was called the *micron*.

micron *See* micrometer.

micronucleus *See* macronucleus.

micropropagation *See* tissue culture.

micropyle *See* chorion.

microsatellite DNA *See* satellite DNA.

microscope An instrument designed to magnify objects and thus increase the resolution with which one can view them. *Resolution* is the ability to distinguish between two separate adjacent objects. Radiation (light or electrons) is focused through the specimen by a *condenser lens*. The resulting image is magnified by further lenses. Since radiation must pass through the specimen, it is usual to cut larger specimens into thin slices of material (sections) with a

microtome. Biological material has little contrast and is therefore often stained. If very thin sections are required the material is preserved and embedded in a supporting medium.

The LIGHT MICROSCOPE uses light as a source of radiation. With a *compound microscope* the image is magnified by two lenses, an *objective lens* near the specimen, and an *eyepiece*, where the image is viewed, at the opposite end of a tube. Its maximum magnification is limited by the wavelength of light. Much greater resolution became possible with the introduction of the ELECTRON MICROSCOPE, which uses electrons as a source of radiation, because electrons have much shorter wavelengths than light.

microsomes Fragments of ENDOPLASMIC RETICULUM and GOLGI APPARATUS in the form of vesicles formed during homogenization of cells and isolated by high-speed centrifugation. Microsomes from rough endoplasmic reticulum are coated with ribosomes and can carry out protein synthesis in the test tube.

microtome An instrument for cutting thin sections (slices a few micrometers thick) of biological material for microscopic examination. The specimen is usually embedded in wax for support and cut by a steel knife. Alternatively it is frozen and a *freezing microtome*, which keeps the specimen frozen while cutting, is used. For electron microscopy, extremely thin (20–100 nm) sections can be cut by an *ultramicrotome*. Here the specimen is embedded in resin or plastic for support and mounted in an arm that advances slowly, moving up and down, toward a glass or diamond knife. As sections are cut they float off on to the surface of water contained in a trough behind the knife. The sections are then picked up on a carbon-coated copper grid and dried.

microtubule A thin cylindrical unbranched tube of variable length found in eukaryotic cells, either singly or in groups. Its walls are made of the protein tubulin. Microtubules are one of the three main components of the CYTOSKELETON, along with microfilaments and intermediate filaments, and help cells to maintain their shape and provide the motive force for some forms of cell movement. They are part of the structure of centrioles, basal bodies, and undulipodia (cilia and flagella); and form the spindle during cell division, bringing about chromosome movement. Microtubules help to orientate materials and structures in the cell, for example cellulose fibrils during the formation of plant cell walls. They also serve as tracks along which cargoes (e.g. vesicles) are translocated within cells, for example along the axons of neurons.

A microtubule is typically 24 nm in diameter, with a wall consisting of 13 longitudinally arranged protofilaments. The protofilaments comprise tubulin subunits, each of which has two similar parts (monomers): α-tubulin, which binds GTP (guanosine triphosphate) irreversibly; and β-tubulin, which binds GTP reversibly, and can split it to GDP (guanosine diphosphate), thereby yielding energy. Microtubule formation is initiated by one or more MICROTUBULE-ORGANIZING CENTERS within the cell (the centrosome in animals cells). The microtubules are rapidly assembled from a pool of tubulin subunits, and can be quickly disassembled, according to the requirements of the cell.

Various proteins bind to microtubules and give them particular properties. Motor proteins include KINESIN and DYNEIN; both use ATP to generate the force required to transport vesicles, proteins, or organelles along microtubules. There are several classes of MICROTUBULE-ASSOCIATED PROTEINS (MAPS), which regulate the stability of microtubules, and can also bind them to membranes, to each other, or to intermediate filaments.

microtubule-associated protein (MAP)

Any of a group of proteins that bind to MICROTUBULES. They form cross-links that join microtubules to each other, and also to other components, such as cell membranes and intermediate filaments. By coating the surface of microtubules, MAPs also affect the assembly and disassembly of the microtubules, and hence control their length.

microtubule-organizing center (MTOC)

A point in a cell from which microtubules are assembled and radiate. It also helps to determine the arrangement of organelles inside the cell, including the mitochondria, Golgi complex, and endoplasmic reticulum. Most nondividing animal cells have a single MTOC, called the CENTROSOME, which typically lies close to the nucleus at the center of the cell. It consists of a lattice of microtubule-associated proteins (MAPs), and sometimes a pair of centrioles, arranged at right angles to each other. The ends of the radiating microtubules are adjacent to the centrioles. In contrast to animal cells, plant cells and fungal cells have hundreds of MTOCs, distributed throughout the cell. These produce a series of girdle-like microtubule arrays around the cell's inner periphery (cortex), and also a mesh of microtubules in the growing ends of plant cells.

microvillus (*pl.* **microvilli**) An elongated slender projection of the plasma membrane found especially in secretory and absorptive cells. Microvilli are generally 0.5–10 μm long, and provide an increased surface area over which membrane proteins can be arrayed for the transport of molecules into or out of the cell. The closely packed arrangement of microvilli on the free surface of epithelial cells constitutes a *brush border*. Numerous microvilli occur on the epithelial cells of the intestine and also in the kidney tubules. In addition to their presence in secretory and absorptive cells, they are commonly observed in many other cells although they may not be permanent structures. Each microvillus is supported internally by a bundle of actin filaments. These are cross-linked to each other, and to the surrounding plasma membrane. *Compare* filopodium.

middle lamella A thin cementing layer holding together neighboring plant cell walls. It consists mainly of PECTIC SUBSTANCES (e.g. calcium pectate). The middle lamella is laid down at the CELL PLATE during cell division.

Milstein, César (1927–2002) Argentinian-born British biochemist and immunologist. He was awarded the Nobel Prize for physiology or medicine in 1984 jointly with N. K. Jerne and G. J. F. Köhler for theories concerning the specificity in development and control of the immune system and the discovery of the principle for production of monoclonal antibodies.

minisatellite DNA *See* satellite DNA.

mismatch repair *See* DNA repair.

missense mutation *See* mutation.

Mitchell, Peter Dennis (1920–92) British biochemist. He was awarded the Nobel Prize for chemistry in 1978 for his contribution to the understanding of biological energy transfer.

mitochondrial DNA **(mtDNA)** A small circular DNA molecule found in mitochondria that carries the genes for certain proteins required for the respiratory functions of mitochondria. The size and number of mtDNAs vary between different organisms, but each mitochondrion usually contains several or numerous copies of mtDNA, often grouped to form nucleoids. For example, human mitochondria each have 2 to 10 copies of a 17 kb DNA with 37 genes. Besides protein-coding genes, there are also genes encoding transfer RNAs (tRNAs) and ribosomal RNAs needed for protein synthesis within the mitochondrion. The genetic code used by mitochondrial-derived mRNAs differs in some respects from the nuclear code. In humans and other organisms where the female gamete contains relatively large amounts of cytoplasm compared to the male gemete, mtDNA is inherited exclusively from the mother, i.e. maternal inheritance.

mitochondrial theory of aging *See* senescence.

mitochondrion An organelle of all eukaryotes that is the main site of ATP production in aerobically respiring cells. It is surrounded by two membranes separated by an intermembrane space; the inner membrane forms finger-like processes called *cristae*, which project into the gel-like *matrix*. Mitochondria are typically sausage-shaped, but may assume a variety of forms, including irregular branching shapes. The diameter is always about 0.5–1.0 μm. They contain the enzymes and cofactors of AEROBIC RESPIRATION and therefore are most numerous in active cells (up to several thousand per cell). They may be randomly distributed or functionally associated with other cell components, for example with the contractile fibrils of muscle cells.

The reactions of the KREBS CYCLE take place in the matrix, while those of electron transport coupled to oxidative phosphorylation (i.e. the respiratory chain) are located on the inner membrane. Within the membrane the components of the respiratory chain are highly organized. The matrix is also involved in amino acid metabolism via Krebs cycle acids and transaminase enzymes, and in fatty acid oxidation. According to the ENDOSYMBIONT THEORY, mitochondria, like chloroplasts, are the descendants of once-independent organisms that early in evolution invaded ancestral eukaryotic cells, leading to an extreme form of symbiosis. Part of the evidence for this is the existence of DNA molecules within the matrix of mitochondria (and chloroplasts) (*see* mitochondrial DNA). These encode various mitochondrial proteins, as well as the transfer RNAs and ribosomal RNAs required for their synthesis. It is assumed that most of the other genes, originally required for assembly of the symbiotic bacterium, have subsequently been transferred to the nucleus. *See illustration at* cell.

mitosis (karyokinesis) The ordered process by which the cell nucleus divides in two during the division of somatic (i.e. nongermline, or body) eukaryotic cells. It is the final phase (M phase) in the CELL CYCLE, and follows replication of the chromosomes during the S phase (synthetic phase). The replicated chromosomes are separated during mitosis in such a way that each daughter cell inherits a genetic complement identical to that of the parent cell. Although mitosis is a continuous process it is divided into four phases. During *prophase*, the chromosomes condense by coiling on themselves (*see* chromatin), and in late prophase the nuclear envelope disintegrates and the nucleolus disappears. During *metaphase* the chromosomes become aligned in the cell's equatorial plane. Each chromosome can be seen to consist of two closely associated chromatids. Microtubules extend from either pole of the cell, forming the mitotic SPINDLE, to which the chromatids attach via their kinetochores. During *anaphase*, the daughter chromatids of each chromosome are pulled to opposite poles by the spindle microtubules, and in *telophase*, the chromosomes aggregate and begin to uncoil. A nuclear envelope reforms around them, and the cell returns to

G_1 of the cell cycle. Before the end of telophase, the cytoplasm begins to divide in the process called CYTOKINESIS; this brings about the cleavage of the parent cell to form two new daughter cells. *Compare* meiosis; amitosis; endomitosis. *See illustration.*

mitosis-promoting factor (MPF) *See* cell cycle.

mobile element (mobile DNA) Any segment of genomic DNA that can copy and insert itself into new locations within the genome – a process called transposition (hence the alternative name transposable genetic element). Such elements apparently have no function in the cell, and can be regarded as 'molecular parasites' (*see* selfish DNA). In many eukaryotic organisms they have accumulated over evolutionary time and now account for a sizeable proportion of the genome. There are various types. Some transpose directly as DNA: these in-

Prophase

Chromosomes appear and shorten and thicken. Nuclear membrane disintegrates and nucleolus disappears

Metaphase

Individual chromosomes, each comprising two chromatids, become aligned along the equator of the nuclear spindle

Anaphase

Chromosomes split at the centromere and the daughter chromatids move to opposite poles of the spindle

Telophase

A nuclear membrane forms around each group of daughter chromatids, or new chromosomes

Mitosis

clude relatively small mobile elements called insertion sequences, found in prokaryotes, and TRANSPOSONS, found in both prokaryotes and eukaryotes. Others, called RETROTRANSPOSONS, form an RNA intermediate. Mobile elements have had an important influence on evolution. For example, they can bring about spontaneous mutations by inserting into genes, and also cause duplications and rearrangements of genes through homologous recombination between two identical such elements.

molarity A measure of the concentration of solutions based upon the number of molecules or ions present, rather than on the mass of solute, in any particular volume of solution. The molarity (M) is the number of moles of solute in one cubic decimeter (litre). Thus a 0.5M solution of hydrochloric acid contains $0.5 \times (1 + 35.5)$ g HCl per dm^3 of solution.

mole Symbol: mol The SI base unit of amount of substance, defined as the amount of substance that contains as many elementary entities as there are atoms in 0.012 kilogram of carbon-12. The elementary entities may be atoms, molecules, ions, electrons, photons, etc., and they must be specified. One mole of a compound has a mass equal to its RELATIVE MOLECULAR MASS in grams.

molecular chaperone See chaperone.

molecular systematics A branch of biology that compares functionally equivalent macromolecules from different organisms as a basis for classification. Sequences of amino acids in proteins (e.g. enzymes) or of nucleotides in nucleic acids (e.g. ribosomal RNA) are determined using automated techniques, and compared statistically using computer programs. Essentially, how closely two organisms are related in evolutionary terms is reflected in the degree of similarity of their macromolecules.

molecular weight See relative molecular mass.

molybdenum See trace element.

Monera An alternative name for the kingdom Prokaryotae. It was originally used by the German biologist Ernst Haeckel to refer to bacteria and blue–green algae (now called cyanobacteria) as a group within the kingdom Protista, in his three-kingdom classification scheme. These prokaryotic organisms were later transferred to their own kingdom, in recognition of the fundamental differences between PROKARYOTES and all other organisms. See also eukaryote.

monoclonal antibody A specific antibody produced by a cell clone (i.e. one of many identical cells derived from a single parent). The parent cell is obtained by the artificial fusion of a normal antibody-producing lymphocyte, a B-cell taken from a mouse spleen, with a transformed lymphocyte obtained from cancerous lymphoid tissue (myeloma) of a mouse. This hybrid cell, or *hybridoma*, is immortal, like the myeloma cell, and multiplies rapidly *in vitro* to yield large amounts of antibody, which comprises only a single type of immunoglobulin molecule. Monoclonal antibodies bind to a specific antigen within a mixture, and are used widely as reagents in immunoassays, and for locating particular substances within cells or tissues. They also have potential for treating certain diseases, for example by binding specifically to tumor cells to deliver anticancer drugs.

monocyte The largest type of white blood cell (see leukocyte). It has nongranular cytoplasm and a large kidney-shaped nucleus. Making up 4–5% of all leukocytes, monocytes are actively phagocytic, devouring foreign particles (such as bacteria). They leave the blood and settle in tissue as MACROPHAGES.

Monod, Jacques Lucien (1910–76) French microbiologist and molecular biologist. He was awarded the Nobel Prize for physiology or medicine in 1965 jointly with F. Jacob and A. M. Lwoff for their discoveries concerning genetic control of enzyme and virus synthesis.

mononuclear phagocyte system (reticuloendothelial system) The system of MACROPHAGE cells, which are scattered throughout the body and are capable of engulfing foreign particles. The cells are found in connective tissue and in the lining of small blood vessels, and are concentrated especially in the lung, liver, spleen, and lymph nodes. The mononuclear phagocyte system is important in defending the body against disease and in destroying worn-out erythrocytes.

mononucleotide *See* nucleotide.

monoploid *See* haploid.

monosaccharide A SUGAR that cannot be hydrolyzed to simpler carbohydrates of smaller carbon content. Glucose and fructose are examples.

monosomy *See* aneuploidy.

monotrichous Describing bacteria that possess one flagellum (undulipodium), for example *Vibrio*.

monozygotic twins *See* identical twins.

Moore, Stanford (1913–82) American biochemist. He was awarded the Nobel Prize for chemistry in 1972 jointly with W. H. Stein for their contribution to the understanding of the connection between chemical structure and catalytic activity of the active center of the ribonuclease molecule. The prize was shared with C. B. Anfinsen.

Morgan, Thomas Hunt (1866–1945) American geneticist and zoologist who developed *Drosphilia* mutants and for techniques of gene mapping. He was awarded the Nobel Prize for physiology or medicine in 1933 for his discoveries concerning the role played by the chromosome in heredity.

morphogen A substance that during embryonic development affects the fate of cells in different ways depending on its concentration. Different morphogens are produced at successive stages of develop-

ment, and in different regions of the embryo. A morphogen released from a particular point sets up a concentration gradient. Along the gradient, cells are exposed to a number of threshold concentrations. Below a certain threshold, the morphogen commits a cell to, say, fate A, whereas above that same threshold level, the cell is committed to fate B. Morphogens play a crucial role in establishing the basic body plan of an embryo, and in the INDUCTION of organs. For example, the anteroposterior axis of an early fly embryo, such as that of *Drosophila*, is determined by gradients of morphogens encoded by maternal mRNAs localized to particular regions of the oocyte (egg cell). These morphogens are transcription factors, which activate or repress zygotic genes in a spatially restricted manner, so that distinct anterior and posterior zones appear. Many morphogens, such as the TGFb (*see* transforming growth factor) and HEDGEHOG PROTEIN families, are signal molecules that bind to cell-surface receptors, and activate signaling pathways that ultimately regulate gene transcription. *See also* differentiation.

morphogenesis The development of form and structure.

morula A loose aggregation of blastomeres resulting from cleavage of the egg of mammals. It develops into the blastocyst.

mosaic *See* chimera.

motif A three-dimensional structure within a protein molecule created by a specific arrangement between distinct secondary structures (e.g. coiled and noncoiled regions of the polypeptide). Motifs generally serve a characteristic function, determined not only by their shape but also by the chemical nature of the amino acids found within the motif. For example, the *helix-loop-helix motif* consists of two α helices linked by a noncoiled loop of the polypeptide chain that can form a ring around a calcium ion. It occurs in many calcium-binding proteins. The *zinc finger motif* is common in proteins that bind to

DNA or RNA. It consists of a single α helix and two β strands, between which is held a zinc atom.

motor neuron A nerve cell (*see* neuron) that transmits impulses from the brain or spinal cord to a muscle or other effector.

motor protein Any protein that can hydrolyze ATP and use the resultant energy to propel itself along a protein fiber or filament. For example, the motor protein in muscle is MYOSIN, which actively slides along actin filaments. KINESIN and DYNEIN are other examples that actively 'walk' along microtubules.

mRNA *See* messenger RNA.

MRNA *See* messenger RNA.

MSH *See* melanocyte-stimulating hormone.

mtDNA *See* mitochondrial DNA.

mucin The main constituent of mucus. It is a glycoprotein.

mucopolysaccharide *See* glycosaminoglycan.

mucoprotein *See* proteoglycan.

mucosa (*pl.* mucosae) *See* mucous membrane.

mucous membrane (mucosa) The tissue, in vertebrates, that lines many tracts (e.g. the intestinal and respiratory tracts) that open to the exterior. It consists of surface epithelium containing goblet cells, which secrete mucus, and is underlaid by connective tissue.

mucus A slimy substance produced by goblet cells in mucous membranes of animals. It is viscous and insoluble, consisting mainly of glycoproteins. Its function is to protect and lubricate the surface on which it is secreted.

Muller, Hermann Joseph (1890–1967) American geneticist. He was awarded the Nobel Prize for physiology or medicine in 1946 for the discovery of the production of mutations by means of x-ray irradiation.

Mullis, Kary B. (1944–) American biochemist who invented the polymerase chain reaction. He was awarded the Nobel Prize for chemistry in 1993. The prize was shared with M. Smith.

multicellular Consisting of many cells.

multifactorial inheritance *See* quantitative inheritance.

Murad, Ferid (1936–) American biochemist who was awarded the 1998 Nobel Prize for physiology or medicine jointly with R. F. Furchgott and L. J. Ignarro for work on the action of nitric oxide as a signaling molecule in the cardiovascular system.

muscle Tissue consisting of elongated cells (*muscle fibers*) containing fibrils that are highly contractile. *See* cardiac muscle; skeletal muscle; smooth muscle.

muscle contraction The process by which excitation of muscle cells results in contraction of a muscle. The initiation of contraction varies according to the type of muscle. In SKELETAL MUSCLE contraction is triggered by nerve impulses arriving at neuromuscular junctions. The resulting depolarization spreads along infoldings of the muscle cell membrane, the transverse tubule (or T) system, which activates the SARCOPLASMIC RETICULUM to release calcium ions into the sarcoplasm. The calcium ions act by removing the effect of an inhibitory protein (tropomyosin in skeletal muscle) so that the muscle filaments (actin and myosin) can repeatedly form crossbridges and slide over one another (the SLIDING FILAMENT MODEL of contraction). Skeletal muscle is organized into numerous contractile units called SARCOMERES. The energy for contraction is derived from the breakdown of ATP. On cessation of stimulation (when the T system is no longer depolarized) calcium ions are resorbed into the sarcoplasmic reticulum and relaxation occurs. CARDIAC MUSCLE and SMOOTH MUSCLE each have distinct structural features, related to the particular functions they perform. For example, cardiac muscle and certain types of smooth muscle are self-excitatory and have an inherent rhythmicity of contraction. There are junctions between the individual muscle cells, and the sarcoplasmic reticulum is less well developed than in skeletal muscle. However, contraction still involves the ATP-powered sliding of muscle filaments.

mutagen Any physical or chemical agent that induces MUTATION or increases the rate of spontaneous mutation. Chemical mutagens include ethyl methanesulfonate, which causes changes in the base pairs of DNA molecules, and acridines, which cause base pair deletions or additions. Physical mutagens include ultraviolet light, x-rays, and gamma rays. carcinogens are a class of mutagens whose effects on DNA result in cancer.

mutant *See* mutation.

mutation (**gene mutation**) A change in the nucleotide sequence of DNA, which results in a change in the protein encoded by the affected gene, or altered expression of the gene (usually with reduction or complete loss of synthesis of the corresponding protein). The affected organism is called a *mutant*, and is characterized by some difference in appearance, physiology, behavior, etc. compared to its normal, or wild-type, counterparts. Mutations are inherited only if they occur in the cells that give rise to the gametes; somatic mutations (occurring in body cells) may give rise to chimeras and cancers. Mutations result in new allelic forms of a gene and hence new variations upon which natural selection can act. Most mutations are deleterious but are often retained in the population because they also tend to be recessive and can thus be carried in the genotype without affecting the viability of the organism. The natural rate of mutation is low, but the mu-tation frequency can be increased by mutagens.

Mutations can involve a change in a single base pair (called a *point mutation*) or deletion or addition of one or more base pairs. In terms of their effect on translation, such mutations can be one of three kinds. A *missense mutation* causes the codon for another amino acid to be inserted in the messenger RNA, resulting in substitution of the corresponding amino acid in the protein. A *nonsense mutation* changes an amino acid codon to a stop codon, so causing a premature halt to translation. A *frameshift mutation* shifts the READING FRAME during translation, resulting in the assembly of incorrect amino acids and usually premature termination of translation. Rearrangements or deletions of larger segments of DNA are generally termed CHROMOSOME MUTATIONS.

Mutations can lead to cancers. Ones in the germline cells can cause inherited cancers or predisposition to cancer. The accumulation of mutations in somatic cells can cause nonheritable cancer, which tends to develop in later life. The loss or rearrangement of chromosomal material can, for example, mean the loss or disruption of TUMOR SUPPRESSOR GENES, which help to prevent potentially cancerous changes building up in cells. Alternatively, these mutations might bring a proto-oncogene under the influence of a more powerful promoter, which leads to overproduction of the gene product. *See* cancer; oncogene.

mycelium (*pl.* **mycelia**) A filamentous mass comprising the body of a fungus, each filament being called a HYPHA. The mycelium often forms a loose mesh as in *Mucor*, but the hyphae may become organized into definite structures, for example the fruiting body of a mushroom. The mycelium produces the reproductive organs of a fungus.

mycoplasmas (PPLO; **pleuropneumonia-like organisms**) A group of extremely small bacteria that naturally lack a rigid cell wall. They often measure less than 200 nm in diameter and their cells are delicate and plastic. Mycoplasmas can cause pneumonias in humans, bovine tuberculosis, and other mammalian diseases, but they may exist harmlessly in mucous membranes. Mycoplasmas are resistant to penicillin and related antibiotics that work by inhibiting wall growth.

mycoprotein   Any protein produced by a fungus or bacterium.

mycorrhiza (*pl.* **mycorrhizas** or **mycorrhizae**)   The association between the hyphae of a fungus and the roots of a higher plant. Two main types of mycorrhiza exist, *ectotrophic* in which the fungus forms a mantle around the smaller roots, as in trees, and *endotrophic* in which the fungus grows around and within the cortex cells of the roots, as in orchids and heathers.

In ectotrophic mycorrhizas the fungus, which is usually a member of the Agaricales, benefits by obtaining carbohydrates and possibly B-group vitamins from the roots. The trees benefit in that mycorrhizal roots absorb nutrients more efficiently than uninfected roots, and it is common forestry practice to insure the appropriate fungus is applied when planting seedling trees. In endotrophic mycorrhizas the fungus is generally a species of *Rhizoctonia* and again both partners benefit nutritionally from the relationship.

myelinated nerve fiber   A nerve fiber that is surrounded by a fatty (myelin) sheath. Most nerves of vertebrates consist of thousands of medullated fibers, which appear white because of the fatty sheaths. *See* myelin sheath.

myelin sheath   An insulating covering that surrounds the axon of a neuron in the peripheral nerves of vertebrates. It is composed of multiple layers (some 50–100) of the plasma membrane of SCHWANN CELLS wound tightly in a spiral around the axon. The membranes consist of a fatty material (myelin). The myelin sheath of a single long axon is divided into segments, each of which is produced by a single Schwann cell. In between each segment is a short region of bare axon (*node of Ranvier*). The electrical insulation provided by the myelin

sheath enables SALTATORY CONDUCTION of action potentials along the axon, which greatly increases the speed of conduction of nerve impulses.

myelocyte   A cell in the myeloid tissue of red bone marrow that is a precursor of a GRANULOCYTE. Myelocytes are formed by division of precursor cells (myeloblasts and progranulocytes) and they change into the immediate precursors of the granulocytes, i.e. basophils, eosinophils, and neutrophils, which are released into the bloodstream.

myeloid tissue   Tissue in the red bone marrow that manufactures red blood cells (ERYTHROCYTES) and certain white blood cells. It contains proerythroblasts, cells that produce the precursors of erythrocytes, and also myeloblasts, which divide continuously to give myelocytes (which develop into GRANULOCYTES). Most lymphocytes and monocytes are produced in lymphoid tissue.

myofibers   *See* skeletal muscle.

myofibril   A very fine fiber (1–2 μm in diameter) many of which are embedded in the SARCOPLASM of a muscle fiber (muscle cell). In SKELETAL MUSCLE these fibrils are striated, being divided along their length into a great number of sarcomeres, which constitute the contractile apparatus of the muscle fiber.

myoglobin   A conjugated protein found in muscles (sometimes referred to as 'muscle hemoglobin'). It is similar to hemoglobin in being a heme protein capable of binding oxygen but is structurally simpler, having only one polypeptide chain combined with the heme group. Each molecule of myoglobin can attach one molecule of oxygen.

myoneme   A contractile fibril found in certain protists (e.g. in the retractable stalk of *Vorticella*).

myosin   A protein that is able to move along actin filaments and is involved in muscle contraction and other types of movement within cells. Myosin can hydrolzye ATP to cause conformational changes in the myosin molecule that enable it to 'walk' along an actin filament. Hence myosin is a MOTOR PROTEIN, able to convert chemical energy (in the form of ATP) into mechanical energy. There are several types of myosin, the most common being myosin I, involved in cytoskeletal interactions with actin, and myosin II, which occurs mainly in muscle. All myosins comprise a globular 'head' domain, a flexible 'neck' domain, and an extended 'tail' domain. The head domain contains actin-binding sites and ATP-binding sites, and it is this region that walks along the actin filament, with each 'step' coupled to the hydrolysis of one ATP molecule. *See* actin; sliding filament model.

myotome The part of each SOMITE of a vertebrate embryo that differentiates as a muscle block. The muscle myotomes remain segmental in fishes, but in terrestrial vertebrates they lose much of their original pattern and buds from myotomes (with their ventral root nerves) form muscles in limbs, etc.

myxobacteria (slime bacteria) A group of bacteria in which individual cells are typically rod-shaped and covered in slime, but which may congregate to form gliding swarms or various upright reproductive structures under certain conditions. For example, the reproductive body of *Stigmatella aurantiaca* consists of a stalk bearing several cysts. These open when wetted to release masses of individual gliding bacteria that move together over the substrate as discrete colonies. In some species the reproductive structures are brightly colored and just visible to the naked eye. Myxobacteria are common in soil, animal dung, and decaying plant matter.

Myxomycetes *See* slime molds.

Myxomycota *See* slime molds.

myxovirus One of a group of RNA-containing viruses that cause such diseases as influenza, mumps, measles, and rabies.

NAD

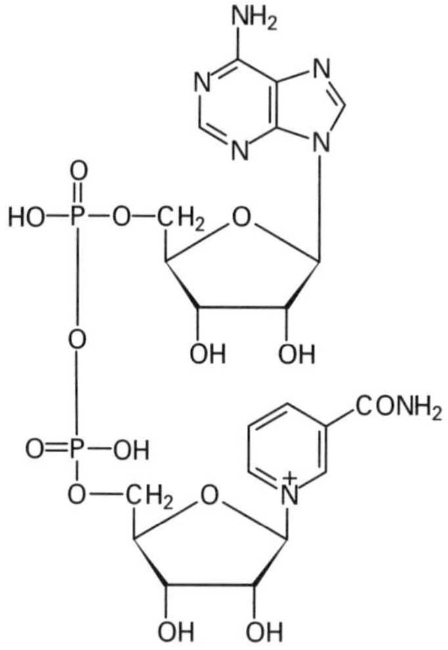

additional phosphate group on the ribose of the adenosine (*see illustration at* NAD). As with NAD, the oxidized form of NADP (NADP⁺) accepts two electrons and a proton to become the reduced form, NADPH. During PHOTOSYNTHESIS, NADP⁺ generally serves as the electron acceptor that receives electrons from the electron transport chain in the thylakoid membrane of the chloroplast. The NADPH generated in this way then provides electrons for the fixation of carbon dioxide and synthesis of carbohydrates. NADP participates in a different set of reactions than those involving NAD, and tends to take part more in synthetic reactions than energy-yielding ones. *See* NAD.

Na⁺/H⁺ antiporter A transport protein found in the plasma membrane that im-

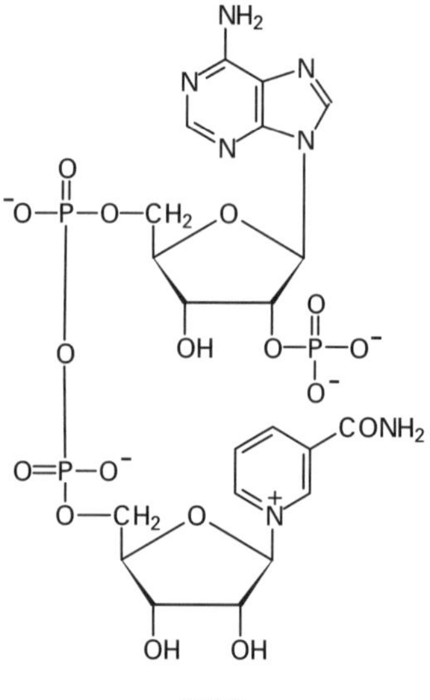

NAD (nicotinamide adenine dinucleotide) A derivative of nicotinic acid that acts as a coenzyme in many biological oxidation–reduction reactions, most notably in the ELECTRON-TRANSPORT CHAIN. Its role is to carry electrons. It consists of nicotinamide, ribose, and adenosine groups linked via two phosphate groups (*see illustration*). The oxidized form, denoted NAD⁺, accepts a pair of electrons and one proton to become the reduced form, denoted as NADH. For example, during aerobic respiration, the main role of the Krebs cycle reactions is to generate NADH, which then gives up its electrons to the mitochondrial electron transport chain, thereby driving the generation of ATP. *See also* NADP.

NADP (nicotinamide adenine dinucleotide phosphate) A coenzyme that acts similarly to NAD as an electron carrier, and is identical in structure apart from an

NADP

ports sodium ions (Na⁺) in exchange for protons (H⁺). It helps to maintain cellular pH by removing excess H⁺ ions produced by cellular metabolism. For every Na⁺ ion that enters via the antiporter down the Na⁺ concentration gradient, the antiporter exports one H⁺ ion.

Na⁺/K⁺ ATPase A transport protein, present in the plasma membrane of all animal cells, that is responsible for the coupled movement of sodium ions (Na⁺) and potassium ions (K⁺) out of and into the cell, respectively, using energy derived from the hydrolysis of ATP. It is the principal means by which animal cells establish their resting membrane potential (*see* resting potential), and thus plays a key role in the life of such cells. Three Na⁺ are exported in exchange for two K⁺ entering the cell, with both types being transported against their concentration gradients. The two-way ion transport involves successive changes in the protein's conformation, which affect the affinity for binding of Na⁺ and K⁺ to specific sites on the interior and exterior faces of the protein. These changes are linked to binding of ATP and transfer of the phosphate group to form a high-energy intermediate with an aspartic acid residue on the protein.

nano- Symbol: n A prefix denoting one thousand-millionth, or 10⁻⁹. For example, 1 nanometer (nm) = 10⁻⁹ meter. *See* SI units.

narrow-spectrum antibiotic *See* antibiotic.

Nathans, Daniel (1928–99) American molecular biologist. He was awarded the Nobel Prize for physiology or medicine in 1978 jointly with W. Arber and H. O. Smith for the discovery of restriction enzymes.

native state The three-dimensional form in which a protein or nucleic acid normally occurs within living cells. The polypeptide chains of proteins are usually folded into their most stable form, and under appropriate conditions will resume this native state spontaneously following unfolding due to physical or chemical treatment (i.e. DENATURATION).

natural killer cell (NK cell) A large lymphocyte with granular cytoplasm that attaches to the surface of virus-infected cells and secretes various substances that destroy the target cell. Unlike cytotoxic T-cells, NK cells are not primed to attack specific target cells, but are thought to recognize certain chemical groups that occur on the surface of infected cells. Recognition triggers the release from granules in the NK cell of proteins called *perforins*, which make pores in the target cell membrane. Further cytotoxic substances found in the granules include tumor necrosis factor β and protease enzymes called granzymes. Their action might induce apoptosis of the target cell, prompting it to 'commit suicide'. *See also* T-cell.

nebulin *See* sarcomere.

negative contrast technique *See* negative staining.

negative staining (negative contrast technique) A method of preparing material for electron microscopy when studying three-dimensional and surface features, notably of viruses, macromolecules (e.g. enzyme complexes), and the cristae of mitochondria. A stain is used that covers the background and penetrates the spaces within the specimen, but leaves the specimen itself unstained. Hence, the specimen appears transparent against an opaque background.

Neher, Erwin (1944–) German biophysicist. He was awarded the Nobel Prize for physiology or medicine in 1991 jointly with B. Sakmann for their work on the function of single ion channels in cells.

nematocyst *See* cnidocyte.

nephrotome *See* somite.

nerve A bundle of nerve fibers surrounded by a protective covering of connective tissue. *Mixed nerves*, such as the spinal nerves, contain both sensory and motor fibers. *See also* neuron.

nerve cell *See* neuron.

nerve cord An enclosed cylindrical tract of nerve fibers that forms a central route for the conduction of nerve impulses within the body; for example, the spinal cord in vertebrates and other chordates.

nerve fiber The axon of a NEURON.

nerve impulse The signal transmitted along neurons. For any particular neuron, the impulses are identical in form and strength, and are transmitted in one direction only. An impulse consists of changes in permeability of the axon membrane followed by flows of ions into and out of the cell. This produces a brief and reversible change in electrical potential across the axon membrane, which can be detected as an ACTION POTENTIAL. The energy required to pass the impulse is derived from the neuron itself, not from the stimulus.

In myelinated neurons, action potentials are generated only at the nodes of Ranvier; the myelin sheath insulates the axon so that the action potential is conducted, more rapidly, from one node to the next (*saltatory conduction*). There is a refractory period following the passage of an impulse, during which another impulse cannot be transmitted. In living organisms the impulse is triggered by local depolarization at a synapse or a receptor cell, but in isolated axons almost any disturbance of the membrane will set off an impulse. Since the impulse is an all-or-nothing event, the strength of the stimulus is signaled by the frequency and number of identical impulses. *See also* node of Ranvier.

nerve net A netlike layer of interconnecting nerve cells that is found in the body wall of certain groups of invertebrate animals; the most primitive type of nervous system. It occurs in cnidarians and ctenophorans.

nervous system A ramifying system of cells, found in all animals except sponges, that forms a communication system between receptors and effectors and allows varying degrees of coordination of information from different receptors and stored memory, producing integrated responses to stimuli. The system consists of neurons, supportive glial cells, and various fibrous tissues surrounding the softer matter. Impulses are transmitted through the neurons, which communicate with each other at specialized one-way junctions, the synapses. The impulse is a propagated change in electrical potential on either side of the neuron membrane, the ACTION POTENTIAL, which travels at between 1 and 120 m/s, depending on the animal and the type of neuron.

At its simplest (e.g. in cnidarians) the nervous system is merely a diffuse net with little concentration of function, but higher animals possess groups of neurons (ganglia), within which integration can take place. The major ganglion develops in the head, as the brain, and becomes increasingly important as a control center in more advanced types. The brain communicates with the body through one or more nerve cords (e.g. the spinal cord of vertebrates); the neurons of the nerve cord(s) connect to the peripheral nervous system, which contains sensory or motor neurons running from receptors or to effectors. Proper functioning of the nervous system depends on the correct pattern of connections between individual nerve cell (neurons). This 'wiring' is established during the organism's development by the process called NEURONAL OUTGROWTH. *See also* autonomic nervous system; central nervous system; nerve impulse.

neural crest A ridge of cells lying above the neural tube in a vertebrate embryo. Derived from the outermost of the three germ layers (i.e. ectoderm), the neural crest forms following apposition and fusion of the neural folds, which run longitudinally either side of the neural plate in the dorsal surface. The cells of the neural crest form part of a population of wandering cells that differentiate into various cell types, including pigment cells, Schwann cells, gill arches, postganglionic autonomic neurons, teeth, and muscles and skeleton of the face.

neural groove *See* neural plate.

neural plate The longitudinal indentation that forms in the outer layer (ectoderm) of the dorsal surface of an early vertebrate embryo and gives rise to the neural tube. As development proceeds, a groove (*neural groove*) forms along the midline of the neural plate, while along either side a neural fold arises. Eventually the folds arch over and meet along the midline, forming a tubelike structure.

neural tube The first formed element of the spinal cord and brain of vertebrate embryos. It is usually formed by deepening of the neural groove that runs along the midline of the NEURAL PLATE, and by arching over and closure of the neural folds from either side of the neural plate. Failure of the neural tube to close completely gives rise to such congenital abnormalities as spina bifida.

neurilemma *See* axon.

neuroblast A cell of an animal embryo that gives rise to a neuron. Such cells lack axons and dendrites, and typically migrate to specific sites within the developing nervous system, forming cell clusters called ganglia, from where they produce neurons. Although many of the cells of the neural tube of vertebrates will produce neurons, the term neuroblast is usually restricted to cells whose immediate progeny will differentiate as neurons.

neuroglia (**glia**) A specialized tissue, found in the central nervous system of vertebrates, that supports, protects, and nourishes the nerve cells. It consists of various types of cells, including *astrocytes*, which have many fine processes, smaller *oligodendrocytes*, and *ependymal cells*, which may be ciliated and line the brain cavities and spinal canal.

neurohormone A hormone that is produced by a neuron and released into the bloodstream. Examples are norepinephrine, serotonin, vasopressin (antidiuretic hormone), and oxytocin.

neurohypophysis The posterior lobe of the pituitary gland in higher vertebrates. It is derived from a fold in the floor of the brain and stores and releases into the blood the hormones oxytocin and vasopressin (antidiuretic hormone). These are manufactured in the hypothalamus by neurosecretory cells that have their endings in the neurohypophysis.

neuromast One of numerous groups of sensory cells that occur in pits or canals, scattered or arranged in rows over the head and along the body (lateral line) of fish and aquatic amphibians. The sensory cells (hair cells) bear hairlike processes, which detect low-frequency vibrations, water currents, and pressure changes.

neuromuscular junction The specialized junction between a nerve ending and a muscle, which is a type of synapse. The end of the axon of a motor neuron flattens out to form a motor END PLATE, which lies very close to an extensively folded area of the muscle cell membrane. The end plate and muscle cell membrane (sarcolemma) remain separated by a narrow gap, the synaptic cleft, across which neurotransmitter molecules diffuse when nerve impulses reach the end of the axon. The neurotransmitter at neuromuscular junctions is acetylcholine. This binds to receptors in the sarcolemma, and triggers a localized change in membrane potential (*see* endplate potential), which may lead to contraction of the muscle fibrils within the cell (*see* muscle contraction).

neuron (**nerve cell**) A cell that is specialized for the transmission of NERVE IMPULSES. It consists of a CELL BODY, which contains the nucleus and Nissl granules and has numerous branching extensions (DENDRITES), and a single long fine AXON (nerve fiber), which has few branches and may be surrounded by a myelin sheath.

Dendrites carry nerve impulses toward the cell body and the axon carries them away from the cell body. The end of the axon forms a junction with another neuron at a SYNAPSE or with an effector (e.g. a muscle or gland). *Sensory neurons* carry impulses from sense organs to the central nervous system and usually have rounded cell bodies; *motor neurons* carry impulses from the central nervous system to muscles and usually have star-shaped cell bodies. *Interneurons* relay impulses between sensory and motor neurons. In more advanced animals the cell bodies are located within the brain, spinal cord, or in ganglia and the fibers collectively form nerves.

neuronal outgrowth The process by which nerve cells (neurons) grow axons and connect to the appropriate target cells during development of the nervous system. A newly born neuron extends one or more axons; at the end of each is a *growth cone* bearing receptors for extracellular signal molecules. It is thought that each neuron has a specific set of growth-cone receptors that respond to particular signals in the extracellular matrix or on adjacent cells. These signals control motility of the growing axon and guide it to the correct target, for example, a certain muscle or another neuron. Often, pioneer axons establish distinct tracts, which serve as pathways for the subsequent growth of secondary axons.

neuropeptide *See* neurotransmitter.

neurotransmitter A chemical that is released from neuron endings to cause either excitation or inhibition of an adjacent neuron (*see* synapse) or an effector cell, such as a muscle cell (*see* neuromuscular junction) or gland. It is stored in minute vesicles near the synapse and released when a nerve impulse arrives. In mammals the main neurotransmitters are acetylcholine, found throughout the nervous system, and norepinephrine, occurring in the sympathetic nervous system. However, over 30 other compounds can act as neurotransmitters. They can be classified into four main groups according to their chemical structure: *acetylcholine*, which is derived from choline and is chemically unique among neurotransmitters; *amines*, such as serotonin, histamine, dopamine, epinephrine, and norepinephrine; *amino acids*, such as gamma-aminobutyric acid (GABA); and *neuropeptides*, such as the enkephalins, endorphins, and vasoactive intestinal peptide (VIP). Particular neurotransmitters can be excitatory or inhibitory, depending on the site in the nervous system. For example, acetylcholine is excitatory at neuromuscular junctions with skeletal muscle, but has an inhibitory effect on cardiac muscle.

Neurotransmitters produce their effect on the postsynaptic cell (neuron or effector cell) by acting as ligands and binding to receptors on the postsynaptic cell membrane. There are two main classes of these receptors. In LIGAND-GATED ION CHANNELS, the receptor is part of the channel protein, and binding of the neurotransmitter causes an immediate opening of the ion channel. With G-PROTEIN-linked receptors, binding of the neurotransmitter activates a G-protein, which may itself then bind to a channel protein causing it to open; alternatively, the G-protein may activate enzymes that increase the concentration of a second messenger (e.g. cyclic AMP or calcium ions), which then opens the relevant ion channels. Neurotransmitters binding to G-protein-linked receptors have slower but longer-lasting effects compared to ones binding to ligand-gated ion channels.

neurula The stage of vertebrate embryos, following gastrulation, when the neural tube is formed (the process is called *neurulation*).

neutrophil A white blood cell (*see* leukocyte) containing granules that do not stain with either acid or basic dyes. Neutrophils have a many-lobed nucleus and are therefore called *polymorphonuclear leukocytes* or *polymorphs*. Comprising about 65% of all leukocytes in humans and certain other mammals, they engulf and digest foreign particles, such as bacteria, using enzymes from their granules. This is the body's first line of defense against disease. They can pass out of capillaries by an ameboid process (see DIAPEDESIS) and wander in the tissues, gathering in large numbers at the site of an infection, where they may die, forming pus. *Compare* macrophage.

nicotinamide adenine dinucleotide *See* NAD.

nicotinamide adenine dinucleotide phosphate *See* NADP.

nidation *See* implantation.

ninhydrin A reagent used to test for the presence of proteins and amino acids. An aqueous solution turns blue in the presence of alpha amino acids in solution. When dissolved in an organic solvent it is used as a developer to color amino acids on chromatograms. If a chromatogram treated with ninhydrin is heated strongly the amino acids appear as purple spots that can be identified by measuring the R_f value. Ninhydrin is carcinogenic.

Nirenberg, Marshall Warren (1927–) American biochemist and molecular biologist. He was awarded the Nobel Prize for physiology or medicine in 1968 jointly with R. W. Holley and H. G. Khorana for their interpretation of the genetic code and its function in protein synthesis.

Nissl granules Densely staining granules found in the cell bodies of neurons. They consist of endoplasmic reticulum covered by ribosomes, plus many free ribosomes. The granules are stained by the same basic dyes that stain nuclei.

nitrification The conversion of ammonia to nitrite, and nitrite to nitrate, carried out by certain *nitrifying bacteria* in the soil. The chemosynthetic bacteria *Nitrosomonas* and *Nitrobacter* carry out the first and second stages respectively of this conversion. The process is important in the NITROGEN CYCLE.

nitrifying bacteria *See* nitrification.

nitrogen An essential element found in all amino acids and therefore in all pro-

teins, and in various other important organic compounds, for example, nucleic acids. Gaseous nitrogen forms about 80% of the atmosphere but is unavailable in this form except to a few species of nitrogen-fixing bacteria. Nitrogen is therefore usually incorporated into plants as the nitrate ion, NO_3^-, absorbed in solution from the soil by roots. In animals, the nitrogen compounds urea and uric acid form the main excretory products. *See also* nitrogen cycle.

nitrogen cycle The circulation of nitrogen between organisms and the environment. Atmospheric gaseous nitrogen can only be used directly by certain nitrogen-fixing bacteria (*see* nitrogen fixation). They convert nitrogen to ammonia, nitrites, and nitrates, which are released into the soil by excretion and decay. Some atmospheric nitrogen is also fixed by lightning. When plants and animals die, the organic nitrogen they contain is converted back into nitrate in the process termed *nitrification*. Apart from uptake by plants, nitrate may also be lost from the soil by *denitrification* and by leaching. The use of nitrogen fertilizers in agriculture and the emission of nitrous oxides in car exhaust fumes are also important factors in the nitrogen cycle.

nitrogen fixation The formation of nitrogenous compounds from atmospheric nitrogen. In nature this may be achieved by electric discharge in the atmosphere or by the activities of certain microorganisms. For example, the symbiotic bacteria, *Rhizobium* species, are associated with leguminous plants, forming the characteristic nodules on their roots. The bacteria contain the nitrogenase enzyme that catalyzes the fixation of molecular nitrogen to ammonium ions, which the plant can assimilate. In return the legume supplies the bacteria with carbohydrate. Free-living bacteria that can fix nitrogen include members of the genera *Azotobacter* and *Clostridium*. Some sulfur bacteria (e.g. *Chlorobium*), some Cyanobacteria (e.g. *Anabaena*), and some yeast fungi have also been shown to fix nitrogen. In industry the most important method for fixing nitrogen is the Haber process, which is used to make ammonia from nitrogen and hydrogen.

NK cell   See natural killer cell.

NMR   See nuclear magnetic resonance.

NMR spectroscopy   See nuclear magnetic resonance.

node of Ranvier   A region of bare axon that occurs at intervals of up to 2 mm along the length of myelinated nerve axons. See also myelin sheath; nerve impulse; saltatory conduction.

nondisjunction   The failure of homologous chromosomes to move to separate poles during cell division. This can occur during either of the divisions of meiosis, or during mitosis. For example, nondisjunction during anaphase I of meiosis results in two of the four gametes formed at telophase missing a chromosome (i.e. being $n - 1$). If these fuse with normal haploid (n) gametes then the resulting zygote is *monosomic* (i.e. $2n - 1$). The other two gametes formed at telophase have an extra chromosome (i.e. are $n + 1$) and give a *trisomic* zygote (i.e. $2n + 1$) on fusion with a normal gamete. If two gametes deficient for the same chromosome fuse then *nullisomy* ($2n - 2$) will result, which is almost always lethal, and if two gametes with the same extra chromosomes fuse, *tetrasomy* ($2n + 2$) results. All these abnormal chromosome conditions are collectively referred to as ANEUPLOIDY.

nonessential amino acid   See amino acids.

nonsense mutation   See mutation.

noradrenaline   See norepinephrine.

norepinephrine (noradrenaline)   A catecholamine, secreted as a hormone by the adrenal medulla, that regulates heart muscle, smooth muscle, and glands. It causes narrowing of arterioles and hence raises blood pressure. It is also secreted by nerve endings of the sympathetic nervous system in which it acts as a neurotransmitter. Norepinephrine binds to alpha and beta ADRENERGIC RECEPTORS in the various effector organs whose activity it influences. In the brain, levels of norepinephrine are related to mental function; lowered levels lead to mental depression.

Northern blotting   A technique used to isolate and estimate the amount of particular RNA molecules in a sample. The RNA is chemically denatured to ensure that it consists solely of linear single strands, and then subjected to gel electrophoresis to separate the RNAs according to size. The RNAs are then blotted onto a nitrocellulose filter, to which they adhere, and are incubated with a labeled DNA probe. After washing away excess probe, the location of any bound probe is revealed as a pattern of bands by autoradiography. The density of such bands is a measure of the amounts of corresponding RNAs present in the original sample. The technique is named, somewhat humorously, by analogy to SOUTHERN BLOTTING, which is a similar technique used for DNA.

Northrop, John Howard (1891–1987) American biochemist. He was awarded the Nobel Prize for chemistry in 1946 jointly with W. M. Stanley for their preparation of enzymes and virus proteins in a pure form. The prize was shared with J. B. Sumner.

nucellus   The nutritive and protective tissue enclosing the megaspore or egg cell inside a plant ovule.

nuclear basket   See nuclear pore complex.

nuclear envelope   The double membrane that forms the boundary between the nucleus and the cytoplasm in eukaryotic cells. It consists of an inner and an outer nuclear membrane with a lumen between them. Perforating the envelope are numerous pores, each surrounded by a NUCLEAR PORE COMPLEX, which controls the flow of substances through the pore.

nuclear magnetic resonance (NMR) A property of atomic nuclei used in the analysis of molecular structure. Small changes in the resonance of certain atomic nuclei (e.g. ^{1}H, ^{13}C) can be induced by irradiating them with radio waves in the presence of a strong magnetic field. This property is utilized in a form of spectroscopy called *NMR spectroscopy* to provide information about the composition and structure of complex molecules. It is also exploited in medicine as the basis of *NMR imaging*, which is used to detect tumors, etc. in the body.

nuclear pore *See* nucleus.

nuclear pore complex The structure that surrounds each of the many NUCLEAR PORES that perforate the nuclear envelope in eukaryotic cells. It is a large and elaborate complex consisting of multiple copies of maybe 50–100 different proteins. Electron micrographs suggest that there is a nuclear ring and a cytoplasmic ring, in the inner and outer nuclear membranes respectively, linked by a cylindrical array of radially projecting 'spokes' surrounding a central plug. Extending into the nucleoplasm from the nuclear ring are eight filaments, each about 100 nm long, joined to a terminal ring and forming an array called the nuclear basket. This resembles an inverted basketball hoop. Ions and small molecules can diffuse through water-filled channels in the pore complex. Larger molecules, such as the messenger ribonucleoproteins (mRNPs) that represent the mature products of gene transcription, are actively transported through the central plug. Similarly, proteins destined for the nucleus are imported from their sites of synthesis in the cytosol via the pore complexes.

nuclear transfer A method used in cloning vertebrates that involves transferring the nucleus from a donor cell to direct the development of a recipient egg cell into an embryo. There are several variations, depending on the species of organism involved, type of material available, aim of the procedure, etc. The feasibility of nuclear transfer was established in 1968, when John Gurdon (1933–) and colleagues removed the nucleus from the skin cell of an adult frog and transplanted it into an egg cell from which the nucleus had been removed. Some of the treated egg cells developed into embryos, but none reached adulthood.

The first successful attempts at cloning vertebrates used whole donor embryo cells, instead of body cell nuclei. In 1996 a research team at the Roslin Institute, led by Ian Wilmut (1945–), produced the first sheep clones, called Morag and Megan, using donor cells taken from a nine-day-old sheep embryo. These were injected into unfertilized egg cells from which the chromosomes had been removed using an ultrafine micropipette. Pulses of electric current stimulated donor and recipient cells to fuse, and begin developing like a normal embryo. A successfully fused embryo was implanted in the uterus of a surrogate mother sheep to continue development. In 1997, work by Wilmut's team resulted in the birth of Dolly, the world's first mammal to be successfully cloned from an adult body cell. Donor cells taken from a sheep's udder were cultured and transferred to a low-nutrient medium to starve them into a state of quiescence, in which the cells stop dividing and 'switch off' all but nonessential genes. Individual such cells were then fused with recipient egg cells and stimulated to begin development by the established route, in a surrogate mother.

The birth of Dolly showed that it was possible to 'reprogram' fully differentiated tissue cells so that they could direct the formation of an entirely new organism from an embryo. Donor cells derived from body tissues have several advantages. They are relatively easy to harvest, maintain in culture, and work with. Also, it is easier to genetically engineer cultured donor cells, and to select successfully engineered cells before nuclear transfer takes place. Cloning can be undertaken when embryo cells are scarce or difficult to obtain, such as with endangered species. However, the technique has a high failure rate.

nuclease *See* endonuclease; exonuclease.

nucleic acid hybridization (DNA hybridization) The pairing of a single-stranded DNA or RNA molecule with another such strand, forming a DNA–DNA or RNA–DNA hybrid. In order to achieve hybridization the base sequences of the strands must be complementary. This phenomenon is exploited in many techniques, notably in DNA PROBES, which are designed to bind to particular complementary base sequences among a mass of DNA fragments.

nucleic acids Linear polymers of NUCLEOTIDES, consisting of chains of alternating sugar and phosphate groups, with nitrogenous bases attached to the sugar units. They are crucial constituents of the cells of all organisms, occurring as DNA (in which the sugar is deoxyribose) and RNA (in which the sugar is ribose). *See* DNA; RNA.

nucleocapsid *See* capsid.

nucleoid The region of a bacterium containing aggregated DNA but not enclosed by membranes. *Compare* nucleus.

nucleolar organizer *See* nucleolus.

nucleolus (*pl.* **nucleoli**) A more or less spherical structure found in nuclei of eukaryote cells, and easily visible with a light microscope. One to several per nucleus may occur. Rich in RNA and protein, the nucleolus is the site of synthesis of ribosomal RNA and is thus most conspicuous in cells making large quantities of protein. Some ribosomal proteins are also packaged in the nucleolus to form ribosomal subunits, which are exported from the nucleus. Nucleoli disappear during cell division. The nucleolus forms around particular loci of one or more chromosomes called *nucleolar organizers*. These loci contain numerous tandem repeats of the genes coding for ribosomal RNA.

nucleoplasm *See* nucleus.

nucleoprotein A compound consisting of a protein associated with a nucleic acid. Examples of nucleoproteins are the chromosomes, made up of DNA, some RNA, and histones (proteins), and the ribosomes (ribonucleoproteins), consisting of ribosomal RNA and proteins.

nucleoside A molecule consisting of a purine or pyrimidine base linked to a sugar, either ribose or deoxyribose. Adenosine, cytidine, guanosine, thymidine, and uridine are common nucleosides.

nucleosome A beadlike particle that is the basic structural unit of chromosomes in their condensed state (*see* chromatin). Each comprises a disk-shaped core of interlocking histone proteins, around which the DNA double helix coils twice, equal to a length of about 146 bp. Consecutive nucleosomes are joined by uncoiled linker DNA of variable length, ranging from some 15 to 55 bp. This 'string of beads' is itself coiled in a spiral some 30 nm in diameter – the solenoid model of condensed chromatin.

nucleotide The compound formed by condensation of a nitrogenous base (a purine, pyrimidine, or pyridine) with a sugar (ribose or deoxyribose) and phosphoric acid; i.e. a NUCLEOSIDE having one, two, or three phosphate groups esterified to the sugar at the 5′ position. For example, ATP is a *mononucleotide* with three phosphate groups. The coenzymes NAD and FAD are *dinucleotides* (consisting of two linked nucleotides) whereas the NUCLEIC ACIDS are *polynucleotides* (consisting of chains of many linked nucleotides).

nucleus An organelle of eukaryote cells containing the genetic information (DNA) and hence controlling the cell's activities. It is found in virtually all living cells (exceptions include mature sieve tube elements and mature mammalian red blood cells). It is the largest organelle, typically spherical and bounded by a double membrane, the *nuclear envelope* or *nuclear membrane*, which is perforated by many pores (*nuclear* *pores*) that allow exchange of materials with the cytoplasm. The outer nuclear membrane is an extension of the rough endoplasmic reticulum (RER), and the space between the two membranes is an extension of the lumen of the RER. The DNA is packaged with proteins into several or many CHROMOSOMES, of which there may be one, two, or more sets. In the nondividing (interphase) nucleus the chromosomal material, called CHROMATIN, is highly diffuse, and the most visible structures in the nucleus are the one or more nucleoli (*see* nucleolus), which are concerned with RNA synthesis. The nonnucleolar region of the nucleus is termed the *nucleoplasm*, and the inner surface of the nuclear membrane is covered with an array of structural proteins called lamins. During nuclear division (mitosis or meiosis) the nuclear envelope and nucleoli disappear, and the chromatin condenses so that the individual chromosomes can be readily stained and identified under a microscope. *Compare* nucleoid. *See also* macronucleus.

nullisomy   *See* aneuploidy.

Nurse, (Sir) Paul M. (1949–   )   British cell biologist who was awarded the 2001 Nobel Prize for physiology or medicine jointly with R. T. Hunt and L. H. Hartwell for discoveries of the key regulators of the cell cycle.

Nüsslein-Volhard, Christiane (1942–   ) German biochemist who was jointly awarded the 1995 Nobel Prize for physiology or medicine with E. B. Lewis and E. F. Wieschaus for work on the genetic control of early embryonic development.

OAA *See* oxaloacetic acid.

Ochoa, Severo (1905–93) Spanish-born American biochemist. He was awarded the Nobel Prize for physiology or medicine in 1959 jointly with A. Kornberg for their discovery of the mechanisms in the biological synthesis of ribonucleic acid and deoxyribonucleic acid.

odontoblast A cell that secretes the dentine of a tooth. Odontoblasts lie around the outside of the pulp cavity and their long fine processes extend into fine canals (canaliculi) in the dentine.

oil body *See* spherosome.

oil immersion A microscopic technique using special high-powered objective lenses. A drop of immersion oil (e.g. cedarwood oil) is placed on the coverslip of a microscope slide and the objective lens carefully lowered into it. The oil has the same refractive index as the lens glass and increases the resolving power obtainable by letting a wider angle of rays enter the objective lens.

Okazaki fragment A short segment of nucleic acid that is assembled during discontinuous replication of the lagging strand of DNA (*see* DNA replication). Each consists of a short RNA primer, which is extended by the addition of 1000–2000 nucleotides to its 3′ end, forming a segment of DNA complementary to the lagging strand. When each fragment is complete, the RNA primer is removed, the gap between the new fragment and the previous fragment is filled, and the ends are joined together producing a continuous new DNA strand. It is named after the Japanese molecular biologist, Reiji Okazaki (1930–75), who discovered the fragments.

oleoplast *See* elaioplast.

oleosome *See* spherosome.

olfactory organ The organ involved in the detection of smells, which consists of a group of sensory receptors that respond to air- or water-borne chemicals. Vertebrates possess a pair of olfactory organs in the mucous membrane lining the upper part of the nose, which opens to the exterior via the external nares (nostrils). Olfactory organs are found on the antennae in insects and in various positions in other invertebrates.

oncogene A gene that is capable of transforming a normal cell into a cancerous cell. Oncogenes are typically mutated forms of normal cellular genes (*proto-oncogenes*). The products of proto-oncogenes are generally involved in some aspect of controlling cell growth and differentiation, such as regulating the cell cycle, or cell signaling and gene expression. The oncogenic mutaton can cause the inappropriate expression or over-expression of a normal protein, or give rise to a defective protein (*oncoprotein*) that behaves in a faulty manner. For example, the *ras* oncogene produces a faulty signal transduction protein that cannot 'switch off', and so continuously activates components downstream in its signal pathway. Oncogenes arise from 'gain-of-function' mutations, and are well placed to disrupt the normal activities of cells and cause the changes that lead to cancer.

Oncogenes were originally discovered in RETROVIRUSES. In 1910 the US pathologist Peyton Rous (1879–1970) showed that a cell-free filtrate of a particular tumor (called a sarcoma) found in chickens could induce similar tumors in healthy chickens. The agent causing the tumors was later identified as a virus – now called the *Rous sarcoma virus*. One of the three genes carried by this virus is the oncogene *src* (short for sarcoma). When the virus infects a host cell, the oncogene is inserted into the host's genome. Certain other viruses carry oncogenes that cause cancers in their animal hosts, including the Abelson murine leukemia virus (which affects mice) and the avian erythroblastosis virus (which affects chickens).

oncogenic Causing the production of a tumor, especially a malignant tumor (i.e. cancer). The term is usually applied to viruses or genetic mutations known to be implicated in causing cancer.

oncoprotein *See* oncogene.

one gene–one enzyme hypothesis The theory that each gene controls the synthesis of one enzyme. Thus by regulating the production of enzymes, genes control the biosynthetic reactions catalyzed by enzymes and ultimately the character of the organism. The theory was advanced in the 1940s by G.W. Beadle (1903–89), E.L. Tatum ((1909–75), and their colleagues following studies of nutritional mutants of fungi. Genes also code for proteins, or polypeptides that form proteins, other than enzymes, so the idea is more accurately expressed as the *one gene–one polypeptide hypothesis*.

ontogeny The course of development of an organism from fertilized egg to adult.

oocyte A reproductive cell in the ovary of an animal that gives rise to an ovum. The primary oocyte develops from an oogonium, which has undergone a period of multiplication and growth. It divides by meiosis and the first meiotic (or reduction) division produces a secondary oocyte, con-taining half the number of chromosomes, and a small polar body. The secondary oocyte forms the ovum, and undergoes the second meiotic division to form an ovum and a second polar body. In many species the second meiotic division is not completed until after FERTILIZATION. *See also* oogenesis.

oogamy Sexual reproduction involving the fusion of two dissimilar gametes. The male gamete is usually motile and smaller than the female gamete, which is usually nonmotile, contains a food store, and is retained by the parent. The term is generally restricted to descriptions of plants, particularly those that produce female gametes in oogonia. It is an extreme form of ANISOGAMY.

oogenesis The formation of sex cells (ova) within the ovary of female animals. In mammals, this process begins during fetal development, when precursor cells (oogonia) in the germinal epithelium of the fetal ovary start to divide by mitosis to form the potential egg cells, or primary oocytes. In female humans there are some 250 000 primary oocytes in each ovary at the time of birth. Before birth, these start to divide by meiosis, but further development is halted in prophase I until puberty. During childhood, granulosa cells develop around each primary oocyte, forming primary follicles. Then, during each menstrual cycle, about a thousand follicles begin to mature and migrate toward the ovary surface. However, usually only a single follicle survives to become a fully mature (Graafian) follicle, at the ovary surface. As the follicle matures, the primary oocyte resumes meiosis I and divides to produce a large secondary oocyte, containing the bulk of the cytoplasm, and a much smaller cell called the first polar body. Meiosis again pauses, and at ovulation, the secondary oocyte is released to form the ovum. The ovum enters the second meiotic division only if triggered by entry of a sperm, when it produces two haploid sets of chromosomes – one set is discarded as the second polar body, while the nuclear membrane re-forms around the other set to form the female pronucleus. *See also* fertilization; ovarian follicle.

oogonium (*pl.* **oogonia**) **1.** A precursor cell in the ovary of an animal that gives rise to a primary oocyte. *See* oogenesis. **2.** A unicellular female reproductive organ of certain protists.

oosphere A large, nonmotile female gamete formed by certain protists.

oospore A diploid zygote of certain protists, especially water molds (oomycetes), produced by fertilization of the female gamete, the oosphere, by the male gamete, the antherozoid. The zygote may form a thick coat and go through a resting period before germination.

OP Osmotic pressure. *See* osmosis.

open reading frame (**ORF**) A long stretch of cloned DNA whose base sequence does not contain a stop codon (ATT, ATC, or ACT). Such a sequence, typically of the order of hundreds or thousands of bases long, can potentially represent a coding sequence, such as an exon or other portion of a complete gene. Computerized searches for ORFs enable researchers to pinpoint candidate genes in sequenced cloned fragments of genomic DNA that have been isolated using DNA probes.

operator *See* operon.

operator site *See* operon.

operon A genetic unit found in prokaryotes and comprising a group of closely linked genes that are expressed in a coordinately controlled fashion. Expression frequently results in a single messenger RNA molecule containing a transcript of all the structural genes in the operon. Typically, an operon encodes the various enzymes of a particular biochemical pathway; an example is the *lac* operon, which encodes enzymes involved in lactose metabolism. At its simplest, an operon has an *operator site* lying at one end; under certain conditions this site is repressed by a protein (repressor) encoded by another gene outside the operon, the *regulator gene*. When the repressor binds to the operator site, it blocks transcription of the operon's structural genes, and hence enzyme production. In the presence of the relevant substrate, the substrate binds to the repressor so that the latter cannot bind to the operator. Hence transcription can proceed, enzyme production commences, and the substrate is metabolized by the cell. Another site in the operon, the *promoter*, binds the RNA polymerase enzyme that transcribes the structural genes into messenger RNA.

opsin The protein component of the retinal pigment RHODOPSIN, which is localized in the rod cells of the retina. Opsin is released from rhodopsin when light strikes the retina.

opsonin Any of various blood proteins that bind to microorganisms or other foreign material and enhance their susceptibility to engulfment by phagocytic cells. The main types of opsonins are immunoglobulin G antibodies and certain complement proteins.

opsonization The process by which a host coats the surface of an invading cell with opsonins to render it susceptible to phagocytosis.

optical isomer (**enantiomer**) Either of a pair of molecules that are mirror images of each other; i.e. they contain the same combination of atoms but the bonds can be arranged in two different ways to form structures that cannot be superimposed on each other. Any pair of such isomers can be distinguished by their optical activity – the ability to rotate plane-polarized light as it passes through crystals or solutions containing them. The dextrorotatory isomer, or D-isomer, rotates light one way, whereas the levorotatory or L-isomer, rotates light the other way. Many organic compounds contain an asymmetric carbon atom that gives rise to optical isomers, and this quality is usually crucial to their biological occurrence and activity. For example, both D and L isomers of the amino acid alanine are found in nature, although only L-alanine occurs in proteins. Similarly, glucose and other sugars can exist in both forms, although L isomers of these are virtually unknown in nature.

optical rotation Rotation of the plane of polarization of plane-polarized light by a solution of an optically active substance. *See* optical isomer.

optical trapping *See* micromanipulation.

organ A part of an organism that is made up of a number of different tissues specialized to carry out a particular function. Examples include the lung, stomach, wing, and leaf.

organelle A discrete subcellular structure with a particular function. The largest organelle is the nucleus; other examples are chloroplasts, mitochondria, Golgi apparatus, vacuoles, and ribosomes. Organelles allow division of labor within the cell. Prokaryotic cells have very few types of organelle compared with eukaryotic cells.

organizer A part of an embryo that causes neighboring tissue to develop in a particular way. Examples are the eye-cup of vertebrates, which causes lens, and later cornea, to be produced; the gut of snail embryos, which organizes shell gland and mantle; the dorsal lip of the blastopore of frogs, which becomes notochord and organizes all the axial structures of the embryo; and the dermal papilla of a hair, feather, or tooth, which organizes local epidermis and dermis to form the follicle and appendages. The term *primary organizer* is restricted to the first or most important initiator at gastrulation; for example, the dorsal lip of the amphibian blastopore or Hensen's node in mammals and birds. Determinants of major systems (e.g. notochord) are called *secondary organizers*; local centers of developmental activity (e.g. dermal papillae) are *tertiary organizers*. *See also* inducer; induction.

ornithine *See* amino acids.

orthotropism *See* tropism.

osmiophilic globules *See* plastoglobuli.

osmium tetroxide A stain used in electron microscopy because it contains the heavy metal osmium. It also acts as a fixative, i.e. preserves material in a lifelike condition, often being used in conjunction with the fixative glutaraldehyde. It stains lipids, and therefore membranes, particularly intensely.

osmoregulation *See* osmosis.

osmosis The movement of solvent from a dilute solution to a more concentrated solution through a membrane. For example, if a concentrated sugar solution (in water) is separated from a dilute sugar solution by a membrane, water molecules can pass through from the dilute solution to the concentrated one. A membrane of this type (which allows the passage of some kinds of molecule and not others) is called a *semipermeable membrane*. Osmosis between two solutions will continue until they have the same concentration. If a certain solution is separated from pure water by a membrane, osmosis also occurs. The pressure necessary to stop this osmosis is called the *osmotic pressure* (OP) of the solution. The more concentrated a solution, the higher its osmotic pressure.

Most cell membranes are not perfectly semipermeable, but are more permeable to water than to ions or other solutes; they are called *differentially permeable membranes*. Some contain channel proteins (e.g. AQUA-PORINS) that enable the bulk flow of water across the membrane. If an animal cell is immersed in a hypotonic solution (i.e. a solution less concentrated than the cell's cytosol), water will enter the cell by osmosis and may eventually cause the cell to burst (*osmotic lysis*). Conversely, if a cell is immersed in a hypertonic solution (i.e. one more concentrated than the cell's cytosol), water will leave the cell and the cell will shrink. To avoid either of these consequences, a cell must perform *osmoregulation*, i.e. maintain its solute concentration so that it is roughly the same as its surroundings – a state described as isotonic. In cells that have cell walls (e.g. in plants, fungi, and algae), osmotic influx of water causes a rise in intracellular pressure, because the cell wall maintains a fixed cell volume. In plant cells the vacuole contents are generally more concentrated than the cytosol, so water flows through the cell wall, plasma membrane and cytosol into the vacuole. Consequently, the protoplast is pushed against the cell wall. As a plant cell loses water, the protoplast shrinks away from the cell wall – a condition called PLASMOLYSIS. Physiologists now describe the tendency for water to move in and out of plant cells in terms of WATER POTENTIAL. *See also* turgor.

osmotic lysis   *See* osmosis.

osmotic potential   *See* water potential.

osmotic pressure   *See* osmosis.

ossification (**osteogenesis**)   The transformation of embryonic or adult connective tissue (*intramembranous ossification*) or cartilage (*endochondral ossification*) into bone. Bone is produced by the action of special cells (*see* osteoblast), which deposit the bone matrix and its network of collagen fibers; they eventually become enclosed in the matrix as bone cells (*see* osteocyte).

osteoblast   Any of the cells that secrete the matrix and collagen fibers of bone during OSSIFICATION. Initially they occupy sites on the outside of the embryonic cartilage or membrane, but after the embryonic tissue has been eroded by OSTEOCLASTS the osteoblasts accompany the ingrowing blood vessels and form temporary trabeculae of bone. Later the osteoblasts lay down the permanent structure of bone, and those that become trapped between the lamellae are called OSTEOCYTES.

osteoclast   Any of the cells that erode bone minerals. In conjunction with OS-TEOBLASTS, osteoclasts are vital for the erosion and remodeling of the calcified cartilage or membrane in the early stages of bone formation (*see* ossification). Blood vessels, preceded by osteoclasts, invade the tissue, and then osteoblasts lay down the permanent structure of bone. Even in adult life, osteoclasts are required for the continual remodeling of bone.

osteocyte   Any of the cells that secrete the hard matrix of bone. They are found in small spaces (lacunae) between the concentric lamellae of bone that form the *Haversian systems*. Each osteocyte has many fine cytoplasmic processes that pass, in fine channels (canaliculae), through the matrix and connect with each other and with blood vessels to maintain supplies of food and oxygen to the living cells. *See also* osteoblast.

osteogenesis   *See* ossification.

otolith   One of numerous tiny granules composed of protein and calcium carbonate that are contained in a gelatinous matrix within the utriculus and sacculus of the vertebrate inner ear. The matrix is attached to hairlike processes of sensory cells, and the otoliths give the matrix appreciable mass so it responds to changes in the position of the head, thereby stimulating the sensory cells.

ovarian follicle   A fluid-filled ball of cells within the mammalian ovary, in which an ovum develops. A follicle matures periodically during the active reproductive years from one of the enormous number of follicles present in the ovary at birth (*see* oogenesis). In humans about 1000 follicles begin to mature and migrate through the ovary tissue in each round of the menstrual cycle, but usually only one survives to reach the surface, as a *Graafian follicle*. Finally it bursts and releases the ovum (OOCYTE) to the Fallopian tube. The follicle then becomes a solid body, the *corpus luteum*. The growth of follicles is under the influence of a hormone (*follicle-stimulating hormone*, *FSH*) released by the pituitary gland.

ovulation The release of an egg (ovum) from a Graafian follicle at the surface of a vertebrate ovary. In humans it first occurs at the onset of sexual maturity and a single ovum is released about every 28 days from alternate ovaries until menopause, at the age of about 45. In some other species, several ova can be released at one ovulation. Ovulation is stimulated by luteinizing hormone (LH) produced by the pituitary gland in the presence of estrogen. *See also* oogenesis.

ovule Part of the female reproductive organs in seed plants. It consists of the nucellus, which contains the embryo sac, surrounded by the integuments. After fertilization the ovule develops into the seed. In angiosperms the ovule is contained within an ovary and may be orientated in different ways being upright, inverted, or sometimes horizontal. In gymnosperms ovules are larger but are not contained within an ovary. Gymnosperm seeds are thus naked while angiosperm seeds are contained within a fruit, which develops from the ovary wall.

ovum (egg cell; *pl.* ova) **1.** (*Zoology*) The immotile female sex cell (gamete) produced in the ovary of an animal. Although variable in form, it essentially contains a haploid complement of chromosomes, which may or may not be bounded by a nucleus. The cytoplasm contains a variable amount of yolk, and is surrounded by a vitelline membrane and often additional protective layers. Size varies between species; in humans, the mature is about 0.15 mm in diameter. In chickens it is about 30 mm in diameter and further enlarged by a layer of albumen, more membranes, and a shell to become a true egg. *See also* fertilization; oogenesis; ovulation. **2.** (*Botany*) In flowering plants, the female gamete produced by mitotic division of a haploid megaspore within the embryo sac of an ovule. At fertilization it fuses with a sperm cell from the pollen tube to form the zygote, from which the embryo develops.

oxaloacetic acid (**OAA**) A water-soluble carboxylic acid, structurally related to fumaric acid and maleic acid. Oxaloacetic acid forms part of the KREBS CYCLE, it is produced from L-malate in an NAD-requiring reaction and itself is a step towards the formation of citric acid in a reaction involving pyruvate ion and coenzyme A.

oxidative phosphorylation The production of ATP from phosphate and ADP in aerobic respiration. Oxidative phosphorylation occurs in mitochondria, and is the cell's principal means of storing the energy released during the complete oxidation of foodstuffs and other compounds. The energy released is transferred as electrons to the ELECTRON-TRANSPORT CHAIN, and according to the CHEMIOSMOTIC THEORY, is harnessed to establish an electrochemical proton (H^+) gradient across the mitochondrial inner membrane. This gradient then drives the conversion of ADP to ATP by an enzyme complex called ATP SYNTHASE, located in the inner membrane.

oxonium ion *See* acid.

oxygen An element essential to living organisms both as a constituent of carbohydrates, fats, proteins, and their derivatives, and in aerobic respiration. It enters plants both as carbon dioxide and water, the oxygen from water being released in gaseous form as a by-product of photosynthesis. Plants and other oxygen-evolving organisms (e.g. algae, cyanobacteria) are the main if not the only source of gaseous oxygen and as such are essential in maintaining oxygen levels in the air for aerobic organisms.

oxyhemoglobin *See* hemoglobin.

oxytocin A peptide hormone, produced by the hypothalamus and posterior pituitary gland, that promotes labor and also the release of milk from the mammary gland.

P

pachytene In MEIOSIS, the stage in midprophase I that is characterized by the contraction of paired homologous chromosomes. At this point each chromosome consists of a pair of chromatids, and the two associated chromosomes are termed a tetrad.

pairing *See* synapsis.

pair-rule gene A gene in the fruit fly (*Drosophila*) whose product establishes the repeating pattern of segments during development of the early embryo. There are several such genes, and their expression along the anterior–posterior (head-to-tail) axis of the embryo defines a series of stripes, which become the 14 PARASEGMENTS of the embryo. Each pair-rule gene is expressed in only 7 of the 14 parasegments – either the even-numbered ones or the odd-numbered ones.

Palade, George Emil (1912–) Romanian-born American cell biologist. He was awarded the Nobel Prize for physiology or medicine in 1974 jointly with A. Claude and C. R. M. J. de Duve for their discoveries concerning the structural and functional organization of the cell.

palisade mesophyll *See* mesophyll.

palmella A stage formed under certain conditions in various unicellular algae in which, after division, the daughter cells remain within the envelope of the parent cell and are thus rendered nonmotile. The cells may continue dividing giving a multicellular mass, which is contained in a gelatinous matrix. Palmelloid forms may develop undulipodia (flagella) and revert to normal motile cells at any time. Members of the algal genus *Palmella* typically exist in the palmella condition.

pancreas A gland lying between the spleen and the duodenum, and having a duct that enters the duodenum. It secretes pancreatic juice, which contains various enzymes, including: trypsin for breaking down proteins to amino acids, amylase for converting starch to maltose, and lipase for changing emulsified oils to glycerin and fatty acids. The gland also contains endocrine tissues and produces insulin and glucagon. *See* pancreatic islets.

pancreatic islets (islets of Langerhans) Clusters of endocrine cells within the pancreas (described by Langerhans in 1869) that produce various hormones, notably insulin, which controls blood-sugar level. The islets consist of four types of cells: A-cells, which secrete glucagon; B-cells, which secrete insulin; D-cells, which secrete somatostatin; and PP cells, which secrete pancreatic polypeptide. The B-cells also secrete amylin, an insulin antagonist.

paper chromatography A chromatographic method using absorbent paper by which minute amounts of material can be analyzed. A paper strip with a drop of test material at the bottom is dipped into the carrier liquid (solvent) and removed when the solvent front almost reaches the top of the strip. Two-dimensional chromatograms can be produced using square paper and two different solvents. The paper is removed from the first carrier liquid, turned at right angles and dipped into the second. This gives a two-dimensional 'map' of the constituents of the test drop.

papovavirus One of a group of small

DNA-containing viruses, about 50 nm in diameter, that cause tumors in animals. The group includes SV40 (simian virus 40), a much-studied virus originally isolated from the African green monkey (*see* SV40).

Paramecium A genus of ciliated protists belonging to the phylum CILIOPHORA and common universally in fresh water containing decaying vegetable matter. *Paramecium* is slipper-shaped and covered with cilia, the beating of which produces rapid locomotion. It reproduces asexually by binary fission and sexually by conjugation. There are two contractile vacuoles for osmoregulation and food is taken in through the oral groove and cytopharynx and digested in food vacuoles. There are two nuclei, the meganucleus controlling the vegetative functions and the smaller micronucleus controlling sexual reproduction.

parasegment A developmental unit within an embryo, such as the larva of a fruit fly (*Drosophila*). The parasegments form a repeating pattern along the the the anterior–posterior (head-to-tail) axis of the embryo, and regional body structures develop in the appropriate parasegments. In *Drosophila* the parasegments are defined primarily by the expression of PAIR-RULE GENES, and are subsequently subdivided by the expression of SEGMENT-POLARITY GENES. The development of regional body structures (e.g. antennae, legs, wings) is determined by patterns of SELECTOR GENE expression unique to each parasegment. The *Drosophila* larva has 14 parasegments, each of which corresponds with two halves of adjoining segments in the adult fly. For example, parasegment 5 corresponds with the posterior compartment of thoracic segment 2 (T2) plus the anterior compartment of thoracic segment 3 (T3).

parasympathetic nervous system One of the two divisions of the AUTONOMIC NERVOUS SYSTEM, which supplies motor nerves to the smooth muscles of the internal organs and to cardiac muscles. Parasympathetic fibers emerge from the central nervous system via cranial nerves, especially the vagus nerve, and a few spinal nerves in the sacral region. Their endings release acetylcholine, which slows heart rate, lowers blood pressure, and promotes digestion, thereby antagonizing the effects of the SYMPATHETIC NERVOUS SYSTEM.

parathyroid glands Four small oval-shaped structures embedded in the thyroid gland. The glands are composed of columns of cells with vascular channels between the columns. They produce parathyroid hormone, which antagonizes the effects of calcitonin in controlling blood calcium level.

parenchyma 1. (*Botany*) Tissue made up of cells that are not differentiated for any specific function, but perform important metabolic processes, including photosynthesis, secretion, and storage. Parenchyma cells typically have thin primary cell walls, but can also have secondary walls, sometimes thickened with lignin. The leaf mesophyll and the stem medulla and cortex consist of parenchyma. The vascular tissue is also interspersed with parenchyma; for instance the medullary rays of secondary vascular tissue.
2. (*Zoology*) Spongy tissue made of loosely packed cells as occurs, for example, between the outer skin (ectoderm) and the lining of the gut (endoderm) in flatworms.

parthenogenesis Development of unfertilized eggs to form new individuals. It occurs regularly in certain plants (e.g. dandelion) and animals (e.g. aphids). Animals produced by parthenogenesis are always female and, if diploid, look exactly like the parent. Parthenogenesis produces haploid or diploid individuals depending on the genetic state of the ovum when development of the embryo begins. Genetic recombination cannot occur in parthenogenesis and so sexual reproduction occurs occasionally. *Artificial parthenogenesis* can be induced by pinpricks or treatment with, for instance, cold or acid, especially in eggs shed in water.

passage cells Thin-walled cells found in the endodermis of the roots of some plants that might act as an entry point to the stele for water and solutes from the cortex. The walls of such cells have Casparian strips (*see* endodermis) but remain otherwise unthickened after deposits of lignin and cellulose have been laid down elsewhere in the endodermis.

passive immunity *See* immunity.

Pasteur, Louis (1822–95) French chemist, microbiologist, and immunologist. Pasteur is remembered for his recognition of the role of microorganisms in fermentation and for devising methods of immunization against certain diseases. He also did pioneering work in stereochemistry.

pathogen Any organism that is capable of causing disease or a toxic response in another organism. Many bacteria, viruses, fungi, and other microorganisms are pathogenic.

patterning The establishment of repeated or corresponding parts during the development of an organism. Patterning defines the head-to-tail (anterior–posterior) axis, by subdiving the embryo into a series of segments. It also defines the back-to-front (dorsoventral) axes, and other basic reference points, so that tissues and organs develop in the appropriate parts of the body, according to the body plan of the species concerned. Patterning has been best studied in experimental organisms, notably the fruit fly *Drosophila*. It results from the products of numerous maternal and embryonic genes, expressed in a coordinated fashion over time and in different parts of the embryo. In *Drosophila*, proteins called MORPHOGENS, encoded by maternal genes of follicle cells, diffuse into the developing early embryo where they lay the foundations of the general body plan. Concentration gradients of the various morphogens cause genes in different zones of the embryo to be activated to different extents, creating a rudimentary pattern of body segments. This segmentation pattern is reinforced and refined by the sequential expression of three groups of the embryo's own genes– pair-rule genes, segment-polar-

ity genes, and selector genes. The latter specify the structures of each segment, such as legs or wings, and are described as HOMEOTIC GENES – mutation or misexpression can result in homeosis, in which one body part is transformed into another. Homeotic genes are also responsible for tissue patterning in vertebrates. *See also* induction (def. 2).

Pauling, Linus Carl (1901–94) American chemist noted for his early work on the nature of the chemical bond. He also studied a number of complex biochemical systems. He was awarded the Nobel Prize for chemistry in 1954. Pauling was also awarded the Nobel Peace Prize in 1962.

Pavlov, Ivan Petrovich (1849–1936) Russian physiologist and experimental psychologist. He was awarded the Nobel Prize for physiology or medicine in 1904 for his work on the physiology of the saliva glands.

PCR *See* polymerase chain reaction.

pectin (pectic substance) Any of a group of complex polysaccharides that, together with hemicelluloses, form the matrix of plant cell walls. They are rich in galacturonic acid, and serve to cement the cellulose fibers together. Pectins are especially abundant in the middle lamella, which cements adjoining cell walls together. Under suitable conditions pectins form gels with sugar and acid, and they are used commercially as gelling agents, for example, in preserves.

pellicle A thin flexible transparent outer protective covering of many unicellular organisms, especially undulipodiated protozoan protists, for example, *Euglena* and *Paramecium*. It is made of protein, and maintains the shape of the body. It may thus be regarded as an exoskeleton. In *Paramecium*, the pellicle is perforated by fine pores through which the cilia emerge. Pellicles are also found in many parasitic organisms, e.g. *Monocystis* and *Trypanosoma*.

pentose A SUGAR that has five carbon atoms in its molecules.

pentose phosphate pathway (**hexose monophosphate shunt**) A pathway of glucose breakdown occurring in the cytosol of most organisms, including plants and animals. It is an alternative to GLYCOL-YSIS, with which it shares several intermediates, and has two main functions. The first is to generate reducing power in the form of NADPH, which is required to drive biosynthetic reactions, for example, fatty acid synthesis. The second is to produce pentose phosphates (e.g. ribulose-5-phosphate), which serve as precursors for the ribose and deoxyribose needed in the synthesis of nucleic acids.

pepsin Any of a family of enzymes that catalyze the partial hydrolysis of proteins to polypeptides. Pepsins are secreted by the gastric glands in inactive forms, *pepsinogens*, and are activated by hydrogen ions.

pepsinogen *See* pepsin.

peptidase An enzyme that is responsible for catalyzing the hydrolysis of certain peptide bonds. Peptidases help break down peptides into amino acids.

peptide Any compound comprising two or more amino acids linked together by *peptide bonds*. Such bonds are formed by reaction between the carbonyl group of one amino acid and the amino group of another amino acid with the elimination of water. According to the number of amino acids linked together, peptides are called di-, tri-, oligo-, or polypeptides. In general, peptides have an amino group at one end of the chain and a carbonyl group at the other. They can be produced by the partial hydrolysis of proteins. *See also* polypeptide.

perforation plates The remains of the cross walls between the VESSEL MEMBERS in the xylem vessels of a plant's water-conducting tissue. The cross wall may have disintegrated completely (making effectively one long cylinder), or partially so

that remnants of the cross wall form bars across the cavity of the vessel. Most commonly only the centers of the cross walls disappear leaving a distinct rim.

perforin *See* natural killer cell.

periclinal Describing a line of cell division parallel to the surface of the organ. *Compare* anticlinal.

periclinal chimera *See* chimera.

periosteum The connective tissue membrane that surrounds a bone. It is tough and fibrous, with many interlacing bundles of white collagen fibers. It contains OS-TEOBLASTS, important in the formation of bone.

peripheral nervous system The system of nerves and their ganglia that run from the CENTRAL NERVOUS SYSTEM to the organs and peripheral regions of the body. It constitutes all parts of the nervous system not included in the central nervous system. In vertebrates it comprises the cranial and spinal nerves with their many branches. These convey impulses from sense organs for processing by the central nervous system and transmit the consequent motor impulses to muscles, glands, etc.

peritrichous Describing bacteria that possess flagella at many places or all over the cell surface. An example is *Proteus*.

peroxisome *See* microbody.

Perutz, Max Ferdinand (1914–) Austrian-born British molecular biologist who worked on hemoglobin and myoglobin. He was awarded the Nobel Prize for chemistry in 1962, the prize being shared with J. C. Kendrew for their studies on the structures of globular proteins.

petri dish A shallow circular glass or plastic container, fitted with a lid, that is used for tissue culture or for growing such microorganisms as bacteria, molds, etc., on nutrient agar or some other medium. It is named after the German bacteriologist J. R. Petri (1852–1921).

PG *See* prostaglandin.

pH A measure of the acidity or alkalinity of a solution on a scale 0–14. A neutral solution has a pH of 7. Acid solutions have a pH below 7; alkaline solutions have a pH above 7. The pH is given by $\log_{10}(1/[\,H^+])$, where [H^+] is the hydrogen ion concentration in moles per liter.

phage (bacteriophage) A virus that infects bacteria. Some phages are small and relatively simple, whereas others have complex capsids composed of a polyhedral head (containing the nucleic acid), and a helical tail through which nucleic acid is injected into the host. After infection with a VIRULENT PHAGE, the viral nucleic acid is reproduced and the host cell usually undergoes lysis and is killed. However, TEMPERATE PHAGES can enter a state called LYSOGENY, in which the phage genes are not expressed and the phage genome is replicated with the host's genetic material. Various phages are used in genetic engineering as cloning VECTORS. Nonviral DNA can be inserted into a phage, which is then used to infect a host cell, thereby introducing the foreign DNA. *See also* lambda phage; T phage.

phagocyte A cell that is capable of engulfing particles from its surroundings by a process termed PHAGOCYTOSIS. Examples are the neutrophils and macrophages in vertebrates, which play an important role in protecting the organism against infection. Many other cells are capable of phagocytosis, for example, amebas and certain other protists.

phagocytosis The process in which a cell's plasma membrane surrounds and engulfs solid particles so that they can pass inside the cell. In a cell (phagocyte) performing phagocytosis, the plasma membrane extends around the particles due to the controlled assembly and disassembly of actin microfilaments of the cytoskeleton, and accompanying changes in viscosity of the cytoplasm (*see* ameboid movement). *Compare* endocytosis, pinocytosis.

phase-contrast microscopy A form of microscopy that exploits slight differences in the phase of light after it has passed through transparent specimens, such as single cells or thin layers of cells. These phase differences are caused by tiny differences in refractive index and thickness, and are converted by the phase-contrast microscope into differences in light and dark in the image. This gives the image greater contrast compared to that obtained with an ordinary compound microscope. This form of microscopy is useful for studying organelles within unstained living cells.

phenotype 1. All the observable characteristics of an organism, which are determined by the interaction of the organism's GENOTYPE with the environment. Many genes present in the genotype do not show their effects in the phenotype because they are masked by dominant alleles. Genotypically identical organisms, such as identical twins, may have very different phenotypes in different environments. 2. One of two or more distinct forms of a particular character, for example, albino or pigmented skin. Individuals having such clearly separable phenotypes usually have different alleles at a gene determining the character.

phenylalanine *See* amino acids.

phloem Plant vascular tissue that transports food from areas where it is made to where it is needed or stored. It consists of sieve tubes, which are columns of living cells (*see* sieve elements) with perforated end walls, that allow passage of substances from one cell to the next. Accompanying the sieve tubes are COMPANION CELLS, FIBERS, and PARENCHYMA.

phosphatide A glycerophospholipid; any derivative of glycerol phosphate containing an acyl, alkyl, or alkenyl group attached to the glycerol residue. *See* lipid.

phosphatidylcholine (**lecithin**) One of a group of phospholipids that contain glycerol, fatty acid, phosphoric acid, and choline and are found widely in higher plants and animals, particularly as a component of cell membranes.

phosphatidylinositol (**PI**) A membrane phospholipid that is the precursor for several important SECOND MESSENGERS – the *phosphoinositides* – including inositol 1,4,5-trisphosphate (IP_3). PI has an inositol group that protrudes into the cytosol, and this is phosphorylated at different positions by enzymes to yield the various products.

phosphoglyceride *See* phospholipid.

phospholipid Any LIPID containing phosphate. The most common are the *phosphoglycerides*, which occur as key components of cell membranes. They consist of two fatty (acyl) chains esterified to two of the hydroxyl groups of glycerol, with the third hydroxyl group esterified to phosphate. Often the phosphate is itself esterified to another group, such as choline (in phosphatidylcholine), ethanolamine, or inositol. These groups tend to be charged, and interact strongly with water, forming a hydrophilic (water-loving) 'head' to the molecule, whereas the acyl chains form a hydrophobic (water-hating) 'tail'. This dual, or amphipathic, nature of phosphoglycerides enables them spontaneously to form organized structures in aqueous solutions, including double (bilayer) sheets, spherical micelles, and bilayered liposomes. Phospholipid bilayers form the basis of virtually all biological membranes.

phosphoprotein A conjugated protein containing one or more phosphate groups bound directly to amino acids. Casein is an example.

phosphorus One of the essential elements in living organisms. In vertebrates, calcium phosphate is the main constituent of the skeleton. Phospholipids are important in cell membrane structure, and phosphates are necessary for the formation of the sugar–phosphate backbone of nucleic acids. Phosphates are also necessary for the formation of high-energy bonds in compounds such as ATP. Phosphate compounds are important in providing energy for muscle contraction in vertebrates (creatine phosphate) and invertebrates (arginine phosphate). Phosphorus has many other important roles in living tissues, being a component of certain coenzymes. The phosphate ion, PO_4^{3-}, is an important buffer in cell solutions.

phosphorylation The introduction of a phosphate group (PO_4^{3-}) into a molecule. The phosphorylation of biomolecules, notably proteins and nucleotides, is crucial to many aspects of cellular life, including energy metabolism, uptake and transport of substances, the cell cycle, and cell signaling. High-energy phosphoanhydride bonds are formed by the phosphorylation of certain nucleotides, notably in the formation of ATP from ADP, and provide a means of trapping chemical energy in a usable form. Phosphate groups are transferred from these phosphorylated nucleotides to a host of different proteins by enzymes called PROTEIN KINASES. This protein kinase-catalyzed phosphorylation enables the cell to 'switch on' enzymes and other proteins, for example, by greatly increasing their ability to bind substrates. Conversely, removal of phosphate groups (dephosphorylation), which is catalyzed by phosphorylase enzymes, tends to 'switch off' proteins. *See also* oxidative phosphorylation, SUBSTRATE-LEVEL PHOSPHORYLATION.

photomicrograph *See* micrograph.

photophore *See* bioluminescence.

photophosphorylation (**photosynthetic phosphorylation**) The conversion of ADP to ATP using light energy. *See* photosynthesis.

photoreceptor Any light-sensitive organ or organelle. The eyes of vertebrates and the ocelli and compound eyes of insects are photoreceptors, as are the EYESPOTS of such protists as *Euglena*.

photorespiration A light-dependent metabolic process of most green plants that resembles respiration only in that it uses oxygen and produces carbon dioxide. It effectively 'competes' with photosynthesis, and wastes fixed carbon and energy. Photorespiration occurs because of the dual activity of the key enzyme in carbon fixation, RIBULOSE 1,5-BISPHOSPHATE CARBOXYLASE. Besides its carboxylase activity, incorporating carbon dioxide into ribulose bisphosphate (RuBP) as part of the Calvin cycle, this enzyme also has oxygenase activity, splitting RuBP into one molecule of phosphoglycerate and one of phosphoglycolate (a two-carbon compound). Enzymes located in peroxisomes and mitochondria, as well as the chloroplast, constitute a *glycolate cycle*, by which the cell recovers some of the carbon from the excess glycolate by converting it to glycerate. However, some carbon is lost as CO_2. It is estimated that in C_3 plants (*see* C_3 pathway) 40% of the potential yield of photosynthesis is lost through photorespiration. It is therefore economically important and ways of inhibiting the process are sometimes employed. For example, artificially raising the CO_2:O_2 ratio in the air is effective, and CO_2 enrichment of greenhouses is often used for high-value crops such as tomatoes. Yields are increased 30–100%. C_4 plants (*see* C_4 pathway) are more efficient at photosynthesis, and photorespiration is not detectable.

photosynthesis The conversion of light energy to chemical energy. Various types of living organisms, notably green plants and algae, perform photosynthesis, using chlorophyll pigments to absorb photons of light. This energy is stored in the form of ATP and reduced coenzymes, usually NADPH. Most photosynthetic organisms can manufacture all their organic constituents (carbohydrates, proteins, lipids, etc.) with carbon dioxide as the sole source of carbon. In this process, the carbon dioxide is chemically reduced. ATP supplies the energy, and NADPH the electrons for the reduction. The formation of NADPH depends on reduction of NADP+ by an electron donor. In plants, algae, and certain bacteria (cyanobacteria) this electron donor is water, which is split by light energy to yield electrons, protons, and molecular oxygen. In contrast, photosynthetic purple and green bacteria use various reduced compounds in their environment as electron donors; examples include hydrogen sulfide, elemental sulfur, hydrogen gas, certain organic compounds, and even ferrous iron (Fe^{2+}). These bacteria do not evolve oxygen. Directly or indirectly, photosynthesis is the source of carbon and energy for all except chemoautotrophic organisms.

In green plants, photosynthesis takes place in CHLOROPLASTS, mainly in leaves. The mechanism is complex and involves several stages, sometimes divided into two sets called *light-dependent reactions* and *light-independent reactions* (or 'dark' reactions), although the latter two terms are misnomers. The overall reaction in green plants can be summarized by the equation:

$$CO_2 + 4H_2O \rightarrow [CH_2O] + 3H_2O + O_2$$

Light energy is absorbed by chlorophyll (and other pigments), setting off a chain of chemical reactions in which water is split, electrons are transferred along an electron-transport chain (ultimately to reduce NADP+), oxygen is evolved, and ADP is converted to ATP (in a process known as *photophosphorylation*). The NADPH is used to reduce carbon dioxide to carbohydrates in the subsequent carbon fixation reactions.

The enzymes and other components responsible for the absorption of light are located in the thylakoid membrane of the CHLOROPLAST, where they constitute an ELECTRON-TRANSPORT CHAIN, analogous to that involved in aerobic respiration in mitochondria. The key components of each chain are complexes of chlorophylls, proteins, and carotenoids called PHOTOSYSTEMS. A chain comprises two such photosystems, designated PSII and PSI, linked by a CYTOCHROME complex. Photons of light are trapped by each photosystem, in clusters of chlorophyll molecules called an *antenna*, and channeled to another cluster of pigment molecules and proteins, the *reaction center*. The process starts at PSII, where electrons enter the electron transport chain, and water is split into molecular oxygen, electrons, and protons (H^+) – a process called *photolysis*. Electrons are transferred through the chain to PSI, via a group of cytochrome proteins, called the b_6-f complex, and certain electron-carrier molecules (plastoquinones and plastocyanins). At PSI, absorbed photons generate electrons that reduce ferredoxin, which in turn reduces $NADP^+$ to NADPH. According to the CHEMIOSMOTIC THEORY, as electrons flow along the chain they cause protons to be pumped into the thylakoid lumen, creating a chemiosmotic gradient. Protons also enter the lumen due to the photolysis of water by PSII. As these protons diffuse back through the thylakoid membrane into the stroma of the chloroplast they drive ATP production by the F_0F_1 enzyme complex (see ATP SYNTHASE), also located in the thylakoid membrane.

There are two patterns of electron flow. *Non-cyclic electron flow* involves all components of the electron-transport chain, there is photolysis of water yielding oxygen, and both ATP and NADPH are generated. *Cyclic electron flow* involves only PSI, ferredoxin, the b_6-f complex, and plastocyanin. It produces the extra ATP needed for the light-independent reactions, but not NADPH.

Carbon fixation uses ATP and NADPH from the light-dependent reactions to reduce carbon dioxide to carbohydrate. These reactions also require light, and in eukaryotes they take place in the chloroplast stroma. Carbon dioxide is first fixed by combination with the 5-carbon sugar ribulose bisphosphate (RuBP) to form two molecules of phosphoglyceric acid (PGA), a reaction catalyzed by the enzyme RIBULOSE 1,5-BISPHOSPHATE CARBOXYLASE (rubisco). PGA is then reduced to phosphoglyceraldehyde (triose phosphate) using the NADPH and some of the ATP. Some of the triose phosphate and the rest of the ATP is used to regenerate the carbon dioxide acceptor RuBP in a complex cycle involving 3-, 4-, 5-, 6-, and 7-carbon sugar phosphates, usually called the *Calvin cycle*. The rest of the triose phosphate is transported out of the chloroplast into the cytosol, where it can be used to make sucrose, and also amino acids and lipids. *See* C_3 pathway; C_4 pathway; photosynthetic pigments.

photosynthetic bacteria   Bacteria that can perform PHOTOSYNTHESIS through possession of various light-absorbing pigments, notably bacteriochlorophylls (which are slightly different from the chlorophylls of plants) and carotenoids. *Anoxygenic* photosynthetic bacteria, including the green sulfur bacteria, purple sulfur bacteria, and purple nonsulfur bacteria, do not use water as an electron donor and thus do not produce oxygen as a product of photosynthesis. In contrast, the CYANOBACTERIA do use water as as electron donor, and are *oxygenic*.

photosynthetic phosphorylation   *See* photophosphorylation.

photosynthetic pigments   Pigments that absorb the light energy required in photosynthesis. They are located in the chloroplasts of plants and algae, whereas in most photosynthetic bacteria they are located in thylakoid membranes, typically distributed around the cell periphery. All photosynthetic organisms contain chlorophylls and carotenoids; some also contain phycobilins. Chlorophyll *a* is the *primary pigment* since energy absorbed by this is used directly to drive the light-dependent reactions of photosynthesis. The other pigments (chlorophylls *b*, *c*, and *d*, and the carotenoids and phycobilins) are *accessory pigments* that pass the energy they absorb on to chlorophyll *a*. They broaden the spectrum of light used in photosynthesis. *See* absorption spectrum.

photosystem   A complex of proteins and pigment molecules responsible for absorbing light in photosynthesis. The photosystems are located in the thylakoid membranes, and drive the entire photosynthetic process by passing electrons to the electron-transport chain. Each photosystem consists of one or more *antennae*, which capture the photons and transfer the absorbed energy to a *reaction center*, where the energy is converted to electron flow. The antenna comprises pigment molecules organized into arrays called *light-harvesting complexes*. The antenna pigments include chlorophyll *a*, chlorophyll *b* (in plants), and carotenoids. The reaction center consists of proteins and chlorophyll *a* molecules; the latter donate electrons to a primary electron acceptor, which transfers them to the electron-transport chain. In oxygenic photosynthesis there are two different photosystems, named PSI and PSII (*see* photosynthesis).

phototactic movement   *See* phototaxis.

phototaxis (phototactic movement)   A TAXIS in response to light. Many motile algae and photosynthetic bacteria are positively phototactic.

phototropic movement   *See* phototropism.

phototropism (heliotropism; phototropic movement)   A directional growth movement of part of a plant in response to light. The phenomenon is clearly shown by the growth of shoots and coleoptiles towards light (*positive phototropism*). According to the *Cholodny–Went hypothesis*, the stimulus is perceived in the region just behind the shoot tip. If light falls on only one side of the apex then auxins produced in the apex tend to diffuse towards the shaded side. Thus more auxin diffuses down the stem from the shaded side of the tip. This results in greater elongation of cells on the shaded side thus causing the stem to bend towards the light source. Most roots are light-insensitive but some (e.g. the adventitious roots of climbers such as ivy) are negatively phototropic. *See also* tropism.

phragmoplast   A barrel-shaped body appearing in dividing plant cells during late anaphase and telophase between the two separating groups of chromosomes. It consists of microtubules associated with the spindle, and transports vesicles that coalesce to form the early CELL PLATE. The vesicles' membranes form the plasma membranes of the new daughter cells, and their contents contain material for the future cell walls.

phylloquinone   *See* vitamin K.

physical map   *See* chromosome map.

physiological saline   A solution of sodium chloride and various other salts in which animal tissues are bathed *in vitro* to keep them alive during experiments. It must be isotonic with, and of the same pH as, body fluids. One of the most commonly used is *Ringer's solution*, which contains (in addition to sodium chloride) calcium, magnesium, and potassium chlorides. Other solutions may also contain a food supply, e.g. glucose. *See also* tissue culture.

phytochrome   A proteinaceous pigment found in low concentrations in most plant organs, particularly meristems and dark-grown seedlings. It exists in two interconvertible forms. P_r (or P_{660}) has an absorption peak at 660 nm (red light) and P_{fr} (or P_{730}) at 730 nm (far-red light). Natural white light favors formation of P_{fr}, the physiologically active form. Light intensities required for conversion are very low and it occurs within seconds. Phytochrome plays a vital role as a photoreceptor in a wide range of light-induced physiological processes (e.g. photoperiodic responses); developmental changes, including stem expansion, leaf expansion, leaf unrolling in grasses and cereals, and greening; and in germination of light-sensitive seeds such as lettuce. Phytochrome is thought to act by controlling gene expression, although how it does this is unclear.

phytohormone   *See* plant hormone.

pico-   Symbol: p   A prefix denoting one million-millionth, or 10^{-12}. For example, 1 picogram (pg) = 10^{-12} gram. *See* SI units.

picornavirus   One of a group of small RNA-containing viruses including those responsible for influenza, the common cold, poliomyelitis, and foot-and-mouth disease.

pilus (*pl.* **pili**) A fine straight hairlike protein structure, one or several of which emerge from the walls of certain bacteria (e.g. *E. coli*). They are associated with bacterial CONJUGATION, pulling the conjugating cells together, and are involved in attachment of certain pathogenic bacteria to human tissues. Some viruses specifically adhere to pili. Shorter and more numerous extensions, called *fimbriae* (singular: fimbria), confer the property of 'stickiness' whereby bacteria tend to adhere to one another.

pinocytosis *See* endocytosis.

pit (*Botany*) A gap in the secondary cell wall of certain plant cells that enables communication between thickened cells, for example, tracheids. According to whether or not the secondary wall forms a lip over the pit, pits are described as *bordered* or *simple*, respectively. Usually pits occur in pairs so that the only barrier separating adjacent cells is the middle lamella and the respective primary cell walls. If a pit occurs singly it is termed a *blind pit*.

pith (**medulla**) The central region of a plant stem and, occasionally, root that is normally composed of parenchymatous tissue.

pituitary gland An endocrine gland in the vertebrate brain that secretes various hormones. It consists of two separate parts. The *adenohypophysis* (anterior pituitary) secretes somatotropin (growth hormone), corticotropin, thyrotropin, prolactin, and gonadotropins. The *neurohypophysis* (posterior pituitary) secretes antidiuretic hormone (vasopressin) and oxytocin.

plagiotropism *See* tropism.

plant Any of a large group of multicellular organisms that make their own food by building simple inorganic substances into complex molecules using PHOTOSYNTHESIS. This process uses light energy, absorbed by a green pigment called chlorophyll. One major characteristic that distinguishes plants from other plantlike organisms, such as ALGAE or FUNGI, is the possession of an embryo that is retained and nourished by maternal tissue. Fungi and algae lack embryos and develop from spores. Plants are also characterized by having cellulose cell walls, and by the inability to move around freely except for some mobile microscopic plants. Plants also generally respond to stimuli very slowly, the response often taking a matter of days and only occurring if the stimulus is prolonged.

plant hormone (**phytohormone**) One of a group of essential organic substances produced in plants. They are effective in very low concentrations and control growth and development by their interactions. The main groups are auxins, gibberellins, cytokinins, abscisic acid, and ethylene. Terms sometimes used instead of plant hormone include *plant growth substance* and *plant growth regulator*.

plasma *See* blood plasma.

plasma cell A mature B-CELL (white blood cell) that is differentiated to secrete just one particular type of antibody. A single plasma cell lives for several days, and can secrete up to 2000 antibody molecules per second.

plasmagel *See* ectoplasm.

plasmalemma *See* plasma membrane.

plasma membrane (**cell membrane; plasmalemma**) The MEMBRANE that surrounds all living cells, and acts as the chief interface between the cytosol and the cell's exterior. Its main role is to control the entry and exit of substances to and from the cell, for which it is equipped with various proteins within the basic lipid bilayer membrane structure. These proteins serve, for example, as ION CHANNELS or TRANSPORT PROTEINS, enabling the selective uptake or export of ions and small molecules, sometimes against their concentration gradient. RECEPTOR proteins in the plasma membrane bind different signaling molecules arriving at the cell, and relay the signals to the cytosol, thereby influencing the activity of the cell. The plasma membrane is also an anchorage for various internal cytokeletal fibers, and the site of junctions between neighboring tissue cells (*see* cell junctions). For tissue cells embedded in EX-TRACELLULAR MATRIX, such as epithelial cells, proteins in the plasma membrane attach to external matrix components to provide strength and rigidity. In plants, fungi, bacteria, and certain protists, the plasma membrane is bounded externally by a CELL WALL.

plasmasol *See* endoplasm.

plasmid An extrachromosomal genetic element found within a cell that replicates independently of the chromosomal DNA. Plasmids occur in bacteria, yeasts, and some other eukaryotic cells, and typically consist of circular double-stranded DNA molecules ranging in size from a few thousand base pairs to over 100 kb. Numbers of a particular plasmid range from just one or two up to about 100 or per cell. They can carry a variety of genes, including ones for antibiotic resistance, toxin production, and enzyme formation. Plasmids are duplicated by the host cell before cell division, and at least one copy is transmitted to each daughter cell, thereby ensuring inheritance of plasmid-encoded characters.

Various natural and genetically modified plasmids are widely used as *cloning vectors* in GENETIC ENGINEERING. A simple engineered plasmid vector consists of two different antibiotic *resistance genes* and a specific DNA sequence called the *replication origin* site, which is essential for replication of the plasmid within the host cell. The plasmid must also contain *restriction sites*, where the plasmid DNA can be cut by restriction enzymes in order to insert a foreign DNA fragment. The DNA is cut within one of the resistance genes, thus inactivating the gene. The antibiotic resistance genes provide the means of identifying: (a) which cells contain the plasmid (i.e. cells containing the active resistance gene, which are resistant to the first antibiotic), and (b) which of the latter contain a successfully engineered plasmid

(i.e. cells with the inactivated second antibiotic gene, which are susceptible to the second antibiotic).

plasmodesma (*pl.* **plasmodesmata**) A type of plant cell junction that connects the cytosol of adjacent cells, traversing the cell walls. It consists of a channel, typically about 60 nm in diameter, that is lined with plasma membrane and filled with cytosol. Running through the center is an extension of endoplasmic reticulum, called a *desmotubule*. Plasmodesmata permit the passage of various substances, ranging in size from ions, sugars, and amino acids up to proteins, nucleic acids, and even viruses. Soluble molecules pass in the cytosol, whereas membrane-bound molecules can travel via the desmotubule. The permeability is regulated according to developmental, physiological, or environmental factors. The continuity of cytoplasm between cells created by plasmodesmata is called the symplast.

plasmodium (*pl.* **plasmodia**) A multinucleate mass of cytoplasm surrounded by a plasma membrane. Such structures are formed during the life cycles of certain SLIME MOLDS.

Plasmodium A genus of parasitic protists, some species of which are the cause of malaria in humans. *Plasmodium* is spread by mosquitoes of the genus *Anopheles*, and has a complicated life cycle involving asexual reproduction in humans and sexual reproduction in the mosquito. *See also* Apicomplexa.

plasmogamy Fusion of the cytoplasm of different cells without fusion of their nuclei. Plasmogamy in the absence of karyogamy (fusion of nuclei) occurs in the life cycles of fungi and certain fungus-like protists (e.g. slime molds and chytrids). Plasmogamy between fungal mycelia of different strains leads to the formation of a heterokaryon.

plasmolysis Loss of water from a walled cell (e.g. of a plant or bacterium) to the point at which the PROTOPLAST shrinks away from the cell wall. The point at which this is about to happen is called *incipient plasmolysis*. Here the cell wall is not being stretched; i.e. the cell has lost its turgidity or become *flaccid* (wall pressure is zero). Wilting of herbaceous plants occurs here. As plasmolysis proceeds parts of the protoplast may remain attached to the cell wall, giving an appearance characteristic of the species. Plasmolysis occurs when a cell is surrounded by a more concentrated solution, i.e one with a lower WATER POTENTIAL.

plastid An organelle enclosed by two membranes (the envelope) that is found in plants and certain protists (e.g. algae), and develops from a PROPLASTID. Various types exist, but all contain DNA and ribosomes. They include CHLOROPLASTS, LEUKOPLASTS, and CHROMOPLASTS.

plastocyanin An electron carrier in the electron-transport chain of PHOTOSYNTHESIS. It receives electrons (ultimately derived from photosystem II) from cytochrome *f*, and transfers these to reduce the reaction center of photosystem I.

plastoglobuli (osmiophilic globules) Spherical lipid-rich droplets found in varying numbers inside CHLOROPLASTS. They stain intensely with osmium tetroxide and so appear black under the electron microscope.

plastoquinone A lipid-soluble compound that acts as an electron carrier in PHOTOSYNTHESIS. Each molecule removes two electrons from the reaction center of photosystem II. In the process it picks up two protons, becoming reduced quinone (QH_2), which diffuses through the thylakoid membrane and releases electrons to the cytochrome b_6-f complex.

platelet (thrombocyte) A tiny irregular body found in blood plasma, that is important in stopping blood flow (hemostasis) at wound sites, and in blood coagulation (clotting). Platelets are 2–4 μm in diameter and number on average about 250 000 per cubic millimeter of blood. They are made in red bone marrow, from the fragmenta-

tion of large cells (called *megakaryocytes*). When they come into contact with a rough surface, such as a damaged tissue, platelets aggregate to form a plug. They also start the chain of reactions leading to the formation of a blood clot. Platelets release *serotonin*, which causes constriction of blood vessels, so reducing capillary bleeding; and *platelet-derived growth factor*, which stimulates tissue cells to grow and repair the wound.

platelet-derived growth factor *See* platelet.

pleated sheet (beta-sheet) A form of SECONDARY STRUCTURE occurring within certain proteins. It consists of a planar array of multiple polypeptide chains (called β strands) aligned side by side and joined by hydrogen bonds. The peptide bonds lie in the plane of the sheet, and side chains of the constituent amino acids project on either side of the sheet. Pleated sheets can be found within the substrate-binding sites of certain proteins, and when stacked together they lend flexibility and toughness to certain structural proteins, as in silk fibers, for example.

pleiotropism The situation in which one gene is involved in the production of several characters. For example, a mutant allele at the gene responsible for juvenile-onset diabetes gives rise to a host of disorders, ranging from constipation to heart failure.

pleuropneumonia-like organisms *See* mycoplasmas.

pluripotent 1. (*Cell Biology*) Describing a cell, particularly a stem cell, that is capable of giving rise to two or more different types of fully differentiated progeny cells. *Compare* totipotent.
2. (*Biochemistry*) Describing a molecule that can be converted into two or more different types of active product.

point mutation *See* mutation.

polar body A minute cell produced by

the unequal division of the primary and secondary oocytes during formation of an ovum. *See* oogenesis.

polar nuclei The two nuclei found midway along the EMBRYO SAC in the ovule of a plant. They may fuse to form the diploid *definitive nucleus*. The endosperm is formed from the fusion of one or both polar nuclei with one of the male gametes from the pollen tube.

pollen tube A filamentous outgrowth of the pollen grain that in most seed plants transports the male gamete to the ovule.

polyamine An aliphatic compound having two or more amino and/or imino groups. Polyamines are found in virtually all organisms, and have a range of biological properties. Examples of polyamines include spermine, spermidine, cadaverine, and putrescine. The latter, synthesized from ornithine by the enzyme ornithine decarboxylase, is the precursor of other polyamines. At normal cellular pH they carry multiple positive charges, and bind to nucleic acids and membrane phospholipids. They play crucial roles in cell division and differentiation, and cell growth ceases if polyamine synthesis is inhibited. Hence, drugs aimed at disrupting polyamine synthesis or metabolism have potential as anti-cancer agents. In plants, polyamines might act as plant hormones.

polygene A hypothetical gene with an individually small effect on the phenotype that interacts with numerous other polygenes controlling the same character to produce the continuous quantitative variation typical of such traits as height, weight, and skin color. Because such characters are generally subject to considerable variation due to environmental factors, the existence of polygenes remains unproven.

polygenic inheritance *See* quantitative inheritance.

polymerase An enzyme that regulates the synthesis of a polymer. Examples include RNA POLYMERASES and DNA POLY-MERASES. There is only one type of RNA polymerase in prokaryotes, but in eukaryotes there are three different types: type I makes ribosomal RNA, type II makes messenger RNA precursors, and type III makes transfer RNA and 5S ribosomal RNA. DNA polymerases are involved either in the synthesis of double-stranded DNA from single-stranded DNA or in the repair of DNA by scanning the DNA molecule and removing damaged nucleotides. *See also* reverse transcriptase.

polymerase chain reaction (PCR) A technique for amplifying small samples of DNA rapidly and conveniently. Invented in 1983 by the US molecular biologist Kary Mullis (1944–), it is now used widely in research, genetic testing, and forensic science, e.g. to produce a suitable quantity of DNA for genetic fingerprinting from the minute amounts present in traces of blood or other tissue. To amplify a particular segment of DNA (the target DNA) it is necessary first to know the sequence of bases flanking it at either end. This enables the construction of short single DNA strands (primers) that are complementary to and will bind with these flanking regions. Then the sample is incubated in a water bath with the primers, nucleotides, and enzymes, especially a temperature-resistant DNA polymerase called *Taq polymerase* (after the thermophilic bacterium, *Thermus aquaticus*, from which it was isolated). By varying the temperature precisely and rapidly, amplification proceeds in cycles of DNA denaturation, annealing of primers, and replication of new DNA strands, each lasting about 20 seconds. After 30 cycles, some 10^9 copies of the original DNA are produced.

polymorph (**polymorphonuclear leukocyte**) *See* neutrophil.

polymorphonuclear leukocyte *See* polymorph.

polynucleotides *See* nucleotide.

polypeptide A compound that contains

many amino acids (>10) linked together by peptide bonds. *See* peptide; protein.

polyploid The condition in which a cell or organism contains three or more times the HAPLOID number of chromosomes. Polyploidy is far more common in plants than in animals and very high chromosome numbers may be found; for example in octaploids and decaploids (containing eight and ten times the haploid chromosome number). Polyploids are often larger and more vigorous than their DIPLOID counterparts and the phenomenon is therefore exploited in plant breeding, in which the chemical colchicine can be used to induce polyploidy. Polyploids may contain multiples of the chromosomes of one species (*autopolyploids*) or combine the chromosomes of two or more species (*allopolyploids*). Polyploidy is rare in animals because the sex-determining mechanism is disturbed. For example, a tetraploid XXXX would be sterile.

polyribosome (polysome) A complex consisting of several RIBOSOMES simultaneously translating a single messenger RNA (mRNA) molecule. This increases the rate of TRANSLATION, and hence the rate of protein synthesis. The two ends of the mRNA are bridged by binding proteins, effectively creating a circular mRNA. After each round of translation, the ribosomal subunits dissociate from the 3′ end and readily find the nearby 5′ end, where they reassociate to initiate another round of translation.

polysaccharide A polymer of MONOSACCHARIDES joined by glycosidic links (*see* glycoside). They contain many repeated units in their molecular structures and are of high molecular weight. They can be broken down to smaller polysaccharides, DISACCHARIDES, and monosaccharides by hydrolysis or by the appropriate enzyme. Important polysaccharides are inulin (hydrolyzed to fructose), starch (hydrolyzed to glucose), glycogen (also known as animal starch), and cellulose (hydrolyzed to glucose but not metabolized by humans).

Polysaccharides are now classified as GLYCANS. *See also* carbohydrates; sugar.

polysome *See* polyribosome.

polysomy *See* aneuploidy.

polyspermy The penetration of several sperm into one ovum at fertilization; only one sperm actually fuses with the ovum nucleus. It occurs normally in a few animals with yolky eggs (e.g. birds), but in most animals a fertilization membrane forms around the fertilized ovum, preventing polyspermy.

polytene Describing the chromosome condition caused by chromatids not separating after duplication. It leads to the formation of *giant chromosomes* consisting of numerous identical chromatids lying parallel to each other. Giant chromosomes have characteristic bands visible during interphase, caused by the degree of coiling of the DNA-histone fiber – highly coiled regions stain darker when prepared for microscopy. They are used to study gene activity and make CHROMOSOME MAPS. Polytene chromosomes are a means of amplifying the genes in large or highly active cells, for example, in the salivary gland cells of dipterous insects (e.g. *Drosophila*).

polyteny *See* endomitosis.

porphyrins Cyclic organic structures that have the important characteristic property of forming complexes with metal ions. Examples of such *metalloporphyrins* are the iron porphyrins (e.g. heme in hemoglobin) and the magnesium porphyrin, chlorophyll, the photosynthetic pigment in plants. In nature, the majority of metalloporphyrins are conjugated to proteins to form a number of very important molecules, including hemoglobin, myoglobin, and the cytochromes.

Porter, Rodney Robert (1917–85) British biochemist and immunochemist noted particularly for his isolation of the constituent polypeptide chains of antibody molecules. He was awarded the Nobel

Prize for physiology or medicine in 1972 jointly with G. M. Edelman.

positional cloning   *See* chromosome map.

potassium   One of the essential elements in plants, animals, and other living organisms, found at high concentrations within living cells compared to the external enviornment (e.g. extracellular fluid). It is absorbed by plant roots as the potassium ion, K^+, and in plants is the most abundant cation in the cell sap. Potassium ions are required in high concentrations in the cell for efficient protein synthesis, and for glycolysis in which they are an essential cofactor for the enzyme pyruvate kinase. In animals the gradient of potassium and sodium ions across the plasma membrane is responsible for the potential difference across the membrane, which is important for the transmission of nerve impulses. *See also* Na^+/K^+ ATPase.

poxvirus   One of a group of large DNA-containing viruses that are responsible for smallpox, cowpox, and certain tumors in animals.

PPLO   *See* mycoplasmas.

P$_r$ (P$_{660}$)   *See* phytochrome.

Prelog, Vladimir (1906–98)   Bosnian-born Swiss organic chemist. He was awarded the Nobel Prize for chemistry in 1975 for his research into the stereochemistry of organic molecules and reactions. The prize was shared with J. W. Cornforth.

pre-mRNA (pre-messenger RNA)   Any newly formed RNA transcript of a protein-coding gene in the nucleus of a eukaryotic cell. Pre-mRNAs are assembled by the enzyme RNA polymerase II (*see* transcription), and processed by further enzymes inside the nucleus to form the mature MESSENGER RNA molecules (mRNAs) that are transported to the cytosol for translation and protein synthesis at ribosomes. During this time, the pre-mRNAs associate with various nuclear proteins to form heterogeneous ribonucleoproteins (hnRNPs). Processing of pre-mRNA involves three main stages. Soon after the initiation of transcription, the 5′ end of the RNA transcript is capped with methylguanosine, a process called *5′ capping*. The complete transcript is cleaved at a specific 3′ site just downstream of the final exon and and a string of adenine residues – the poly(A) tail – is added. This stage is called *cleavage and polyadenylation*. Finally, the noncoding introns are removed and the coding regions of the transcript, the exons, are spliced together. This *RNA splicing* stage is performed by a catalytic complex of RNAs and proteins called a SPLICEOSOME.

pressure potential   *See* water potential.

presumptive   Describing embryonic tissues that are presumed to develop in a certain way. For example, presumptive NEURAL PLATE of amphibians lies towards the animal pole of the blastula.

primary structure 1. (of a protein) The linear sequence of amino acids in a polypeptide chain.
2. (of a nucleic acid) The linear sequence of nucleotides along a strand of RNA or DNA – i.e. the base sequence. *Compare* secondary structure; tertiary structure.

primase   *See* DNA replication.

primitive streak   The first sign of embryo formation on the blastoderm of reptiles and birds and in the inner cell mass of mammals (*see* blastocyst). It usually appears in the dorsal midline as a longitudinal wrinkle in the outer layer with a pit (*Hensen's node*) at the anterior end. This appearance is caused by convergence of the outer cells toward the streak and their sinking beneath the surface to become mesoderm; at the anterior end the cells sink into Hensen's node and move anteriorly to become notochord.

primordium (*pl.* **primordia**)   A collection of cells that differentiates into an organ or tissue, such as the apical shoot and apical root primordia of a plant embryo.

prion An infectious protein particle that causes various nervous diseases in humans and other animals, including bovine spongiform encephalopathy ('mad cow disease') in cattle, and a form of Creutzfeldt–Jakob disease (CJD) in humans. Prions are apparently unique in that unlike viruses, virions, and all other infectious agents they lack any form of genetic material (i.e. DNA or RNA). Infectious prion is a variant of a membrane protein normally produced in brain cells. It can be transmitted in food, and induces changes in the folding of the normal proteins, causing them to form rod-shaped aggregates in the central nervous system, which are responsible for the symptoms of disease. Abnormal prion proteins are also produced following mutation of the corresponding gene.

probe *See* DNA probe.

probiotic Any compound produced by a microorganism that promotes growth in other microorganisms. *Compare* antibiotic.

procambial strand The layer of cells that gives rise to the vascular tissue of a plant. It is discernible just below the apex of a shoot or root as a strand of flattened cells which, if traced back, can be seen to give rise to the primary vascular tissues (xylem and phloem).

procambium *See* meristem.

procaryote *See* prokaryote.

processed pseudogene *See* pseudogene.

progesterone A steroid hormone secreted by the *corpus luteum* in the ovary after ovulation. It initiates the preparation of the uterus for implantation of the ovum, the development of the placenta, and the development of the mammary gland in preparation for lactation. *See also* estrogen; ovarian follicle.

progestogen Any hormone whose effects resemble those of progesterone. Synthetic progestogens are used in therapy and oral contraceptives. *See also* estrogen.

programmed cell death *See* apoptosis.

prohormone The inactive form of a hormone: the form in which it is stored. Activation usually involves enzymatic removal of some part of the prohormone; for example, removal of amino acids from the polypeptide prohormone, proinsulin, to form insulin.

prokaryote (**procaryote**) An organism whose genetic material (DNA) is not enclosed by membranes to form a nucleus but lies free in the cytoplasm. Organisms can be divided into prokaryotes and EUKARYOTES, the latter having a true nucleus. This is a fundamental division because it is associated with other major differences. Prokaryotes comprise BACTERIA and constitute the kingdom Monera (Prokaryotae). Eukaryotes comprise all other organisms. Prokaryote cells evolved first and gave rise to eukaryote cells (*see* endosymbiont theory).

prolactin A hormone produced by the anterior pituitary gland that, in mammals, stimulates and controls lactation. In birds prolactin stimulates secretion of crop milk from the crop glands. *See also* gonadotropin.

prolamellar body *See* etioplast.

proline *See* amino acids.

promoter A specific DNA sequence at the start of a gene or group of genes (*see* operon) that initiates TRANSCRIPTION by binding RNA POLYMERASE. In *Escherichia coli* the RNA polymerase has a protein 'sigma factor' that recognizes the promoter; in the absence of this factor the enzyme binds to, and begins transcription at, random points on the DNA strand. The promoter of most eukaryotic genes consists of a *core promoter*, very near the transcription start site, and *promoter-proximal elements*, which are control regions lying further 'upstream' from the start site. In many cases the core promoter contains a region including the base sequence TATA, called the TATA BOX. Various proteins (TRANSCRIPTION FACTORS) bind to the promoter, enabling the correct positioning of RNA polymerase, and leading to the formation of an INITIATION COMPLEX.

promoter-proximal elements   *See* promoter.

proofreading   *See* DNA repair.

prophage   *See* lysogeny.

prophase   The first stage of cell division in meiosis and mitosis. During prophase the chromosomes become visible and the nuclear membrane dissolves. Prophase may be divided into successive stages termed leptotene, zygotene, pachytene, diplotene, and diakinesis. The events occurring during these stages differ in meiosis and mitosis, notably in that bivalents (pairs of homologous chromosomes) are formed in meiosis, whereas homologous chromosomes remain separate in mitosis.

proplastid   A self-duplicating undifferentiated plastid, about 0.5–1 μm in diameter and found in the meristematic regions of plants. They grow and develop into plastids of different types. They consist of an outer and an inner membrane enclosing a small volume of stroma containing DNA and ribosomes (sparse). Exposure to light triggers the formation of chloroplast proteins and membrane lipids, and vesicles begin to bud from the inner membrane. Eventually these form the thylakoid vesicles of the mature chloroplast.

prostaglandin (**PG**)   One of a group of lipid-soluble hormones derived from arachidonic acid. They tend to act locally within tissues, being released from cells into the intercellular space to influence neighboring cells, and are rapidly broken down by enzymes. There are at least 16 types of prostaglandins, in 9 different classes, designated PGA to PGI. Among their many physiological effects, they modulate the effects of other hormones, stimulate smooth muscle contraction in the uterus, influence blood pressure, enhance inflammation, and promote platelet adhesion and blood clotting. Different prostaglandins often have opposing actions; for example, PGE and PGA reduce blood pressure whereas PGF raises it.

prosthetic group   The nonprotein component of a conjugated protein. Thus the heme group in hemoglobin is an example of a prosthetic group, as are the COENZYMES of a wide range of enzymes.

protamine   One of a group of small polypeptides formed chiefly from the amino acids arginine, alanine, and serine. They occur in the sperm of vertebrates, packing the DNA into a condensed form in a role analogous to that of histones.

protease (**proteinase**)   An enzyme that catalyzes the hydrolysis of peptide bonds in proteins to produce peptide chains and amino acids. Individual proteases are highly specific in the type of peptide bond they hydrolyze.

protein   One of a large number of substances that are important in the structure and function of all living organisms. Proteins are POLYPEPTIDES; i.e. they are made up of AMINO ACID molecules joined together by peptide links. Their molecular weight may vary from a few thousand to several million. About 20 amino acids are present in proteins. Simple proteins contain only amino acids. In conjugated proteins, the amino acids are joined to other groups.

The PRIMARY STRUCTURE of a protein is the particular sequence of amino acids present. The SECONDARY STRUCTURE is the way in which this chain is arranged; for example, coiled in an alpha helix or held in beta pleated sheets. The secondary structure is held by hydrogen bonds. The TERTIARY STRUCTURE of the protein is the way in which the protein chain is folded. This may be held by cystine bonds and by attractive forces between atoms. *See also* alpha helix; beta sheet.

proteinase   *See* protease.

protein kinase   Any of a group of enzymes that catalyze the addition of a phosphate group to a protein. Such phosphorylation typically alters the activity or binding properties of the target protein in some way. Protein kinases are important components in cellular signaling pathways, and fall into two main categories, depending on which amino acids they phosphorylate: *tyrosine kinases* are directed toward tyrosine residues, whereas *serine/threonine kinases* are directed toward either serine or threonine residues.

protein sequencing   The determination of the primary structure of proteins, i.e. the type, number, and sequence of amino acids in the polypeptide chain. This is done by progressive hydrolysis of the protein using specific proteases to split the polypeptides into shorter peptide chains. Terminal amino acids are labeled, broken off by a specific enzyme, and identified by chromatography. The first protein to be sequenced was insulin, by Frederick Sanger at Cambridge University in 1954. *See* Sanger method.

protein synthesis   The process whereby proteins are synthesized on the ribosomes of cells. The sequence of bases in messenger RNA (mRNA), transcribed from DNA, determines the sequence of amino acids in the polypeptide chain: each codon in the mRNA specifies a particular amino acid. As the ribosomes move along the mRNA in the process of TRANSLATION, each codon is 'read', and amino acids bound to different transfer RNA molecules are brought to the ribosomes and bound together in their correct positions to form the growing polypeptide chain.

proteoglycan (mucoprotein)   A type of glycoprotein consisting of long branched heterogeneous chains of glycosaminogly-

can molecules linked to a protein core of amino acids. Unlike more typical glycoproteins, they have a greater carbohydrate content, the protein core is rich in serine, and they have a higher molecular weight.

proteolysis   The hydrolysis of proteins into their amino acids, for example, by a PROTEASE enzyme.

proteoplast   A colorless PLASTID (leukoplast) that stores protein.

Protista (Protoctista)   A kingdom of simple eukaryotic organisms that includes the algae, slime molds, fungus-like oomycetes, and the organisms traditionally classified as protozoa, such as flagellates, ciliates, and sporozoans. Most are aerobic, some are capable of photosynthesis, and most possess undulipodia (flagella or cilia) at some stage of their life cycle. Protists are typically microscopic single-celled organisms, such as the amebas, but the group also has large multicellular members, for example the seaweeds and other conspicuous algae.

The name Protista was originally used by Ernst Haeckel for a kingdom of simple organisms including the bacteria, algae, fungi, and protozoans. It was introduced to overcome the difficulties of assigning such organisms, which may show both animal and plantlike characteristics, to the existing animal and plant kingdoms. Today the bacteria and fungi are usually assigned to separate kingdoms, while algae and protozoans are joined by other eukaryotic microorganisms to constitute the kingdom Protista, or Protoctista. *See* Apicomplexa; Ciliophora; Euglenophyta; prokaryote.

Protoctista   *See* Protista.

protoderm   *See* meristem.

protofilament   *See* microtubule.

proton-motive force (pmf)   A force that is capable of moving protons (hydrogen ions, H^+). Such a force is generated across the inner mitochondrial membrane and bacterial plasma membrane during respiration, and across the thylakoid membrane of chloroplasts during photosynthesis. It is the sum of two forces: the proton concentration gradient and the difference in electrical potential (voltage) across the membrane. The pmf results from the active pumping of protons by components of electron transport chains located in the membranes. The pmf then drives the return of protons through the membrane, a process coupled to ATP synthesis. *See also* chemiosmotic theory.

proton pump Any transport protein that moves protons (hydrogen ions, H^+) across a cell membrane against their concentration gradient. There are several different types, all of which power the proton pumping using energy released by the hydrolysis of ATP. In cells of plants, fungi, and bacteria, the export of protons by proton pumps in the plasma membrane (and also in the tonoplast membrane surrounding the vacuole in plant cells) is the principal mechanism for maintaining the RESTING POTENTIAL. Proton pumps also occur in the membranes of lysosomes and endosomes in animal cells, where they import protons to maintain the acidic conditions inside the organelle. Acid secretion by cells lining the mammalian stomach again involves proton pumps. ATP SYNTHASE, found in mitochondria and chloroplast membranes, is a proton pump working in reverse; i.e. it uses the movement of protons down a concentration gradient to power ATP synthesis.

proto-oncogene *See* oncogene.

protoplasm The living contents of a cell, comprising the CYTOPLASM plus NUCLEUS.

protoplast The living part of a cell; i.e. the plasma membrane and protoplast (cytoplasm and nucleus) minus any cell wall. The protoplast of a plant or algal cell, or of a bacterium, is obtained by removing the cell wall. This can be achieved by physical means or by enzymic digestion. Protoplasts can be grown in culture and make possible certain observational or experimental

work such as study of new cell wall formation, pinocytosis, and fusion of cells. Fusion of protoplasts of different species is being investigated by plant breeders as a means of crossing otherwise incompatible plants. Under suitable culture conditions the hybrid cell can develop to form a mature fertile plant. Protoplasts are also used in genetic engineering, because DNA can be introduced through the wall-less cell using the technique of electroporation.

protozoa A group of single-celled heterotrophic often motile eukaryotic organisms, nowadays classified with other single-celled or simple multicelled eukaryotes in the kingdom Protista. Protozoa were traditionally classified as animals and constituted a phylum, or subkingdom, Protozoa. They range from plantlike forms (e.g. *Euglena*, *Chlamydomonas*) to members that feed and behave like animals (e.g. *Amoeba*, *Paramecium*). There are over 30 000 species living universally in marine, freshwater, and damp terrestrial environments. Some form colonies (e.g. *Volvox*) and many are parasites (e.g. *Plasmodium*). Protozoa vary in body form but specialized organelles (e.g. cilia and flagella) are common. Reproduction is usually by binary fission although multiple fission and conjugation occur in some species. The main protozoan phyla are: Rhizopoda (rhizopods); Zoomastigota (flagellates); Apicomplexa (sporozoans); and Ciliophora (ciliates).

provirus A viral chromosome that is integrated in a host chromosome and multiplies with it. Proviruses do not leave the host chromosome to begin a normal cycle of viral replication unless triggered to do so (*see* latent virus). *See also* retrovirus.

proximal Denoting the part of an organ, limb, etc., that is nearest the origin or point of attachment. *Compare* distal.

Prusiner, Stanley B. (1942–) American biochemist who discovered prions. He was awarded the 1997 Nobel Prize for physiology or medicine.

pseudoallele A MUTATION in a gene that produces an effect identical to another mutation at a different site in the same gene locus. The two pseudoalleles thus act as a single gene but do not occupy the same position.

pseudogene A DNA sequence comparable to a gene that does not produce a polypeptide product, and hence is nonfunctional. Pseudogenes contain MUTATIONS that either prevent transcription, or prevent processing of any messenger RNA produced by transcription. They are thought to represent ancestral duplicates of functional genes that have acquired mutations and drifted into nonfunctionality, with no harmful effects to the organism. *Processed pseudogenes* differ in that they lack the intron and flanking sequences characteristic of functional genes. These are thought to represent DNA copies of cellular RNA, produced by reverse transcription, and then integrated at random into the genome, like a RETROTRANSPOSON. Pseudogenes have potential to form new genes by further mutation as they already have useful sequences, such as those signaling transcription.

pseudoplasmodium *See* slime molds.

pseudopodium (*pl.* **pseudopodia**) A temporary finger-like projection or lobe on the body of an ameboid cell. It is formed by a flowing action of the cytoplasm and functions in locomotion and feeding. *See* ameboid movement.

psychrophilic Describing microorganisms that can live at temperatures below 20° C. *Compare* thermophilic.

puff (**Balbiani ring**) A swelling that is seen in certain areas of the giant (POLYTENE) chromosomes found in the salivary glands and other tissues of certain dipterous insects. Puffs originate in different regions of the chromosome in a certain sequence and their occurrence can be correlated with specific developmental events. Others occur only in certain tissues. The puffs are sites of active transcription of

probably just a single gene, albeit present as numerous copies.

pulvinus (*pl.* **pulvini**) A specialized group of cells with large intercellular spaces that are located at the bases of leaves or leaflets in certain plants. They are involved in nongrowth nastic movements, bringing these about by rapid changes in turgor through loss of water to the intercellular spaces.

Purcell, Edward Mills (1912–97) American electrical engineer, physicist, and radioastronomer notable for his discovery (independently of F. Bloch) of the phenomenon of nuclear magnetic resonance. He was awarded the Nobel Prize for physics in 1952 jointly with F. Bloch.

pure culture *See* axenic culture.

Purine

purine A simple nitrogenous organic molecule with a double ring structure. Members of the purine group include adenine and guanine, which are constituents of the NUCLEIC ACIDS, and certain plant alkaloids, for example, caffeine and theobromine.

pyranose A SUGAR that has a six-membered ring form (five carbon atoms and one oxygen atom).

pyrenoid *See* chloroplast.

Pyrimidine

pyrimidine A simple nitrogenous organic molecule whose ring structure is contained in the pyrimidine bases cytosine, thymine, and uracil, which are constituents of the NUCLEIC ACIDS, and in thiamin (vitamin B_1).

pyruvate An intermediate in several metaoblic pathways, including glycolysis and gluconeogenesis, with the formula CH_3COCOO^-.

The last step in glycolysis is the virtually irreversible conversion of phosphoenolpyruvate into pyruvate by pyruvate kinase. The reactions that turn glucose into pyruvate (glycolysis) are common to all cells and do not depend on the presence of oxygen. The subsequent fate of pyruvate is dependent on the amount of oxygen present. If oxygen is available i.e. aerobic conditions pyruvate is converted to acetyl CoA by pyruvate dehydrogenase with the release of NADH and carbon dioxide. Acetyl CoA enters the Krebs cycle and generates more NADH and $FADH_2$, which enter the electron transport chain and transfer electrons to oxygen, leading to the synthesis of ATP. The oxidation of glucose in aerobic conditions leads to the formation of 36 or 38 molecules of ATP. Pyruvate is therefore the link between glycolysis and the Krebs cycle.

If limited amounts of oxygen are available i.e. anaerobic conditions, most microorganisms and the cells of higher organisms convert pyruvate to lactate. Some microorganisms e.g. brewer's yeast convert pyruvate to ethanol (fermentation) under anaerobic conditions. These anaerobic conversions do not generate NADH or FADH2 and therefore do not generate ATP. They regenerate NAD^+, which sustains glycolysis in anaerobic conditions. The only ATP synthesised is the two molecules of ATP made during glycolysis. In higher organisms lactate is made during intense muscle activity and cannot be metabolised further. It diffuses out of the muscles and enters the liver where it is oxidised to pyruvate, which is converted into glucose by the reactions of gluconeogensis. Glucose can then re-enter the blood and be taken up by muscle. See

Pyruvate is also the precursor for the synthesis of the amino acids alanine, valine, and leucine.

quantitative inheritance (polygenic inheritance; multifactorial inheritance) The pattern of inheritance shown by *quantitative traits*, such as height in humans or grain yield in wheat, that show continuous (quantitative) variation within a range of values, rather than distinct qualitative differences. Such traits are typically controlled by several genes – occurring at genetic loci called *quantitative trait loci* (QTLs) – and the interaction of these genes with the environment. The identification and manipulation of genes at QTLs is of great importance in plant and animal breeding, where the goal is often to improve quantitative traits. *See also* polygene.

quaternary structure The number and arrangement of the constituent polypeptides in proteins that comprise two or more subunits (i.e. multimeric proteins). In some proteins the subunits are identical polypeptides, whereas in others they are not. For example, the viral protein hemagglutinin has three identical polypeptide subunits, whereas human hemoglobin A comprises two different pairs of polypeptide chains – two α chains and two β chains.

quiescent center A group of cells in the center of the apical meristem of a plant root in which mitotic divisions are rare or absent. Its role is uncertain. Cells can begin dividing if another part of the meristem is damaged, or if the plant is under stress.

radioimmunoassay (RIA) A type of IM-MUNOASSAY used for finding the concentration of a particular substance, for example a protein, in a biological sample. The substance acts as an antigen, binding to specific antibodies. Also required is a preparation of the substance labeled with a radioisotope. Two series of mixtures are prepared, each using the same known concentrations of antibody and labeled (antigenic) substance; in one series various standard solutions of the substance are added; in the second series, various dilutions of the sample are added. In each case, the unlabeled antigen competes with the labeled antigen for binding sites on the antibody. When the antigen–antibody complexes are separated from the reaction mixture, the ratio of labeled antigen/unlabeled antigen can be measured, and the concentration of substance in the sample found by comparison with the series of standard solutions. The technique is highly sensitive, and is widely used in medicine and research to make accurate determinations of a huge range of enzymes, hormones, drugs, and other substances.

Ramón y Cajal, Santiago (1852–1934) Spanish anatomist and histologist. He was awarded the Nobel Prize for physiology or medicine in 1906 jointly with C. Golgi in recognition of their work on the structure of the nervous system.

ray initial *See* initial.

reading frame A sequence of codons in a messenger RNA (mRNA) molecule that runs from a start codon to a stop codon, as recognized by the ribosome during translation. It thus contains the genetic information required for the assembly of a specific polypeptide. However, a few mRNAs contain two or more overlapping reading frames, in which the same base sequence can be read as different sequences of codons, depending on where the ribosome starts translation. Hence, two or more different polypeptides can be encoded by the same mRNA. *See also* open reading frame.

reagin *See* allergy.

receptor 1. (*Physiology*) A cell or organ that is specialized to receive and respond to stimuli from outside or inside the body of an organism. The eyes, ears, and nose are receptors that respond to light, sound, and airborne chemicals, respectively.
2. (*Cell Biology*) A protein on the surface of a cell that, upon binding to a signal molecule in the cell's environment, initiates a response by the cell. The signal molecule acts as a LIGAND, fitting into a specific site on the receptor protein and thereby causing some change in the shape, or conformation, of the receptor. This conformational change in the receptor leads to the cell's response. The incoming signal is generally relayed and modulated within the cell via various ions or molecules (SECOND MESSENGERS), in a process called TRANSDUCTION. When the signal molecule is removed or degraded, the original conformation of the receptor is restored, and the receptor effectively 'switches off'. Different types of receptors bind all the various signal molecules arriving at a cell, including hormones, neurotransmitters, and growth factors.

Cell-surface receptors fall into four main classes. *G-protein-linked receptors* consist of a receptor protein coupled to a G-PROTEIN, also located in the plasma membrane. *Tyrosine kinase-linked recep-*

tors bind cytosolic protein-tyrosine kinases, thus activating them (*see* protein kinase). The activated protein–tyrosine kinases phosphorylate tyrosine residues in the receptor, which in turn phosphorylate target proteins. Binding of the ligand can directly activate enzymatic activity in the receptor itself; such *enzymatically active receptors* often are protein kinases, which can phosphorylate themselves, or activate other target proteins by phosphorylation. With LIGAND-GATED ION-CHANNELS, binding of ligand to a receptor alters the shape of an ion channel within the receptor protein, enabling ions to flow through the channel.

recessive   An allele that is only expressed in the phenotype when it is in the homozygous condition. *Compare* dominant.

recombinant DNA   1. Any DNA fragment or molecule that contains inserted foreign DNA, whether from another organism or artificially constructed. Recombinant DNA is fundamental to many aspects of GENETIC ENGINEERING, particularly the introduction of foreign genes to cells or organisms. There are now many techniques for creating recombinant DNA, depending on the nature of the host cell or organism receiving the foreign DNA. Particular genes or DNA sequences are cut from the parent molecule using specific RESTRICTION ENZYMES, or are assembled using a messenger RNA template and the enzyme reverse transcriptase (*see* complementary DNA). In gene cloning, or DNA CLONING, using cultures of bacterial or eukaryote tissue cells, the foreign gene is inserted into a vector (e.g. a bacterial plasmid or virus particle), which then infects the host cell. Inside the host cell the recombinant vector replicates and the foreign gene can, in certain circumstances, be expressed. Plasmids are also used to insert foreign DNA into plants. One of the most common is the Ti (tumor-inducing) plasmid of the bacterium *Agrobacterium tumefaciens*. This causes crown gall tumors in plants, and its plasmid has been used on a range of crop plants. Some animal and plant cells can be induced to take up foreign DNA directly from their environment. For example, mouse embryos can be injected with DNA containing specific mutations to produce strains of so-called knockout mice (*see* gene knockout). Another technique, termed *biolistics* and used in transfecting certain plant cells or cell organelles, is to shoot DNA-coated microprojectiles, such as tungsten or gold particles, at the host cell target. Organisms that stably incorporate foreign genes and transmit them to their offspring are called transgenic.
2. DNA formed naturally by RECOMBINATION, for example, by crossing over in meiosis or by conjugation in bacteria.

recombinant DNA technology   *See* genetic engineering.

recombination   The regrouping of genes that regularly occurs during meiosis as a result of the independent assortment of chromosomes into new sets, and the exchange of pieces of chromosomes (*see* crossing over). Recombination results in offspring that differ both phenotypically and genotypically from both parents and is thus an important means of producing variation.

red blood cell (red blood corpuscle)   *See* erythrocyte.

red blood corpuscle   *See* erythrocyte.

reduction division   The first division of MEIOSIS, including prophase, metaphase I, and anaphase I. It results in a haploid number of chromosomes gathering at each end of the nuclear spindle.

refractory period   The period following passage of an impulse along a neuron (nerve cell) when either no stimulus, however large, will evoke a further impulse (the *absolute refractory period*) or only an abnormally large stimulus will evoke further impulse (the *relative refractory period*). The refractory period is a consequence of the nature of the sodium ion channels, which open briefly to allow the influx of sodium ions associated with the passage of the impulse. The refractory period lasts a few milliseconds, and ensures that the impulse is propagated in one direction only along the neuron.

Reichstein, Tadeus (1897–1996) Polish-born Swiss organic chemist and botanist noted for his isolation and study of 29 different steroids from the cortex of the adrenal gland. He also invented an industrial method of synthesizing ascorbic acid from glucose. He was awarded the Nobel Prize for physiology or medicine in 1950 jointly with E. C. Kendall and P. S. Hench for their discoveries relating to the hormones of the adrenal cortex.

relative molecular mass (molecular weight) Symbol: M_r The ratio of the mass of an atom or molecule to one-twelfth of the mass of an atom of carbon-12. It is thus a mass ratio, and has no units. It is commonly used to indicate the size of proteins and other large biomolecules. For example, a protein of M_r 20 000 would have an equivalent molecular mass of 20 000 daltons (or 20 kDa) (*see* dalton).

release factor   *See* translation.

renaturation   *See* denaturation.

repetitive DNA   DNA that consists of multiple repeats of the same NUCLEOTIDE sequences. Unlike prokaryotic cells, eukaryotic cells contain appreciable amounts of repetitive DNA – in mammals it accounts for roughly half of the total DNA. Most repetitive DNA does not encode any product and its function is unknown; it constitutes part of the so-called 'junk DNA' (*see* selfish DNA). It occurs in the noncoding intron sequences within genes, forms part of the 'spacer' DNA between genes, and is especially concentrated around the centromeres of the chromosomes. However, some repetitive DNA does have a function, notably duplicated protein-coding genes, with very similar DNA sequences (*see* gene family). The genes encoding ribosomal RNAs, transfer RNAs, and histones are typically duplicated many times to form *tandemly repeated arrays*. Moreover, the tandem repeats of short DNA sequences that form

telomeres at the ends of chromosomes play a vital role in chromosome replication.

Repetitive DNA falls into two broad categories, according to the number of copies per cell. *Moderately repetitive DNA* is typically repeated tens to thousands of times per genome. It includes families of genes with repetitive elements, tandemly repeated genes, nonfunctional genes (PSEUDOGENES) and various types of mobile DNA segments, including TRANSPOSONS and RETROTRANSPOSONS. *Highly repetitive DNA* is repeated hundreds of thousand of times overall throughout the genome. It consists typically of different sets of short nucleotide sequences (e.g. 5–10 bp), each set comprising arrays of hundreds or thousands of tandem repeats. In mammals, the bulk of these highly repetitive sequences occur near the centromere; each is about 100 kb in extent, and on ultracentrifugation of cellular DNA they form distinct bands called satellite DNA. Other similar sequences, called VARIABLE NUMBER TANDEM REPEATS, consist of tandemly repeated units of 15–100 bp, found dispersed throughout the chromosomes in numbers that vary between individuals. On ultracentrifugation these form minisatellite DNA. Arrays of tandem repeats of just two base pairs also occur, forming microsatellite DNA on ultracentrifugation. *See* satellite DNA.

replica   A thin detailed metal copy of a biological specimen, used in electron microscopy for studying the shape of viruses, large molecules, subcellular components, etc. The specimen is placed in a vacuum and first coated with a thin film of a heavy metal (e.g. gold or platinum), then with a stabilizing film of carbon. The biological material is dissolved by acid, leaving the metal replica of the specimen. *See* shadowing.

replication   The mechanism by which exact copies of the genetic material are formed. Replicas of DNA are made when the double helix unwinds and the separated strands serve as templates along which complementary nucleotides are assembled through the action of the enzyme

DNA polymerase (*see* DNA replication). In RNA viruses an RNA polymerase is involved in the replication of the viral RNA.

replication origin *See* DNA replication.

reporter molecule (**reporter group**) A small molecule or chemical group having a characteristic property (e.g. fluorescence, ultraviolet absorbance, electron spin resonance) that is sensitive to changes in its surroundings. The molecule is introduced into a macromolecule (e.g. a protein) so that changes in the property can be monitored and correlated with changes in the macromolecule or its environment.

repressor A protein molecule that prevents or slows transcription of a gene by binding to a particular DNA sequence. In prokaryotes, where genes are commonly transcribed in groups called OPERONS, the repressor binds to the operator sequence and blocks transcription of the structural genes. The repressor is encoded by a regulatory gene and may act either on its own or in conjunction with a *corepressor*. In some cases another molecule, an *inducer*, may bind to the repressor, weakening its bonds with the operator and derepressing the gene, allowing transcription to proceed. In eukaryotes the actions of repressors are more varied. Some interact with the histone proteins that package the DNA, to prevent the binding of transcription factors to specific control sequences of a gene. In other cases the repressors bind directly to sites on the DNA and interfere with the initiation of transcription, for example by blocking the binding of an activator or other transcription factors, or inhibiting their function, thereby preventing formation of the INITIATION COMPLEX.

resolution (**resolving power**) The ability of an optical system to form separate images of closely spaced objects. Resolution is proportional to the wavelength of the light; and inversely proportional to the refractive index of the transmitting medium (e.g. air or oil) between the specimen and objective lens, and to the angular aperture of light entering the objective lens from the specimen. The most powerful light microscopes can distinguish objects that are more than about 0.2 μm (200 nm) apart, irrespective of the magnifying power of the lenses. Transmission electron microscopes have a maximum resolution in practice of about 0.1 nm. *See* microscope.

resolving power *See* resolution.

respiration The oxidation of organic molecules by organisms to provide usable energy in the form of ATP. Oxidation involves the removal of electrons, and so requires an electron acceptor. In AEROBIC RESPIRATION, organisms use oxygen as the ultimate electron acceptor, reducing it to water. In ANAEROBIC RESPIRATION, organisms can use one of various inorganic substances (e.g. methane, nitrate, sulfur) or organic substances (e.g. pyruvate) as the electron acceptor. In all cases the energy released during oxidation is conserved through the formation of high-energy bonds in the conversion of ADP to ATP. In

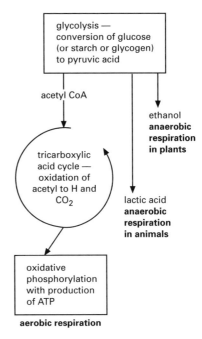

Respiration

heterotrophic organisms (e.g. animals and fungi) food molecules are respired, but autotrophic organisms (e.g. plants) respire molecules that they have themselves synthesized by photosynthesis or other means. *See also* fermentation.

respiratory chain The ELECTRON-TRANS-PORT CHAIN in aerobic respiration.

respiratory pigments Colored compounds that can combine reversibly with oxygen. HEMOGLOBIN is the blood pigment in all vertebrates and a wide range of invertebrates. Other blood pigments, such as hemoerythrin (containing iron) and hemocyanin (containing copper), are found in lower animals, and in many cases are dissolved in the plasma rather than present in cells. Their affinity for oxygen is comparable with hemoglobin, though oxygen capacity is generally lower.

response A change in an organism or in part of an organism that is produced as a reaction to a stimulus.

resting potential The potential difference that exists across the plasma membrane of a cell in a resting, or nonexcited state. Animal cells, including neurons (nerve cells), normally maintain a resting potential of about –mV. In plant cells the value is typically about –120mV in the cytosol; this becomes more positive in the vacuole, by some 20–30mV. The resting potential results from differences in concentrations of ions and electrically charged molecules in the cell's interior compared with the exterior. For example, the concentration of potassium ions (K^+) inside a typical mammalian cell is 20–40 times greater than that in blood, whereas the internal concentration of sodium ions (Na^+) is 8–12 times lower than that in blood. Cells regulate their internal ion concentrations by selectively exporting and/or importing ions. In animal cells, the resting potential is achieved chiefly by the outward diffusion of K^+ through K^+ channels in the plasma membrane, while largely blocking the entry of Na^+ by keeping Na^+ channels closed. The K^+ concentration gradient, on which this outward diffusion depends, is maintained by the active pumping of K^+ ions into the cell, in exchange for Na^+ (*see* Na^+/K^+ ATPase). The chief mechanism in plant cells is the active pumping of H^+, across the plasma membrane out of the cell, and across the tonoplast into the vacuole, by a proton pump. The resting potential is important in various aspects of a cell's life, including the regulation of pH and osmotic pressure; uptake or expulsion of materials; growth and movement (of plant cells); the conduction of nerve impulses by neurons, and initiation of contraction in muscle cells.

restriction endonuclease *See* restriction enzyme.

restriction enzyme (restriction endonuclease) Any of a large group of enzymes, found mainly in bacteria, that can cleave and fragment DNA internally (*see* endonuclease). They are so named because they cut only foreign DNA entering the cell, and so act to 'restrict' the replication of the DNA of an invading virus; hence their natural function is protection of the bacterial cell. Some restriction endonucleases cleave DNA at random, but a particular group of enzymes, known as *class II restriction enzymes*, cleave DNA at specific sites, called *restriction sites*. These sites consist of a sequence of four to eight base pairs. The same sites on the host cell's own DNA are protected from this enzyme activity by methylation, which is controlled by another type of site-specific enzyme.

Two properties of restriction enzymes have made them indispensable tools for producing recombinant DNA molecules for use in genetic engineering. First, digestion of DNA with a particular restriction enzyme produces a characteristic and reproducible set of double-stranded DNA fragments, called *restriction fragments*. Second, many restriction enzymes make a staggered cut across the two strands of the DNA molecule, giving the resulting DNA fragment identical single-stranded 'sticky' ends. The ends readily form hydrogen bonds with the complementary ends on all other fragments generated by the same restriction enzyme. This enables restriction fragments from quite different sources (e.g. different organisms) to be easily joined together, forming chimeric molecules. Over 2500 type II restriction endonucleases have been discovered, with some 200 different cleavage site specificities. They have been named according to the organism in which they occur. For example, *Eco*R1 is obtained from *Escherichia coli*, strain R, and was the first enzyme to be isolated in this bacterium.

restriction fragment length polymorphism (RFLP) Variation among the members of a population in the sites at which restriction enzymes cleave the DNA, and hence in the size of the resulting DNA fragments. It results from differences between individuals in nucleotide sequences at the cleavage sites (restriction sites). The presence or absence of particular restriction sites can be ascertained using DNA probes in the technique called SOUTHERN BLOTTING. Restriction sites vary enormously, and this variation is exploited in analyzing and comparing the genomes of different individuals, for example, to establish how closely related they may be. Restriction sites are also invaluable as genetic markers in mapping mutations in humans and other organisms where experimental crossing is not an option. Called *RFLP mapping*, the restriction sites help to identify and track closely linked genetic loci (e.g. disease-causing genes) and allow investigation of deletions, insertions, or other mutations. They are an essential tool in chromosome mapping. *See also* DNA fingerprinting.

restriction map A map of a segment of DNA showing the cleavage sites of restriction enzymes and their physical distance apart, usually measured in base pairs. Such maps are generally of a fine scale, and can be used for accurately aligning partially overlapping cloned DNA fragments, for example, from a DNA library. Also, detailed knowledge of the restriction sites enables the use of particular combinations of restriction enzymes to obtain a desired re-

striction fragment. *See also* restriction fragment length polymorphism.

restriction point *See* cell cycle.

restriction site *See* restriction enzyme.

reticulocyte An immature red blood cell. It develops in the red marrow in bones from an *erythroblast* (red cell precursor), which accumulates hemoglobin and loses its nucleus to become a fully formed reticulocyte. Once released into the bloodstream, reticulocytes mature into erythrocytes within 24–36 hours.

reticuloendothelial system *See* mononuclear phagocyte system.

retina The innermost light-sensitive layer of the vertebrate eye. It consists of two types of photosensitive cells – RODS and CONES – connected by synapses to bipolar and ganglion nerve cells. From the ganglion cells, nerve fibers pass over the inner surface of the retina to the optic nerve. Light entering the eye through the pupil has to pass through all the layers of the retina before it reaches the sensitive ends of the rods and cones, except at the fovea.

retinal (retinene) An aldehyde derivative of retinol (vitamin A). Retinal is the light-absorbing pigment of the rod cells of the retina, where it occurs covalently bound to the protein opsin, to form RHODOPSIN.

retinene *See* retinal.

retinol *See* vitamin A.

retrotransposon Any of a class of MOBILE ELEMENTS, common in the genomic DNA of eukaryotic organisms, that transpose to new sites via an RNA intermediate. They are so named because this mechanism is analogous to the replication mechanism of RETROVIRUSES. Like retroviruses integrated in the DNA of a host cell, retrotransposons use the RNA polymerase and other enzymes of their host cell to assemble an RNA intermediate. This is converted into DNA by a REVERSE TRANSCRIPTASE enzyme encoded by the retrotransposon. The DNA then inserts at a new site within the host DNA. The main types of retrotransposons in mammalian cells consist of DNA sequences that are repeated typically thousands of times throughout the genome. These are called short interspersed elements (*see* SINES) and long interspersed elements (*see* LINES), according to their length.

retrovirus An RNA-containing virus whose genome becomes integrated into the host DNA (by means of the enzyme REVERSE TRANSCRIPTASE) and then replicates with it. After the retrovirus enters the host cell, viral reverse transcriptase makes a DNA copy of the viral DNA. This is transported into the nucleus of the host cell, where it integrates into the host DNA at one of many possible sites, forming a provirus. This integrated viral DNA is then transcribed into viral RNA, to direct the synthesis of viral proteins, or to be packaged as viral genome to form progeny virions. The virions are released by budding from the host cell membrane. Retroviruses infect only specific host cells bearing surface marker proteins that interact with glycoproteins on the viral envelope. Also, most retroviruses do not kill their host cell, and the integrated proviral DNA is replicated and transmitted to daughter cells, which continue to bud virions from their surface. Some retroviruses can cause cancerous changes in their host cells (i.e. they are oncogenic), either directly by introducing cancer-causing genes (*see* oncogene) within the viral genome, or indirectly by activating host cell oncogenes. For example, the Rous sarcoma virus (RSV), which was discovered in 1911, causes cancer in chickens. The best-known human retrovirus is HIV, which is responsible for AIDS. Retroviruses are also used in experimental work as vectors to introduce desired genes into cultured cells.

reverse transcriptase An enzyme that catalyzes the synthesis of DNA from RNA (i.e. the reverse of transcription, in which mRNA is synthesized from a DNA template). The enzyme occurs in certain RNA viruses (*see* retrovirus) and enables the viral RNA to be 'transcribed' into DNA, which is then integrated into the host DNA and replicates with it. It is also used in genetic engineering to make COMPLEMENTARY DNA (cDNA) from an RNA template.

RFLP *See* restriction fragment length polymorphism.

RFLP mapping *See* restriction fragment.

rhesus factor (**Rh factor**) An antigen attached to human red blood cells, so named because it is also present in the rhesus monkey. The antigen is present in most people, who are therefore described as rhesus-positive (Rh-positive), but absent in others (Rh-negative). The presence of the anti-Rh antibody in a woman pregnant with an Rh-positive fetus poses the risk of a serious condition called *hemolytic disease of the newborn*, in which the red cells of the fetus are destroyed.

Rh factor *See* rhesus factor.

Rhizobium A spherical or rod-shaped bacterium that can live either freely in the soil or symbiotically in the root nodules of leguminous plants and a few other species, such as alder. The bacteria can move slowly through the soil by means of flagella and are attracted to and infiltrate the root hairs of leguminous plants. They produce infection threads that penetrate the cells of the root cortex, which are stimulated to divide rapidly and form a swollen mass of tissue, the root nodule. The central region of the nodule consists of enlarged cells containing large numbers of bacteria. The outer region of the nodule contains vascular strands linking with the vascular bundles of the root. The bacteria in the nodules perform NITROGEN FIXATION – converting atmospheric nitrogen into nitrates, which can be passed to the plant. In return, the plant supplies the bacterium with carbohydrates, such as sugars. *See also* nitrogen cycle.

rhodopsin (**visual purple**) A light-sensitive pigment in the retina. It has a protein component, opsin, linked to a nonprotein molecule, retinal, which is a derivative of vitamin A. It is localized in the ROD CELLS, specifically in the membranous disks of the outer segment. In the dark, the retinal component exists as the *cis*-retinal isomer; absorption of light causes isomerization of this to the *trans*-retinal isomer, forming the unstable meta-rhodopsin (activated opsin). The latter dissociates rapidly, forming opsin and *trans*-retinal. In the dark, enzymes re-form the *cis*-retinal, which recombines with opsin to form rhodopsin. The biochemical mechanism for cone vision is analogous to rod vision, and retinal is used as the chromophore. However, it binds to three types of opsin via different amino-acid side chains, producing blue-, green-, and red-absorbing pigments.

RIA *See* radioimmunoassay.

riboflavin (**vitamin B$_2$**) One of the water-soluble B-group of vitamins. It is found in cereal grains, peas, beans, liver, kidney, and milk. Riboflavin is a constituent of several enzyme systems (flavoproteins), acting as a coenzyme for hydrogen transfer in the reactions catalyzed by these enzymes. Two forms of phosphorylated riboflavin are known to exist in various enzyme systems: FMN (flavin mononucleotide) and FAD (flavin adenine dinucleotide). *See also* vitamin B complex.

ribonuclease (**RNase**) Any enzyme that catalyzes the hydrolysis of the sugar–phosphate bonds of RNA, thereby cleaving adjacent nucleotides. There are several types, each having a specific action. For example ribonuclease T1 degrades RNA to mono- and oligonucleotides terminating in a 3'-guanine nucleotide, while those produced by ribonuclease T2 terminate in a 3'-adenine nucleotide. *Exoribonucleases* cleave nucleotides from one or both ends of the RNA molecule, while *endoribonucleases* cleave bonds within the molecule. *See also* ribozyme.

ribonucleic acid *See* RNA.

ribonucleoprotein (**RNP**) Any protein, or protein complex, that is bound to one or more molecules of RNA. Within the nucleus of eukaryotes there are two main categories: *heterogeneous ribonucleoproteins* (hnRNPs; *see* heterogeneous nuclear RNA) consist of newly formed RNA transcripts (pre-mRNAs) and their associated proteins; *small nuclear ribonucleoproteins* (snRNPs) comprise relatively small RNAs that associate with proteins to form a SPLICEOSOME, which catalyzes the splicing of pre-mRNAs to form mature mRNAs.

ribose A monosaccharide, $C_5H_{10}O_5$; a component of RNA, vitamin B$_{12}$, and various glycosides.

ribosomal RNA (**rRNA**) The RNA found in RIBOSOMES.

ribosome A small organelle found in large numbers in all cells that acts as a site for protein synthesis. During TRANSLATION the ribosome moves along the messenger RNA (mRNA), enabling the peptide linkage of amino acids delivered to the site by transfer RNA molecules according to the code in mRNA. Several ribosomes may be actively engaged in protein synthesis along the same mRNA molecule, forming a POLYRIBOSOME. Ribosomes are assembled partly or fully within the nucleus at the NUCLEOLUS, and then exported into the cytoplasm, where they occur bound to endoplasmic reticulum or free in the cytosol.

In most species ribosomes are composed of roughly equal amounts of protein and RNA. Each ribosome consists of one large subunit and one small subunit, arranged on top of each other. Eukaryotic cells have larger ribosomes than prokaryotic cells but the ribosomes in mitochondria and chloroplasts are about the same size as prokaryotic ribosomes. The subunits, and their component RNA and protein molecules, are designated according to their sedimentation constants as determined by centrifugation, and expressed in svedberg units (S). Hence, in mammals the complete ribosome is an 80S particle, comprising a 60S (large) and a 40S (small) subunit. The large subunit consists of 3 rRNA molecules (28S, 5S, and 5.8S) and 50 proteins; the small subunit has 1 rRNA (18S) and 33 proteins. Pairing between bases in certain regions of the rRNA molecules creates doubled-stranded segments called stems, interspersed with unpaired regions, called loops. This stem–loop formation is important in determining the three-dimensional structure of ribosomes, and is very similar in all types of organisms, indicating that the ancestral rRNA base sequences have been conserved throughout evolution.

ribozyme   Any RNA molecule that acts as an enzyme. During transcription, certain introns (noncoding messenger RNA sequences) can catalyze their own removal from the primary messenger RNA transcript; some can also splice together the cleaved ends, in a process called *self-splicing*. Another crucial ribozyme activity is thought to be performed by RNA in the large subunit of ribosomes in protein synthesis (*see* translation). The rRNA is believed to catalyze the peptidyltransferase reaction, in which a peptide bond is formed between the incoming amino acid and the end of the growing polypeptide chain. Ribozymes have remarkable similarities with VIROIDS, the minute plant pathogens consisting simply of RNA circles, and it has been proposed that viroids are escaped introns.

ribulose bisphosphate (RUBP)   A 5-carbon compound that accepts a molecule of carbon dioxide during PHOTOSYNTHESIS, being converted into two molecules of 3-phosphoglycerate. This reaction is the key step in carbon fixation, and is catalyzed by the enzyme ribulose 1,5-bisphosphate carboxylase (rubisco). The ribulose bisphosphate is subsequently regenerated in the Calvin cycle, when the fixed carbon dioxide is converted into carbohydrate.

ribulose 1,5-bisphosphate carboxylase (**rubisco**)   The enzyme that catalyzes the fixation of carbon dioxide (CO_2) into carbohydrates, the key step of the Calvin cycle during PHOTOSYNTHESIS. One molecule of CO_2 combines with the 5-carbon sugar ribulose bisphosphate (RuBP) to form two molecules of phosphoglyceric acid (PGA).

Rubisco also catalyzes the key reaction of PHOTORESPIRATION, a seemingly wasteful metabolic process that competes with photosynthesis by consuming oxygen and liberating CO_2. Rubisco is a large enzyme, comprising 16 subunits, and is located in the chloroplast stroma. It accounts for some 50% of all chloroplast protein, and is thought to be the most abundant protein in nature.

Ringer, Sydney (1835–1910)   British physician and physiologist who devised Ringer's solution.

Ringer's solution   *See* physiological saline.

RNA (**ribonucleic acid**)   A nucleic acid found mainly in the cytoplasm and involved in protein synthesis. It is a single polynucleotide chain similar in composition to a single strand of DNA except that the sugar ribose replaces deoxyribose and the pyrimidine base uracil replaces thymine. RNA molecules can display various types of folds, turns, and loops due to pairing between complementary bases in different regions of the molecule. These features are important in determining the three-dimensional structure, and hence the function, of the various types of RNA. RNA is synthesized on DNA in the nucleus and exists in three forms (*see* messenger RNA; transfer RNA; ribosome). Some types have catalytic activity (*see* ribozyme). In certain viruses, RNA is the genetic material. *See also* heterogeneous nuclear RNA.

RNA polymerase   Any of a group of enzymes that catalyze the assembly of RNA from its component nucleotides (*see* polymerase) during TRANSCRIPTION of DNA. Essentially, an RNA polymerase binds to a particular initiation site on DNA, temporarily separates, or 'melts', the two DNA strands in that region, and moves along one of the strands, assembling an RNA molecule according to the base sequence of the DNA. In prokaryotes there is only one type of RNA polymerase, responsible for the synthesis of messenger RNAs (mRNAs), ribosomal RNAs (rRNAs), and transfer

RNAs (tRNAs). It is a large protein composed of five subunits: b, b', two a subunits, and a s subunit. In eukaryotes there are three different types of polymerase, each larger and more complex than the prokaryotic one: type I is located in the NUCLEOLUS and makes the bulk of rRNAs; type II makes messenger RNA precursors (pre-mRNAs) that are processed to form the mature mRNA transcript of protein-coding genes; and type III functions outside the nucleolus and makes tRNAs, 5S rRNA, and many other small RNAs whose functions are largely unknown.

In prokaryotes, RNA polymerase binds to the promoter site in the DNA, upstream of the protein-coding region of gene clusters (*see* operon); initiation of transcription depends on the strength of this binding, which is regulated by the s subunit and various other molecules, including repressors and activators. Eukaryotic RNA polymerases also bind to specific start sites of genes, and are controlled by a host of different TRANSCRIPTION FACTORS, many of which bind to distant sites in the DNA. RNA polymerase II forms part of a large complex, along with 60–70 polypeptides, called the INITIATION COMPLEX.

RNase *See* ribonuclease.

Roberts, Richard J. (1943–) British molecular biologist. He was awarded the Nobel Prize for physiology or medicine in 1993 jointly with P. A. Sharp for their discovery of split genes.

Robinson, (Sir) Robert (1886–1975) British organic chemist who worked on plant products, especially the alkaloids. He was awarded the Nobel Prize for chemistry in 1947.

rod One of the two types of light-sensitive cells in the retina of the vertebrate eye. Rods are concerned with vision in dim light, and are found chiefly in the periphery of the retina.They contain a pigment, RHODOPSIN, that is bleached by light energy. This photochemical reaction breaks down rhodopsin into a protein (opsin) and retinal (a derivative of vitamin A) and

causes nerve impulses to pass from the rod cells, via various interneurons, to the brain. Rods cells are long and thin, and are divided into two segments. The outer segment contains a stack of membranous disks, which contain the rhodopsin. The inner segment contains other cellular organelles. Connecting the two segments are the microtubules of a cilium.

The molecules of rhodopsin absorb light and convert into the short-lived activated form, *meta*-rhodopsin, which causes the brief closure of certain ion channels in the plasma membrane of the rod cell. This temporarily changes the membrane potential of the cell, and causes an electrical signal to be transmitted to the interneurons. Rod cells are very sensitive, giving a measurable response to just a single photon of light. The rhodopsin is continually reformed from retinal. In bright light, the reformation does not keep pace with the destruction, so that rods can only function in dim light.

Rodbell, Martin (1925–98) American biochemist. He was awarded the Nobel Prize for physiology or medicine in 1994 jointly with A. G. Gilman for their discovery of G-proteins and the role of these proteins in signal transduction in cells.

root nodule *See* nitrogen fixation.

rough endoplasmic reticulum *See* endoplasmic reticulum.

R plasmid *See* antibiotic.

rRNA *See* ribosomal RNA.

rubisco *See* ribulose 1,5-bisphosphate carboxylase.

RUBP *See* ribulose bisphosphate.

Ružička, Leopold (1887–1976) Croatian-born Swiss chemist who worked on terpenes. He was awarded the Nobel Prize for chemistry in 1939. The prize was shared with A. F. J. Butenandt.

S

saccharide *See* sugar.

Saccharomyces (**budding yeasts**) A genus of unicellular ascomycete fungi that can live in both aerobic and anaerobic conditions. They are important in the brewing and baking industries, respectively, for the alcohol and carbon dioxide they produce by anaerobic respiration. Reproduction is generally asexual by budding although in adverse conditions sexual spores may be formed. *S. cerevisiae* (baker's and brewer's yeast) was the first eukaryotic organism to have its genome sequenced completely. Easy to culture, it is widely used in studies of eukaryotic cell function, especially the events of the cell cycle and cell division. These yeasts are also used in gene cloning as convenient eukaryotic hosts to express inserted foreign genes. *Compare Schizosaccharomyces.*

safranin *See* staining.

Sakmann, Bert (1942–) German physiologist. He was awarded the Nobel Prize for physiology or medicine in 1991 jointly with E. Neher for their work on single ion channels in cells.

saltatory conduction The mode of transmission of a NERVE IMPULSE along a myelinated nerve fiber whereby the impulse leaps between the nodes of Ranvier, considerably speeding up its passage. The fastest nerve impulses known, traveling up to 120 m/s, occur in vertebrate myelinated fibers. The myelin sheath insulates against loss of local currents between the nodes; they are therefore transmitted along the fiber axis to a node, where in the absence of myelin they generate an ACTION POTENTIAL.

Samuelsson, Bengt Ingemar (1934–) Swedish biochemist. He was awarded the Nobel Prize for physiology or medicine in 1982 jointly with S. K. Bergström and J. R. Vane for their discoveries concerning prostaglandins and related biologically active substances.

Sanger, Frederick (1918–) British biochemist. Sanger was awarded the Nobel Prize for chemistry in 1958 for his work on the structure of proteins, especially his work on the structure of insulin. He also shared the Nobel Prize for chemistry in 1980 with W. Gilbert for their work on the determination of base sequences in nucleic acids. The 1980 prize was shared with P. Berg.

Sanger method 1. (of protein sequencing) A technique for determining the overall amino-acid sequence of a polypeptide. It uses selective hydrolysis to break the polypeptide chain into a set of overlapping peptide fragments. These are then sorted using two-dimensional chromatography, and arranged into their correct order. The amino acids in each fragment are identified, and the full amino-acid sequence of the polypeptide can be reconstructed. It is named after its discoverer, the British biochemist Frederick Sanger (1918–). 2. *See* DNA sequencing.

sarcolemma *See* skeletal muscle.

sarcoma *See* cancer.

sarcomere The structural and functional unit of SKELETAL MUSCLE (striated muscle). It consists of a segment of a MYOFIBRIL, typically about 2 μm long in resting muscle. Each sarcomere is joined to the next one by a fibrous lattice called the *Z disk*. Thick filaments of the protein myosin form a dark central *A band*. On either side of this is a light area, the *I band*, in which are thin filaments of another protein, actin. The two types of filament overlap in the dark band except in the center, leaving a slightly lighter *H band* (Hensen's disk). According to the SLIDING FILAMENT MODEL of muscle contraction, projecting parts of myosin molecules form crossbridges with the actin filaments, pulling them along in a ratchet action and thus making the whole sarcomere shorter. The shortening of all the sarcomeres of all the myofibrils results in the contraction of a muscle fiber (muscle cell) when it is stimulated by nerve impulses. Other proteins maintain the arrangement of the filaments and enable the muscle to quickly resume its resting length after being stretched. These include the very long fibrous protein, *titin*, which connects the ends of the myosin filaments to the Z disk, and *nebulin*, which runs along each actin filament.

sarcoplasm   The cytoplasm of the fibers of skeletal muscle, excluding the myofibrils.

sarcoplasmic reticulum (SR)   A modified form of smooth endoplasmic reticulum found in skeletal and cardiac muscle. It forms a network of tubules in the muscle-cell cytosol, into which calcium ions (Ca^{2+}) are continually pumped from the cytosol by CALCIUM ION PUMPS in the membrane of the reticulum. Muscle contraction is initiated by the release of Ca^{2+} from the SR via calcium ion release channels. This release of Ca^{2+} is triggered by the arrrival of a nerve impulse at the muscle cell's plasma membrane. The impulse penetrates deep within the muscle cell, between the myofibrils, via invaginations of the plasma membrane called transverse tubules. Within about 30 milliseconds, the Ca^{2+} ions are pumped back into the SR, thus restoring the resting concentration of Ca^{2+} in the cytosol.

satellite DNA   DNA that can be separated by centrifugation from the main DNA fraction. Such DNA forms distinct 'satellite' bands along the centrifuge tube. The bulk of satellite DNA is from chromosomal regions adjacent to the centromeres, which consists of multiple tandem repeats of short (typically /10 bp) base sequences, stretching to hundreds of kilobases in length. *Minisatellite DNA* consists of another type of repetitive DNA, namely variable number tandem repeats (VNTR). These are slightly longer (15–100 bp) base sequences repeated in tandem arrays extending typically to 1–5 kb. VNTRs are dispersed throughout the genome, and are used to identify individuals in genetic fingerprinting. *Microsatellite DNA* consists of dinucleotide tandem repeats dispersed throughout the genome. The uniformity of base composition of these different types of repetitive DNA gives each a characteristic density on centrifugation, hence their tendency to separate into satellite bands.

scanning electron microscope   *See* electron microscope.

Schally, Andrew Victor (1926–)   Polish-born American biochemist and physiologist. He was awarded the Nobel Prize for physiology or medicine in 1977 jointly with R. Guillemin for their discoveries concerning the peptide hormone production of the brain. The prize was shared with R. Yalow.

schizogeny   The separation of plant cells at the middle lamellae to give intercellular spaces, which may have special functions. An example is seen in the resin ducts of conifers. *Compare* lysogeny (def. 1).

Schizosaccharomyces (**fission yeasts**)   A genus of yeasts that reproduce asexually by fission, dividing into two cells of roughly equal size. *S. pombe* is used as a model experimental organism for genetic studies and other investigations, notably of cell-cycle control mechanisms.

Schwann cell   A cell that makes a section of the MYELIN SHEATH of a medullated nerve fiber. During development, the cell spirally wraps around several axons simultaneously. Both cytosolic and extracellular spaces are progressively reduced, resulting in a sheath of some 50–100 myelin membranes surrounding each axon. The membranes are held together by interacting membrane proteins, which behave rather like a zipper. The cytosol of the Schwann cell becomes confined to the outermost layer of the myelin sheath.

sclereid Any SCLERENCHYMA cell, excluding the fibers. The various forms of sclereid include the star-shaped astrosclereid, the rod-shaped macrosclereid, and the isodiametric stone cell.

sclerenchyma The main supporting tissue in plants, made up of cells that when mature characteristically have heavily thickened, often lignified walls and empty lumina. Sclerenchyma is often found associated with vascular tissue and exists as two distinct types of cell: the fiber and the sclereid.

sclerotome *See* somite.

SCP *See* single-cell protein.

SDS-polyacrylamide gel electrophoresis *See* electrophoresis.

secondary structure 1. (of a protein) A localized arrangement of part of a polypeptide chain due to the formation of stabilizing hydrogen bonds between constituent amino acids. There are two main forms: the spiral rod-shaped ALPHA HELIX, and the planar PLEATED SHEET.
2. (of a nucleic acid) A regular helical structure due to hydrogen bonding between complementary bases along one or two polynucleotide chains. *Compare* primary structure; tertiary structure.

second messenger Any molecule or ion that relays signals from cell-surface receptors to enzymes and other components inside cells. Typically, binding of a hormone to an external binding site on a receptor triggers enzymes that bring about a rapid increase in the concentration of a particular second messenger inside the cell. This acts in turn to activate other components downstream in the signal pathway. Second messengers include CYCLIC AMP, cyclic GMP (cyclic guanosine monophosphate), INOSITOL 1,4,5-TRISPHOSPHATE (IP$_3$), and CALCIUM ions. They mediate a wide range of hormonally induced cellular activities, such as uptake of glucose or secretion of cell products, as well as more general regulation of cell metabolism, including cell proliferation and differentiation. Each has a particular set of targets and mode of action. For example, cyclic AMP often activates PROTEIN KINASES, which in turn activate a cascade of target enzymes, whereas IP$_3$ can stimulate the release into the cytosol of Ca^{2+} stored in the endoplasmic reticulum.

secretion 1. A substance or fluid produced by a cell or group of cells (e.g. a gland) and released to the surrounding medium. The secretion may be a fluid (e.g. sweat) or molecules (e.g. enzymes, hormones, extracellular matrix proteins).
2. The process by which cells release proteins and other molecules to their surroundings. Secretory proteins are sorted and packaged by a sequence of events within cells known as the *secretory pathway*. This pathway deals not only with proteins destined for secretion from the cell surface, but also with ones intended for incorporation in the plasma membrane, lysosomes, and endoplasmic reticulum or Golgi apparatus.

Synthesis of a nascent (new-formed) secretory protein is completed on the endoplasmic reticulum. Then it is encased in a membranous transport vesicle and passes to the *cis* face of the Golgi apparatus. As the protein migrates through the cisternae of the Golgi stack, it is chemically modified to attain its mature form. At the *trans* face of the Golgi stack, proteins are packaged in transport vesicles for *constitutive* (i.e. continuous) *secretion* at the cell surface, or sorting to lysosomes or insertion in the plasma membrane. Some proteins are stored in secretory vesicles for *regulated secretion* (i.e. when triggered by a specific stimulus, such as a hormone). The contents of a vesicle are released by fusion of the vesicle with the plasma membrane, or membrane of the target organelle.

secretory pathway *See* secretion.

sedimentation coefficient (**sedimentation constant**) Symbol: *s* A measure of the rate of sedimentation of a particle in a centrifuge. It is equal to the velocity of sedimentation divided by the centrifugal force, and has the units of time (often measured as *Svedberg units*, symbol: S; 1 S = 10^{-13} second). The sedimentation coefficient of a protein depends on the protein's density and shape, and so can be used to calculate a protein's molecular mass and size under strict analytical conditions.

seed leaf *See* cotyledon.

segment One of a series of repeated parts of the body. For example, an adult fruit fly (*Drosophila*) is divided into 14 segments along its front–rear axis, based on its visible anatomical features. *Compare* parasegment.

segmentation The pattern of SEGMENTS into which the body of an organism is divided. In organisms such as fruit flies (*Drosophila*), the pattern of adult body segments is determined by the arrangement of developmental units (PARASEGMENTS) in the fly embryo. This pattern is established during embryonic development by the expression of three sets of genes: PAIR-RULE GENES; SEGMENT-POLARITY GENES; and SELECTOR GENES. The latter include HOMEOTIC GENES, homologs of which occur in mammals.

segmentation cavity *See* blastocoel.

segment-polarity gene Any of a set of genes in the fruit fly *Drosophila* that help to determine the pattern of cells within each PARASEGMENT during development of the early embryo. Their expression is regulated by the products of the PAIR-RULE GENES, which establish the repeating pattern of parasegments along the anterior–posterior (head to tail) axis of the embryo. The segment-polarity genes encode signal

proteins, which influence the fate of neighboring cells. In particular, they determine which is the most anterior row of cells in each segment. *See also* selector gene.

segregation The separation of the two alleles of a gene into different cells (e.g. gametes) or organisms, brought about by the separation (disjunction) of homologous chromosomes during MEIOSIS. *Mitotic segregation* occasionally occurs in body cells undergoing mitosis. It can result from failure of chromosomes to separate properly, or loss of a chromosome during cell division, or even mitotic crossing-over, which happens rarely in most eukaryotes. *See also* MENDEL'S LAWS.

selectin *See* cell adhesion molecule.

selector gene Any of a set of homeotic genes in the fruit fly *Drosophila* whose products regulate the development of the body parts appropriate to each parasegment of the embryo. Their expression continues through embryonic development into adult life. So, for example, the development of halteres (balancing organs) in the sixth parasegment is determined by expression of the *Ultrabithorax* (*Ubx*) selector gene, which is expressed maximally in the sixth parasegment. The *Drosophila* selector genes are grouped into two clusters on chromosome 3, called the bithorax complex and the antennapedia complex (*see* homeotic gene).

selfish DNA (**junk DNA**) A term sometimes used to describe DNA that is present within the genome of an organism but has no apparent function. This includes much of the REPETITIVE DNA, such as the variable number tandem repeats and the mobile genetic elements (i.e. TRANSPOSONS and RETROTRANSPOSONS), that is dispersed between genes and within introns. Some biologists also regard introns as selfish DNA. The hypothesis is that selfish DNA has somehow evaded the forces of natural selection and become a 'molecular parasite', existing solely to perpetuate itself in the genome. However, it may be that this seemingly functionless DNA is required for some as yet unknown purpose. For example, it might provide the bulk needed for normal chromosome separation during nuclear division, or act as spacer DNA to enable functioning genes to be controlled effectively.

semen A fluid containing spermatozoa and nutritive substances, produced by male mammals. The testes produce the spermatozoa, and the other constituents of the semen are produced by the prostate gland and the seminal vesicles. Semen is placed in the body of the female during mating.

semipermeable membrane *See* osmosis.

senescence 1. (*Biology*) The advanced phase of the aging process of an organism or part of an organism that precedes natural death. It is usually characterized by a reduction in capacity for self-maintenance and repair of cells, and hence deterioration. The degree of senescence varies between groups and its mechanism remains largely uncertain. However, discoveries in cell biology and genetics have prompted certain theories. The *telomere theory of aging* proposes that senescence involves failure of chromosomes to replicate. This is based on the finding that the telomeres, which form the tips of an organism's chromosomes, shorten progressively with successive replication cycles, eventually making replication impossible. The *mitochondrial theory of aging* states that the accumulation of mutations in mitochondrial DNA are linked to a decline in mitochondrial function with age, which is responsible for the age-related decline in oxidative phosphorylation.
2. (*Botany*) The deterioration of the leaves of a deciduous plant toward the end of the growing season, culminating in abscission (leaf fall).

sense organ One or more sensory cells (receptors) and associated structures in an animal that are able to respond to a stimulus from inside or outside. The stimulus is converted into an electrical impulse and sent along nerve fibers to the brain for interpretation and response. In general, a sense organ can only respond to a specific stimulus. Hence there are different organs for touch, heat, pressure, etc.

sensitization The increase in the reaction of an organism or cell to an antigen to which it has been previously exposed. It may occur naturally or be artificially induced, for example, following vaccination.

serine *See* amino acids.

serotonin (5-hydroxytryptamine; 5-HT) A substance that serves as a neurohormone that acts on muscles and nerves, and a neurotransmitter found in both the central and peripheral nervous systems. It controls dilation and constriction of blood vessels and affects peristalsis and gastrointestinal tract motility. Within the brain it plays a role in mood behavior. Many hallucinogenic compounds (e.g. LSD) antagonize the effects of serotonin in the brain.

Sertoli cells Large pillar-like cells in the germinal epithelium of the vertebrate testis, which protect and nourish developing spermatozoa. They also secrete the hormone inhibin, which inhibits secretion of follicle-stimulating hormone by the pituitary. *See also* spermatogenesis.

serum (*pl.* sera) *See* blood serum.

sex chromosome *See* sex determination.

sex determination Any mechanism that causes individuals to develop as either female or male. In species having almost equal numbers of males and females, sex determination is genetic. Very occasionally a single pair of alleles determine sex but usually whole chromosomes, the *sex chromosomes*, are responsible. The 1:1 ratio of males to females is obtained by crossing of the homogametic sex (XX) with the heterogametic sex (XY). In most animals, including humans, the female is XX and the male XY, but in birds, butterflies and some fishes this situation is reversed. In some species sex is determined more by the number of X CHROMOSOMES than by the presence of the Y CHROMOSOME, but in mammals the Y chromosome is important in determining maleness. Many genes are involved in determining all aspects of maleness or femaleness, but it in mammals one particular gene on the Y chromosome acts as a sex switch to initiate male development. In the absence of this male switch gene (i.e. in XX individuals) the fetus develops as a female.

Rarely, sex is subject to environmental control, in which case unequal numbers of males and females develop. In bees and some other members of the Hymenoptera, females develop from fertilized eggs and are diploid while males develop from unfertilized eggs and are haploid, the numbers of each sex being controlled by the queen bee.

sex hormone Any of several hormones responsible for the development and functioning of the reproductive organs. They are also involved in the development of secondary sex characteristics. They are secreted mainly by the gonads and include androgens in males and estrogens and progesterone in females.

sex linkage The coupling of certain genes (and therefore the characters they control) to the sex of an organism because they happen to occur on the X sex chromosome. The heterogametic sex (XY), which in humans is the male, has only one X CHROMOSOME and thus any recessive genes carried on it are not masked by their dominant alleles (as they would be in the homogametic sex). Thus in humans recessive forms of the sex-linked genes appear in the male phenotype far more frequently than in the female (in which they would have to be double recessives). Color blindness and hemophilia are sex linked.

sexual reproduction The formation of new individuals by fusion of two nuclei or sex cells (gametes) to form a zygote. In unicellular organisms whole individuals may unite but in most multicellular organisms only the gametes combine. In organisms showing sexuality, the gametes are of two

types: male and female (in animals, spermatozoa and ova). They are produced in special organs (carpel and anther in plants; ovary and testis in animals), which, with associated structures, form a reproductive system and aid in the reproductive process. Individuals containing both systems are termed monoecious or hermaphrodite.

Generally MEIOSIS occurs before gamete formation, resulting in the gametes being haploid (having half the normal number of chromosomes). At FERTILIZATION, when the haploid gametes fuse, the diploid number of chromosomes is restored. In this way sexual reproduction permits genetic recombination, which results in greater variety in offspring and so provides a mechanism for evolution by natural selection.

shadowing A method of preparing material for electron microscopy enabling surface features to be studied. It can be used for small entire structures, subcellular organelles, or even large molecules (e.g. DNA). The specimen is supported on a plastic or carbon film on a small grid and sprayed with vaporized metal atoms from one side while under vacuum. The coated specimen appears blacker (more electron-opaque) where metal accumulates, and the lengths and shapes of 'shadows' cast (regions behind the objects not coated with metal) give structural information. It is often used in association with FREEZE FRACTURING. See also replica.

Sharp, Phillip Allen (1944–) American molecular biologist. He was awarded the Nobel Prize for physiology or medicine in 1993 jointly with R. J. Roberts for their discovery of split genes.

Shine–Dalgarno sequence A short sequence of nucleotides near the translation start site of a prokaryotic messenger RNA (mRNA) that binds part of the ribosome and so helps correctly to position the ribosome in relation to the start site. The sequence of eight nucleotides contains on average six that are complementary to a sequence on the small (16S) ribosomal subunit. This allows base pairing and binding of the subunit to the mRNA, close to the AUG start codon that signals the fully assembled ribosome to begin translation. It is named after its discoverers, John Shine and Lynn Dalgarno.

short interspersed elements *See* SINES.

sieve elements Elongated cells that, placed end to end, make up the sieve tubes in the vascular tissue of flowering plants. Most of the organelles, including the nucleus, break down during the development of the sieve element, and remaining cytoplasm connects with that in adjacent sieve elements via pores in the perforated end walls. These pores are often grouped to form sieve plates. Sucrose is transported in solution by the flow of water through the sieve tubes (*see* translocation). Sieve elements are intimately associated with neighboring COMPANION CELLS, via numerous cytoplasmic connections (plasmodesmata).

sievert Symbol: Sv   The SI unit of dose equivalent. It is the dose equivalent when the absorbed dose produced by ionizing radiation multiplied by certain dimensionless factors is 1 joule per kilogram (1 J kg^{-1}). The dimensionless factors are used to modify the absorbed dose to take account of the fact that different types of radiation cause different biological effects.

sieve tube   A column of cells formed from SIEVE ELEMENTS, in which food is translocated in plants. *See also* translocation.

signal recognition particle (SRP)   A particle found in the cytosol that helps newly synthesized (nascent) proteins to enter the ENDOPLASMIC RETICULUM (ER). It consists of a single small RNA molecule and six proteins. The SRP binds to the signal peptide of the nascent protein, and also to the ribosome to which the nascent protein is still attached. The complex of SRP, nascent protein, and ribosome then bind to an SRP receptor in the ER membrane. Subseqeunt dissociation of the SRP and SRP receptor then allows the protein to enter the lumen of the ER via a channel (translocon) in the membrane. The SRP is recycled to assist another protein.

silicon   A trace element found in many animals and plants, although not essential for growth in most organisms. It is found in large quantities in the scales or shells of certain aquatic protists (e.g. heliozoans, diatoms, desmids) and horsetails, and in smaller amounts in the cell walls of many higher plants. It forms the skeleton of certain marine animals, for example, the siliceous sponges. Silicon is also found in connective tissue.

SINES (short interspersed elements)   Mobile DNA elements that are found repeated at numerous sites throughout the genome of humans and other mammals. They behave as RETROTRANSPOSONS, but do not encode any proteins, unlike the long interspersed elements (*see* LINES). Many contain a restriction site for the restriction enzyme *Alu*I, and so they are also called *Alu sequences*. Full-length SINES are about 300 bp long, and occur at roughly one million sites in the human genome. However, shorter related sequences are also found, some as small as 10 bp. The cell's RNA polymerases transcribe SINES into RNA copies. It is thought that these are then copied into DNA by a reverse transcriptase enzyme, possibly one encoded by LINES, before being inserted at a new site within the genome.

single-cell protein (SCP)   Protein produced from microorganisms, such as bacteria, yeasts, mycelial fungi, and unicellular algae, used as a protein supplement in food for humans and other animals. Microbial cells contain high proportions of nucleic acids (DNA and RNA), which must be reduced in processing (metabolism of excessive nucleic acids in humans leads to kidney stones and gout).

SI units (Système International d'Unités)   The internationally adopted system of units used for scientific purposes. It has seven base units (the meter, kilogram, sec-

BASE AND DIMENSIONLESS SI UNITS

Physical quantity	Name of SI unit	Symbol for SI unit
length	meter	m
mass	kilogram(me)	kg
time	second	s
electric current	ampere	A
thermodynamic temperature	kelvin	K
luminous intensity	candela	cd
amount of substance	mole	mol
*plane angle	radian	rad
*solid angle	steradian	sr

*supplementary units

DERIVED SI UNITS WITH SPECIAL NAMES

Physical quantity	Name of SI unit	Symbol for SI unit
frequency	hertz	Hz
energy	joule	J
force	newton	N
power	watt	W
pressure	pascal	Pa
electric charge	coulomb	C
electric potential difference	volt	V
electric resistance	ohm	Ω
electric conductance	siemens	S
electric capacitance	farad	F
magnetic flux	weber	Wb
inductance	henry	H
magnetic flux density	tesla	T
luminous flux	lumen	lm
illuminance (illumination)	lux	lx
absorbed dose	gray	Gy
activity	becquerel	Bq
dose equivalent	sievert	Sv

DECIMAL MULTIPLES AND SUBMULTIPLES USED WITH SI UNITS

Submultiple	Prefix	Symbol	Multiple	Prefix	Symbol
10^{-1}	deci-	d	10^{1}	deca-	da
10^{-2}	centi-	c	10^{2}	hecto-	h
10^{-3}	milli-	m	10^{3}	kilo-	k
10^{-6}	micro-	μ	10^{6}	mega-	M
10^{-9}	nano-	n	10^{9}	giga-	G
10^{-12}	pico-	p	10^{12}	tera-	T
10^{-15}	femto-	f	10^{15}	peta-	P
10^{-18}	atto-	a	10^{18}	exa-	E
10^{-21}	zepto-	z	10^{21}	zetta-	Z
10^{-24}	yocto-	y	10^{24}	yotta-	Y

ond, kelvin, ampere, mole, and candela) and two supplementary units (the radian and steradian). Derived units are formed by multiplication and/or division of base units; a number have special names. Standard prefixes are used for multiples and submultiples of SI units.

skeletal muscle (**striated muscle**; **striped muscle**; or **voluntary muscle**) Muscle that moves the bones of the skeleton. Each muscle is made up of many long multinucleated muscle cells (*muscle fibers*, or *myofibers*), each typically 1–40 μm in length and 10–50 μm in diameter. The myofibers are bound together with connective tissue and surrounded by a sheath (*epimysium*). Each muscle fiber has an outer membrane (*sarcolemma*) inside which are up to 100 nuclei. The cytoplasm (*sarcoplasm*) contains many large mitochondria and longitudinal MYOFIBRILS, which contain the contractile elements – the SARCOMERES – giving the striated appearance. The epimysium is continuous with the nonelastic fibers of the tendons, which attach the tapering ends of the muscle to the bones. All skeletal muscles are under the voluntary control of the central nervous system.

Skou, Jens C. (1918–) Danish biochemist noted for his discovery of Na+/K+ ATPase. He was awarded the 1997 Nobel Prize for chemistry. The prize was shared with P. D. Boyer and J. E. Walker.

sliding filament model A model to explain the contraction of skeletal muscles. It proposes that within each functional unit, or SARCOMERE, the thick (myosin) filaments and thin (actin) filaments do not contract but interact so that they slide past each other, thereby shortening the sarcomere. When repeated in each sarcomere, this results in contraction of the muscle as a whole. Although first proposed some 50 years ago, subsequent discoveries have generally supported the model. The myosin filaments have globular heads that 'walk' along the actin filaments by binding to successive actin subunits and then pivoting in a power stroke, which causes the two filaments to slide relative to each other. Each power stroke involves the hydrolysis of an ATP molecule. The myosin filaments lie across the middle of the sarcomere, and their ends interdigitate with the actin filaments at either end of the sarcomere. Because the heads at either end of a myosin filament work in opposite directions, they pull each set of actin filaments toward the center of the sarcomere, thereby making it shorter.

slime bacteria *See* myxobacteria.

slime molds (**slime fungi**) Simple unicellular or multicellular eukaryotic organisms, now usually classified as protists, that display distinct changes in form during their life cycle. They have an ameboid stage, which lacks cell walls and feeds by engulfing bacteria by phagocytosis; and they can form multicellular differentiated reproductive structures that resemble the fruiting bodies of fungi or primitive plants. The ameboid stages aggregate in masses, often visible as slimy masses on rotting logs, vegetation, etc. The slime molds are divided into two groups. The 'true' slime molds, or *plasmodial slime molds*, form a large mass of multinucleate cytoplasm (*plasmodium*), which pulsates internally and moves only as it grows. In contrast, the *cellular slime molds* form a mass of cells, or slug (*pseudoplasmodium*), that moves around leaving a slime trail. In both cases the mass of cells eventually matures to the fruiting stage, in which a stalked fruiting body (*sporophore*) is formed by differentiation. This produces spores, which give rise to individual amebas. In the plasmodial slime molds, haploid amebas of distinct mating types fuse to produce a diploid zygote, whereas sexual reproduction is rare or absent in the cellular slime molds.

Slime molds exhibit features of animals, fungi, and plants, and at some time have been classified as all three. For example, in one scheme they are placed in the phylum Myxomycota as part of the kingdom Fungi. This phylum contains two classes: Myxomycetes (plasmodial slime molds) and Acrasiomycetes (cellular slime molds). However, certain recent classifications place them respectively in the phyla Myxomycota and Rhizopoda in the kingdom Protoctista.

small nuclear ribonucleoprotein (snRNP) *See* spliceosome.

small nuclear RNA (snRNA) *See* spliceosome.

Smith, Hamilton Othanel (1931–) American molecular biologist. He was awarded the Nobel Prize for physiology or medicine in 1978 jointly with W. Arber and D. Nathans for the discovery of restriction enzymes and their application to problems of molecular genetics.

Smith, Michael (1932–) British-born Canadian biochemist. He was awarded the Nobel Prize for chemistry in 1993 for his fundamental contributions to the establishment of oligonucleotide-based, site-directed mutagenesis and its development for protein studies. The prize was shared with K. B. Mullis.

smooth endoplasmic reticulum *See* endoplasmic reticulum.

smooth muscle (involuntary muscle) The muscle of all internal organs (viscera) and blood vessels (except the heart). Usually it is in the form of tubes or sheets, up to several layers thick. The cells are long, narrow, and tapering, with a single nucleus and cytoplasm containing loose bundles of thick and thin contractile filaments, composed of the contractile proteins myosin and actin, respectively. These bundles connect to plaques in the muscle cell plasma membrane and to dense bodies in the cytoplasm, which act like the Z disks of skeletal muscle SARCOMERES. The calcium ions that trigger contraction are stored not in a sarcoplasmic reticulum but in the extracellular space. In vertebrates, smooth muscle is not under voluntary control, being supplied by the autonomic nervous system. It contracts when stretched, may have spontaneous rhythmic contractions, and can remain in a state of continuous contraction (tonus) for long periods without fatigue.

SNAP A membrane protein that assists in the binding and fusion of an intracellular transport vesicle to its correct target, for example, to an organelle or the plasma membrane. The SNAP protein resides in the membrane of the target, and interacts with other proteins (*see* SNARE) both in the target membrane and in the membrane of the incoming vesicle.

SNARE Any of various proteins that function in targeting transport vesicles to their correct destinations inside the cell. A specific SNARE protein, called a *V-SNARE*, is incorporated into the bounding membrane of a vesicle at its formation in the Golgi complex or another organelle. This travels with the vesicle, and recognizes the correct destination membrane by binding to a specific target SNARE (*T-SNARE*) residing in the target membrane, in association with a SNAP protein.

Snell, George Davis (1903–96) American geneticist. He was awarded the Nobel Prize for physiology or medicine jointly with B. Benacerraf and Jean Dausset in 1980 for their discoveries concerning genetically determined structures on the cell surface that regulate immunological reactions.

snRNA (small nuclear RNA) *See* spliceosome.

snRNP (small nuclear ribonucleoprotein) *See* spliceosome.

sodium An element essential in animal tissues, and often found in plants although it is believed not to be essential in the latter. It is found in bones, and is the most abundant ion in the blood and cell fluids, being extremely important in maintaining the osmotic balance of animal tissues. In neurons and other excitable cells, the rapid influx of sodium ions (Na^+) via SODIUM ION CHANNELS is responsible for the depolarization of the cell and generation of an action potential, which marks the passage of a nerve impulse. *See* Na^+/H^+ antiporter, NA^+/K^+ ATPase.

sodium ion channel A transmembrane protein that encloses an aqueous channel for the passage of sodium ions (Na^+) across a cell membrane. Cells generally maintain a low internal Na^+ concentration compared to the exterior. Hence, Na^+ ions tend to flow into the cell down this concentration gradient, and sodium ion channels are gated to regulate this flow. In excitable cells, such as nerve and muscle cells, *voltage-gated sodium channels* are crucial to generating action potentials. The channel protein contains a voltage-sensing region, which responds to slight depolarization of the membrane and triggers opening of the gate. This permits the sudden masssive influx of Na^+ that produces the marked depolarization of an action potential. Each channel is open for roughly 1 ms, during which time about 6000 Na^+ ions pass through. Then a channel-inactivating segment moves to block the channel, and ion flow ceases. The channel remains inactivated as long as the membrane is depolarized – this is the basis of the refractory period, which prevents propagation of the action potential in the reverse direction. Some ligand-gated ion channels also allow the passage of Na^+.

solute potential *See* water potential.

somatic Describing the cells of an organism other than germ cells. Somatic cells divide by MITOSIS producing daughter cells identical to the parent cell. A *somatic mutation* is a mutation in any cell not destined to become a germ cell; such mutations are therefore not heritable.

somatic cell hybridization The fusion of cultured human cells with hamster or mouse cells to create a hybrid cell containing nuclei from both cell types. The nuclei then fuse, and during subsequent divisions, some human chromosomes are lost, eventually resulting in a cell line in which just one or several human chromosomes are stably inherited. The technique is used to assign genes to chromosomes. For example, if the hybrid cell expresses a particular human protein, the corresponding gene must be located on the human chromosome present in the cell. *See also* chromosome map.

somatic embryo A cloned embryo that is produced asexually from cultured plant tissue. Various tissues can be used as the culture material, depending on species, and these are typically grown in successive culture media and under varying conditions to stimulate embryo formation. Somatic embryos are produced experimentally to study plant development, and also commercially as a means of propagating genetically desirable plant lines.

somatomedin *See* insulinlike growth factor.

somatopleure *See* mesoderm.

somatotropin *See* growth hormone.

somite Any of the blocks of tissue into which the mesoderm of vertebrate embryos is divided either side of the notochord and neural tube. Under the influence of signals from surrounding tissue, each somite develops to form a muscle block (*myotome*), a portion of kidney (*nephrotome* or *intermediate cell mass*), and contributions to the axial skeleton (*sclerotome*) and dermis (*dermatome*).

Sonic Hedgehog protein *See* hedgehog protein.

SOS repair *See* DNA repair.

Southern blotting A technique for transferring DNA fragments from an electrophoretic gel to a nitrocellulose filter or nylon membrane, where they can be fixed in position and probed using DNA probes. Named after its inventor, E. M. Southern (1938–), it is widely used in genetic analysis, for example, in making restriction maps and for finding particular DNA fragments in complex mixtures. The DNA is first digested with restriction enzymes and the resulting mixture of fragments separated according to size by electrophoresis on agarose gel. The double-stranded DNA is then denatured to single-stranded DNA using sodium hydroxide, and a nitrocellulose filter pressed against the gel. This transfers, or blots, the single-stranded DNA fragments onto the nitrocellulose, where they are permanently bound by heating. The DNA probe can then be applied to locate the specific DNA fragment of interest, while preserving the electrophoretic separation pattern. *Compare* Northern blotting, Western blotting.

spectrophotometer An instrument for measuring the amount of light of different wavelengths absorbed by a solution. It gives information about the identity or amount of the specimen and can be used to plot absorption spectra.

sperm *See* spermatozoon.

spermatid A reproductive cell resulting from the second meiotic division of a SPERMATOCYTE. It matures and undergoes a series of changes, which transform it into a spermatozoon. *See also* spermatogenesis.

spermatocyte A reproductive cell, within the seminiferous tubules of the testis, that develops during the formation of spermatozoa (*see* spermatogenesis). A *primary spermatocyte* develops from a spermatogonium, which has undergone a period of multiplication and growth. It divides by meiosis and the first meiotic division produces two *secondary spermatocytes* with haploid nuclei. Each secondary spermatocyte undergoes a second meiotic division to produce two spermatids. One primary spermatocyte thus forms four spermatids, which later become spermatozoa.

spermatogenesis The formation of spermatozoa within the testis in male animals. Precursor cells in the germinal epithelium lining the seminiferous tubules begin to multiply by mitosis and form spermatogonia during embryonic development. However, from the onset of sexual maturity some spermatogonia enter meiosis and start to produce huge numbers of spermatozoa. This activity may be confined to distinct breeding seasons, or can be continuous, as in humans.

A spermatogonium destined to form spermatozoa migrates inward towards the lumen of the tubule and enters a growth phase, which results in the formation of a primary spermatocyte. The primary spermatocyte then undergoes the first division of meiosis (meiosis I) resulting in the formation of two secondary spermatocytes, each containing the haploid number of chromosomes. Each secondary spermatocyte undergoes meiosis II and produces two spermatids. Spermatids then become transformed into spermatozoa, during which time they are attached to Sertoli cells. When mature, the spermatozoa pass from the seminiferous tubules into the epididymis for temporary storage.

spermatogonium (*pl.* **spermatogonia**) A reproductive cell in the testis, situated in the germinal epithelium that lines the seminiferous tubules. It undergoes a period of multiplication and growth to give rise to spermatocytes. *See* spermatogenesis.

spermatozoon (**sperm**; *pl.* **spermatozoa**) The small motile mature male reproductive cell (gamete) formed in the testis. It differs in form and size between species; in humans it is about 52–62 μm long and comprises a head region, a middle region (midpiece), and a long tail. The head contains the haploid nucleus and is capped by the ACROSOME, which helps the sperm to penetrate the egg. The midpiece consists of a centriole and a core of microtubules, around which mitochondria are wound; these provide the energy for movement of the tail. The tail is an axoneme (axial filament) made up of a 9 + 2 array of microtubules. Spermatozoa remain inactive until they pass from the testis during coitus, when secretions from the prostate gland and seminal vesicles stimulate undulating movements of the tail and effect locomotion. About 200–300 million spermatozoa may be released in a single ejaculation, although only one fertilizes each ovum.

S phase *See* cell cycle.

spherosome (**oleosome, oil body**) An oil droplet, typically 0.8–1.0 μm in diameter and surrounded by a single layer of phospholipids, that is found especially in the cotyledons or endosperm of seeds as a lipid store.

spindle The structure formed during mitosis and meiosis that is responsible for moving the chromatids and chromosomes to opposite poles of the cell. The spindle consists of a system of microtubules whose synthesis starts late in interphase under the control of a microtubule-organizing center; in plants and animals this is the CENTROSOME. At each pole a centrosome organizes three sets of microtubules. One set form a tuft of microtubules, the aster, that radiates outwards toward the cell perimeter, and helps to orientate the spindle. A second set of microtubules attach to regions of the centromeres of each pair of sister chromatids, the KINETOCHORES. A third set, the polar microtubules, pass from each pole to interact with their counterparts from the opposite pole, in a region of overlap at the *spindle equator*. During anaphase, the ends of the microtubules attached to the kinetochores progressively disassemble and the microtubules shorten, hauling the chromatids towards the spindle pole. Later in anaphase, microtubule motor proteins cause the polar microtubules to slide past each other, elongating the entire spindle.

spindle pole body *See* centrosome.

spirillum (*pl.* **spirilla**) Any helically shaped bacterium. Such cells are usually found singly and possess flagella at one or more poles.

spirochete A long spirally twisted bacterium with a flexible cell wall surrounding a protoplasmic cylinder. Within the wall are several or many internal flagella, extending along the body. These enable a range of body movements, including rotation, torsion, and flexing, with which the organism swims. Many spirochaetes are found in mud and water and can withstand low oxygen concentrations, whereas others are parasites causing diseases such as yaws, syphilis, and relapsing fever.

Spirogyra A genus of filamentous green algae commonly found as pond scum. Species have a characteristic helically wound chloroplast running the length of each cell. They can reproduce asexually by fragmentation of filaments, and sexually by conjugation. In the latter, 'male' and 'female' filaments lie side by side, and extend conjugation tubes. The contents of one cell flows through the tube and fuses with a cell of the opposite filament to form a zygote. This is eventually released into the water, and germinates to form a new filament.

splanchnopleure *See* mesoderm.

spleen A lymphoid organ situated just beneath the stomach in vertebrates. It produces lymphocytes and destroys and stores red blood cells.

spliceosome A complex assemblage of proteins and small RNA molecules that catalyzes the splicing of PRE-MRNA in the nuclei of eukaryotic cells. Splicing is one of the main steps in the processing of pre-mRNA, and involves removal of the noncoding introns and joining together of the coding sequences (exons) to form the mature messenger RNA (mRNA). A spliceosome is roughly the size of a ribosome, and contains five principal components called *small nuclear ribonucleoproteins* (snRNPs). Each of these consists of a particular *small nuclear RNA* (snRNA) molecule plus several associated proteins. The snRNPs sequentially bind to sites bordering each intron, and bend the intervening intron over into a looped structure, or lariat. This is then excised, and the adjoining exons joined (ligated) together. The snRNPs then dissociate in readiness to splice another intron, while the excised intron is degraded.

sporophore *See* slime molds.

SR *See* sarcoplasmic reticulum.

staining A procedure that is designed to heighten contrast between different structures. Normally biological material is lacking in contrast, protoplasm being

COMMON STAINS FOR LIGHT MICROSCOPY		
Stains	*Final Color*	*Suitable for*
aniline (cotton) blue	blue	fungal hyphae and spores
aniline sulphate or hydrochloride	yellow	lignin
borax carmine	pink	nuclei; particularly for whole amounts (large pieces) of animal material
eosin	pink	cytoplasm; *see* hematoxylin
	red	cellulose
Feulgen's stain	red/purple	DNA; particularly to show chromosomes during cell division
hematoxylin	blue	nuclei; mainly used for sections of animal tissue with eosin as counterstain for cytoplasm; also for smears
iodine	blue–black	starch; therefore for plant storage organs
Leishman's stain	red–pink	blood cells
	blue	white blood cell nuclei
light green or fast green	green	cytoplasm and cellulose; *see* safranin
methylene blue	blue	nuclei; suitable as a vital stain
phloroglucinol	red	lignin
safranin	red	nuclei. Lignin and suberin. Mainly used for sections of plant tissue with light green as counterstain for cytoplasm

transparent, and therefore staining is essential for an understanding of structure at the microscopic level. *Vital stains* are used to stain and examine living material. Most stains require dead or nonliving material. Staining is done after FIXING and either during or after dehydration. *Double staining* involves the use of two stains; the second is called the *counterstain*. *Acidic stains* have a colored anion, *basic stains* have a colored cation. Some stains are neutral. Materials can be described as *acidophilic* or *basophilic* depending on whether they are stained by acidic or basic dyes respectively. Basic stains are suitable for nuclei, staining DNA. Stains for light microscopy are colored dyes; those for electron microscopy contain heavy metals, for example osmium tetroxide. *See also* fluorescence microscopy.

Stanley, Wendell Meredith (1904–71) American biochemist. He was awarded the Nobel Prize for chemistry in 1946 jointly with J. H. Northrop for their preparation of enzymes and virus proteins in a pure form. The prize was shared with J. B. Sumner.

starch A polysaccharide found in plants and consisting of linked glucose units. Most plant starches are a mixture of *amylose*, which consists of long straight chains of D-glucose units, and *amylopectin*, which is highly branched with many short glucose chains. During photosynthesis in many plants, a proportion of the assimilated carbon is converted, via the intermediate fructose-6-phosphate, to granules of *leaf starch*. These granules accumulate inside chloroplasts and serve as temporary stores of carbon. The leaf starch is subsequently mobilized and exported from the chloroplast to the cytosol as triose phosphate, which is used to make SUCROSE for transport to other areas of the plant. Some sucrose is subsequently reconverted to starch for storage as granules or grains in plastids (amyloplasts), especially in storage tissue (e.g. tubers). When required by the cell, the starch is broken down by enzymes into glucose, which is transported out of the plastid by specific transport proteins in the plastid membrane.

Starch is a dietary component of animals. In humans it is digested by salivary and pancreatic amylase then further degraded by maltase to yield glucose, which may be stored as glycogen (*animal starch*).

starch–statolith hypothesis   A hypothesis concerning the mechanism of gravity perception in plants. *See* gravitropism.

start codon   *See* translation.

statolith   A collection of starch grains within a membrane (i.e. an AMYLOPLAST) that, according to the starch–statolith hypothesis, acts as a gravity sensor inside plant cells (*see* gravitropism). The statoliths fall to the lower surface of the cell under the influence of gravity, and somehow cause asymmetrical growth of the cell.

statolith hypothesis   *See* gravitropism.

Stein, William Howard   (1911–80) American protein chemist. He was awarded the Nobel Prize for chemistry jointly with s. moore in 1972 for their contribution to the understanding of the connection between chemical structure and catalytic activity of the active center of the ribonuclease molecule. The prize was shared with C. B. Anfinsen.

stem cell   Any cell that remains undifferentiated and capable of unlimited division in order to provide new cells for growth or replacement of tissues. After division it may produce more stem cells or cells that differentiate into specialized tissue cells. An early embryo contains stem cells with the potential to develop into any of the differentiated cells found in the body. Throughout adult life various types of stem cells remain active in many of the body's tissues. For example, intestinal stem cells continually divide to replenish the cells of the gut lining, whereas stem cells in bone marrow give rise to differentiated blood cells, such as lymphocytes and erythrocytes. Stem cells that divide to produce just one type of differentiated cell are called *unipotent* ('having a single capability'). Stem cells that are capable of producing two, three, or more types of differentiated cells are called *pluripotent* ('having many capabilities').

stem–loop formation   *See* ribosomal RNA.

steroid   Any member of a group of compounds having a complex basic ring structure. Examples are CORTICOSTEROID hormones (produced by the adrenal gland), sex hormones (progesterone, androgens, and estrogens), bile acids, and STEROLS (such as cholesterol).

sterol   A steroid with long aliphatic side chains (8–10 carbons) and at least one hydroxyl group. They are lipid-soluble and often occur in cell membranes (e.g. cholesterol and ergosterol).

stigma   *See* eyespot.

stimulus   A change in the external or internal environment of an organism that elicits a response in the organism. The stimulus does not provide the energy for the response.

stoma (*pl.* **stomata**)   One of a large number of pores in the epidermis of plants through which gaseous exchange occurs. In most plants stomata are located mainly in the lower epidermis of the leaf. Each stoma is surrounded by two crescent-shaped GUARD CELLS, which regulate the opening and closing of the pore by changes in their turgidity. The guard cells are connected to adjoining accessory (subsidiary) cells, which assist the guard cells in opening and closing the stoma. The stoma, guard cells, and accessory cells together comprise the *stomatal complex*, or stomatal apparatus.

When the leaf is actively photosynthesizing, ion channels in the guard cell plasma membrane open to allow the influx of ions. This increases the concentration of dissolved solutes inside the guard cell, and so draws water in by osmosis causing the guard cells to swell. As the guard cells expand, they buckle outwards and increase the aperture of the stomatal pore. As they lose water, the guard cells shrink and the pore tends to close.

stone cell (**brachysclereid**) An isodiametric cell – a type of SCLEREID – found either singly or in groups in the parenchyma and phloem of stems and some fruits (e.g. pear).

stop codon *See* translation.

Streptococcus A genus of spherical Gram-positive bacteria that usually occur in pairs or chains; most strains are non-motile. Most species are parasites or pathogens of animals, often occurring in the respiratory or alimentary tracts. Some species are hemolytic (i.e. they destroy red blood cells) and cause such diseases as scarlet fever and rheumatic fever.

striated muscle *See* skeletal muscle.

striped muscle *See* skeletal muscle.

stroma 1. (*Botany*) The space between the inner membrane and the thylakoids (*see* granum) inside a chloroplast. It contains a solution of the enzymes and other components that catalyze the fixation of carbon dioxide and formation of starch (*see* photosynthesis).
2. (*Botany*) A mass of fungal hyphae, sometimes including host tissue, in which fruiting bodies may be produced. An example is the compact black fruiting body of the ergot fungus, *Claviceps purpurea*.
3. (*Zoology*) A tissue that acts as a framework; for example, the connective tissue framework of the ovary or testis that surrounds the cells concerned with gamete production.

stromatolite A layered cushion-like mass of chalk formed by the actions of certain bacteria, notably CYANOBACTERIA. Communities of these organisms trap and bind lime-rich sediments. Modern stromatolite-building communities are confined to salt flats or shallow salty lagoons where bacterial predators cannot survive, but the fossil record demonstrates a much more widespread distribution in the Precambrian period. Some stromatolites date back nearly 4000 million years, making them the oldest known fossils.

suberin A mixture of long-chain fatty acids and alcohols found in the walls of cork tissue in plants. The Casparian band in roots (*see* endodermis) and some stems contain suberin.

subsidiary cell *See* accessory cell.

substrate 1. The substance upon which an enzyme acts.
2. The nonliving material upon which an organism lives or grows.

substrate-level phosphorylation The formation of ATP or a similar high-energy nucleotide through a reaction that does not involve an electron-transport chain. Two of the steps in GLYCOLYSIS are coupled to the phosphorylation of ADP to ATP, and the net yield from glycolysis is 2ATP. This is the sole means of ATP production for certain organisms performing ANAEROBIC RESPIRATION, especially fermentation. Substrate-level phosphorylation also occurs in one of the steps in the Krebs cycle, with the formation of GTP from GDP.

succinic acid A dicarboxylic acid whose succinate anion is an intermediate in the KREBS CYCLE, one of the chief metabolic pathways in living cells.

sucrose (**cane sugar**) A soluble disaccharide consisting of one glucose and one fructose unit that plays a crucial role in plant life. It is the main form in which carbon is transported within plants generally, and is also an important storage carbohydrate in certain plants, such as sugar cane and sugar beet. Sucrose is made primarily in the cytosol of photosynthesizing cells, in a reaction involving fructose 6-phosphate and UDP-glucose; the latter is an activated form of glucose containing the nucleotide uridine diphosphate (UDP). The carbon for sucrose synthesis is exported from chloroplasts in the form of glyceraldehyde 3-phosphate, in exchange for phosphate, by a phosphate–triosephosphate antiporter in the chloroplast inner membrane. The STARCH stored in chloroplasts (as leaf starch) and sucrose in the cytosol represent two interconnected pools of carbon, which fluctuate in order to balance efficient photosynthesis during daylight, with the metabolic needs of the plant through day and night. Sucrose from leaves enters the phloem to form a sugary solution that is translocated to other parts of the plant. Here it can be broken down (hydrolyzed) to fructose and glucose to enter glycolysis for use in respiration, stored temporarily in cell vacuoles, or converted to insoluble starch granules for long-term storage in plastids.

sugar (saccharide) One of a class of sweet-tasting carbohydrates that are soluble in water. Sugar molecules consist of linked carbon atoms with –OH groups attached, and either an aldehyde or ketone group. The simplest sugars are the *monosaccharides*, such as glucose and fructose, which cannot be hydrolyzed to sugars with fewer carbon atoms. They can exist in a chain form or in a ring formed by reaction of the ketone or aldehyde group with an –OH group on one of the carbons at the other end of the chain. It is possible to have a six-membered (*pyranose*) ring or a five-membered (*furanose*) ring. Monosaccharides are classified according to the number of carbon atoms: a *pentose* has five carbon atoms and a *hexose* six. Monosaccharides with aldehyde groups are *aldoses*; those with ketone groups are *ketoses*. Thus, an *aldohexose* is a hexose with an aldehyde group; a *ketopentose* is a pentose with a ketone group, etc. Two or more monosaccharide units can be linked in *disaccharides* (e.g. SUCROSE), *trisaccharides*, etc. *See also* fructose; glucose; polysaccharide.

sugar acid An acid formed from a monosaccharide by oxidation. Oxidation of the aldehyde group (CHO) of the aldose monosaccharides to a carboxyl group (COOH) gives an *aldonic acid*; oxidation of the primary alcohol group (CH_2OH) to COOH yields *uronic acid*; oxidation of both the primary alcohol and carboxyl groups gives an *aldaric acid*. The uronic acids are biologically important, being components of many polysaccharides; for example, glucuronic acid (from glucose) is a major component of gums and cell walls, while galacturonic acid (from galactose) makes up pectin. Ascorbic acid (vitamin C) is an important sugar acid found universally in plant tissues, particularly in citrus fruits.

sugar alcohol (alditol) An alcohol derived from a monosaccharide by reduction of its carbonyl group (CO) so that each carbon atom of the sugar has an alcohol group (OH). For example, glucose yields sorbitol, common in fruits, and mannose yields mannitol.

sulfur An essential element in living tissues, being contained in the amino acids cysteine and methionine and hence in nearly all proteins. Sulfur atoms are also found bound with iron in FERREDOXINS, which occur as components of electron-transport chains, especially in photosynthesis. Plants take up sulfur from the soil as the sulfate ion $SO_4{}^{2-}$. The sulfides released by decay of organic matter are oxidized to sulfur by sulfur bacteria of the genera *Chromatium* and *Chlorobium*, and further oxidized to sulfates by bacteria of the genus *Thiobacillus*. There is thus a cycling of sulfur in nature.

sulfur bacteria Filamentous autotrophic chemosynthetic bacteria that derive energy by oxidizing sulfides, other reduced sulfur compounds, or elemental sulfur, enabling them to build up carbohydrates from carbon dioxide. For example, *Beggiatoa* spp. are common in sulfur springs, where they oxidize hydrogen sulfide to elemental sulfur, and subsequently to sulfate. *See also* photosynthetic bacteria.

summation   1. The additive effect of several impulses arriving at a synapse of a nerve and/or muscle cell, when individually the impulse cannot evoke a response. The impulses either arrive simultaneously at different synapses at the same cell (*spatial summation*) or in succession at one synapse (*temporal summation*). Stimulation of the synapse elicits a graded postsynaptic potential and if the potential exceeds the threshold level, a postsynaptic impulse is triggered. Summation is one of the major mechanisms of integration in the nervous system. *Compare* facilitation.
2. The interaction of two substances with similar effects in a given system, such that the combined effect is greater than their separate effects.

Sumner, James Batcheller (1887–1995) American biochemist who discovered that enzymes can be crystallized. He was awarded the Nobel Prize for chemistry in 1946. The prize was shared with J. H. Northrop and W. M. Stanley.

supergene   A collection of closely linked genes that affect the same character and tend to behave as a single unit because crossing over between them is very rare, for example, due to inversion of part of the chromosome.

suspension culture   A method of growing free-living single cells or small clumps of cells in a liquid medium. Microorganisms or cells of plant CALLUS tissue may be grown in this way; the liquid medium is agitated to keep the cells in suspension. Individual cells from plant suspension cultures can be isolated and grown into entire plants, the process being regulated by hormone treatment. *See also* cell culture.

Sutherland, Earl Wilbur Jr (1915–74) American biochemist who worked on the mechanisms of hormones and discovered cyclic AMP. He was awarded the Nobel Prize for physiology or medicine in 1971.

SV40 (**simian virus 40**)   A DNA-containing virus originally isolated from kidney cell cultures derived from monkeys used to prepare polio vaccine. It has a small genome, and relies heavily on host-cell enzymes to replicate its DNA. The SV40 chromosome can be replicated using an *in vitro* model system consisting of various eukaryotic cell components. This yields useful information about the replication process.

Svedberg, Theodor (1884–1971) Swedish colloid chemist noted for his invention (1923) of the ultracentrifuge. He was awarded the Nobel Prize for chemistry in 1926.

sympathetic nervous system (**thoracolumbar nervous system**)   One of the two divisions of the AUTONOMIC NERVOUS SYSTEM, which supplies motor nerves to the smooth muscles of internal organs and to heart muscle. The endings of sympathetic nerve fibers release mainly norepinephrine, which increases heart rate and breathing rate, raises blood pressure, and slows digestive processes, thereby antagonizing the effects of the PARASYMPATHETIC NERVOUS SYSTEM.

symplast   The living system of interconnected protoplasts extending through a plant body. Cytoplasmic connections between cells are made possible by the plasmodesmata. The symplast pathway is an important transport route through the plant. *Compare* apoplast.

symporter   A type of COTRANSPORTER that couples the transport of a substance across a cell membrane to the movement of ions in the same direction. The substance is transported 'uphill' against its concentration gradient, while the ions move 'downhill' along their concentration gradient. Hence it is a form of active transport, in that the energy for the uphill transport relies on the cell maintaining the ionic concentration gradient. Several types of symporter exploit the concentration gradient of sodium ions (Na^+) across the plasma membrane to import glucose and amino acids into cells. For example, the plasma membrane of kidney tubule cells has a Na^+/glucose symporter that couples the import of each glucose molecule to the movement of two Na^+ ions. *Compare* antiporter.

synapse The junction between two neurons (nerve cells), between a neuron and a muscle cell, or between two muscle cells, across which impulses can be transmitted. There are two main types: *chemical synapses*, which are most common and relay impulses using chemical messengers (NEUROTRANSMITTERS); and *electrical synapses*, which occur between certain neurons in the central nervous system and between cardiac muscle cells.

Chemical synapses are formed between the knoblike axon endings of one neuron and the dendrites or cell body of another. One neuron may have many synapses with other neurons. Each synapse consists of adjacent specialized regions in the plasma membranes of both neurons, separated by a narrow gap (synaptic cleft), typically 20–30 nm wide in vertebrate synapses. A nerve impulse arriving at the axon ending of the presynaptic cell causes small vesicles to fuse with the presynaptic membrane and release neurotransmitter. This diffuses across the cleft and binds to receptors in the plasma membrane of the postsynaptic cell, which triggers the temporary opening of various ion channels in the postsynaptic membrane. Depending on the ion channels involved, this may either start a nerve impulse in the postsynaptic cell (excitation) or prevent impulses from other neurons being transmitted (inhibition). Most chemical synapses will only transmit nerve impulses in one direction.

In electrical synapses, adjoining cells are directly connected by numerous gap junctions, and impulses are transmitted between cells by the flow of ions through channels created by these junctions. Hence, transmission is virtually instantaneous, compared with the typical delay of 0.5ms in chemical synapses. *See* facilitation; summation.

synapsis (**pairing**) The association of homologous chromosomes during the prophase stage of meiosis I. Homologous chromosomes pair point to point so that corresponding regions lie in contact, enabling CROSSING-OVER and recombination of genetic material between chromatids during the ensuing metaphase. Synapsis involves the formation of a SYNAPTONEMAL COMPLEX.

synaptonemal complex A ribbonlike proteinaceous structure that binds together the members of each pair of homologous chromosomes during prophase I of meiosis (*see* synapsis). Assembly begins at the telomeres, which are bound to the nuclear envelope by a synaptonemal attachment plate, and proceeds along the chromosomes rather like a zipper, aligning corresponding regions of the two chromosomes. The complex subsequently breaks down to enable CROSSING OVER and homologous recombination between sister chromatids. The precise details of structure and mechanism of the synaptonemal complex remain uncertain.

synchronous culture A culture of cells in which all the individuals are at approximately the same point in the cell cycle. Cells can be synchronized by a variety of means, e.g. temperature, shock, or drugs. Such cultures are of great value in physiological and biochemical investigations. *See* cell culture.

syncytium (*pl.* **syncytia**) An area of animal cytoplasm containing many nuclei, the whole being bounded by a continuous plasma membrane. This gives rise to a multinucleate condition. The term may be applied to an area of cytoplasm partially divided by membranes into discrete cells but with extensive cytoplasmic continuity. Such structures are to be found in striped and cardiac muscle, insect eggs, and some protists. *Compare* coenocyte; symplast.

synergid cells Two haploid cells located near the egg cell at the micropylar end of the embryo sac, in flowering plants. They do not participate in the fertilization process and abort soon afterwards.

synergism 1. The interaction of two substances, for example, drugs or hormones, such that the effect produced is greater than the sum of their separate effects. *Compare* antagonism, summation. **2.** The coordinated action of muscles to produce a particular movement. *Compare* antagonism.

syngamy   *See* fertilization.

Synge, Richard Laurence Millington (1914–94)   British biochemist and peptide chemist. He was awarded the Nobel Prize for chemistry in 1952 jointly with A. J. P. Martin for their invention of partition chromatography.

syngraft   *See* graft.

Système International d'Unités   *See* SI units.

Szent-Györgyi, Albert (1893–1986) Hungarian-born American biochemist distinguished for his work in a variety of fields, including the isolation of ascorbic acid, the role of dicarboxylic acids in oxidative metabolism, and the biochemistry of muscle contraction. He was awarded the Nobel Prize for physiology or medicine in 1937.

T

tactic movement   *See* taxis.

tandemly repeated array   A series of identical copies of a single gene arranged head to tail along the same chromosome. The genes encoding ribosomal RNAs (rRNAs), transfer RNAs (tRNAs), and histone proteins are often repeated in this manner, sometimes several hundred times or more in each array. This ensures that the genes' products can be supplied in quantities sufficient for the needs of the cell, especially during the repeated cell divisions and rapid growth of early embryonic development. *See also* repetitive DNA.

taste bud   A small bulblike group of chemical receptor cells in vertebrates that is responsible for the sense of taste. In terrestrial vertebrates, taste buds are usually embedded in small projections (papillae) of the epithelium of the throat and mouth, especially the tongue. Chemical substances in solution stimulate the cells to send nerve impulses to the brain for interpretation as taste. Humans are considered to have four kinds of taste buds, which distinguish sweet, sour, salt, and bitter chemicals, using different cellular mechanisms. For example, salt 'receptors' are ungated sodium channels, which allow sodium ions (Na^+) to enter the cell when the extracellular concentration of Na^+ (e.g. due to sodium chloride) rises. This causes depolarization of the plasma membrane, triggering a sensory nerve impulse.

TATA box   *See* promoter.

Tatum, Edward Lawrie (1909–75) American microbiologist. He was awarded the Nobel Prize for physiology or medicine in 1958 jointly with G. W. Beadle for their discovery that genes act by regulating definite chemical events. The prize was shared with J. Lederberg.

taxis (tactic movement)   Movement of an entire cell or organism (i.e. locomotion) in response to an external stimulus, in which the direction of movement is dictated by the direction of the stimulus. Movement towards the stimulus is *positive taxis* and away from the stimulus is *negative taxis*. It is achieved by locomotory appendages, such as cilia and flagella, ameboid movement, etc. *See* aerotaxis; chemotaxis; phototaxis. *Compare* tropism.

TCA cycle   *See* Krebs cycle.

T-cell (T-lymphocyte)   Any of a class of lymphocytes involved primarily in cell-mediated immunity. T-cells originate from stem cells in the bone marrow (like B-CELLS), but during embryological development they migrate to the thymus where they mature and differentiate into thymocytes. These then travel in the bloodstream to reside in lymph nodes as mature T-cells. There are several types of T-cell, which act collaboratively in identifying and destroying virus-infected body cells, cancer cells, and foreign cells (e.g. tissue grafts), and in regulating various aspects of the immune response. All of them possess on their surface T-cell receptors, which are structurally related to the antibody molecules secreted by B-cells, and bind to specific antigens on the surface of target cells. Unlike antibodies, they do not bind soluble antigens. However, a T-cell will generally only recognize its specific antigen if the antigen is bound to a marker protein (an MHC protein; *see* major histocompatibility complex) on the surface of a cell. Binding of the

T-cell receptor to the MHC–antigen complex acts as the stimulus for the T-cell's rapid division, to form a clone of sensitized T-cells.

There are four main subsets of T-cells. *Cytotoxic T-cells* (T_C cells) recognize and bind to cells that display foreign antigens combined with class I MHC markers on their surface. The T_C cell then lyses the target cell by releasing a protein called perforin. This makes a hole in the plasma membrane of the infected cell, causing it to burst. *Helper T-cells* (T_H cells) release substances, called cytokines, that stimulate the growth and maturation of both T_C cells and B-cells. T_H cells bind to antigen–class II MHC complexes on the surface of *antigen-presenting cells*, such as macrophages and B-cells. Antigen-presenting cells take in foreign proteins, process them, and present them on their surface, along with a class II protein, for T_H cells. The latter then stimulate B-cells to make antibodies against the antigen concerned. *Suppressor T-cells* (T_S cells) suppress the activities of T_H cells and thus restrict antibody production by B-cells. *Delayed-type hypersensitivity T-cells* (T_D cells) are responsible for activating macrophages and other phagocytes in a nonspecific manner. The various subsets can be distinguished by the CD MARKERS on their surface. *See* AIDS; killer cell.

telomere The tip of a eukaryotic chromosome. It comprises numerous tandemly repeated noncoding short sequences of DNA For example, in human chromosomes the simple base sequence TTAGGG occurs hundreds or thousands of times at each telomere. Because of the discontinuous replication of the lagging strand of DNA (*see* DNA replication) each replication cycle results in the loss of nucleotides from the tip of the DNA molecule. Hence, although telomeres shorten progressively with age, their presence protects against the loss of coding regions. However, it is suggested that excessive telomere shortening might be the cause of certain age-related degenerative changes (*see* senescence). In humans, the germ-line cells have a *telomerase* enzyme that prevents this pro-

gressive shortening. The enzyme contains a simple RNA template, which it uses to add nucleotides to the 3′ end of the leading DNA strand. This extended strand then acts as the template for completion of the lagging strand.

telomere theory of aging *See* senescence.

telophase The final stage in mitosis and meiosis before cells enter interphase. During this stage chromosomes uncoil and disperse, the nuclear spindle dismantles as its component microtubules depolymerize, and new nuclear envelopes form around each daughter nucleus. The cytoplasm also divides during this phase (*see* cytokinesis).

Temin, Howard Martin (1934–94) American virologist. He was awarded the Nobel Prize for physiology or medicine in 1975 jointly with D. Baltimore and R. Dulbecco for their discoveries concerning the interaction between tumor viruses and the genetic material of the cell.

temperate phage A bacteriophage that becomes integrated into the bacterial DNA and multiplies with it, rather than replicating independently and causing lysis of the bacterium. A prime example is LAMBDA PHAGE, which infects *E. coli*. *See* lysogeny. *Compare* virulent phage.

teratogen Any environmental factor that causes physical defects (teratomas) in a fetus. Tetratogens include various drugs (e.g. thalidomide), infections (e.g. German measles), and irradiation. Teratogens interfere with essential growth mechanisms, causing arrested or distorted growth; the human fetus, for example, is particularly sensitive during the first two months, when rudimentary growth patterns are being established. In later life, when growth patterns are well established, teratogens have no effect.

terminalization The movement of chiasmata (*see* chiasma) to the end of the bivalent arms, a process that may occur during late prophase I of MEIOSIS. The chiasmata can slip off the ends of the bivalents, and thus chiasma frequency may be reduced by terminalization.

tertiary structure The fully folded three-dimensional arrangement of a polypeptide chain. It is stabilized by electrostatic attractions, hydrogen bonds, and in some cases covalent disulfide bonds between certain amino acids. *See also* domain.

tetrad 1. A group of four spores formed as a result of meiosis in a spore mother cell. 2. In meiosis, the association of four homologous chromatids seen during the pachytene stage of prophase.

tetraploid A cell or organism containing four times the haploid number of chromosomes. Tetraploid organisms may arise by the fusion of two diploid gametes that have resulted from the nondisjunction of chromosomes at meiosis. Tetraploid cell lines can also arise in the embryo following fertilization. A cell might go through two successive S phases during one round of the CELL CYCLE, or a cell in metaphase might revert directly to interphase without dividing. Resulting organisms are typically mosaics of tetraploid and diploid cells. *See also* polyploid.

Theorell, Axel Hugo Theodor (1903–82) Swedish biochemist. He was awarded the Nobel Prize for physiology or medicine in 1955 for his discoveries concerning the nature and mode of action of oxidation enzymes.

thermophilic Describing microorganisms, mostly prokaryotes, that require high temperatures (around 60°C) for growth. Some eubacteria are thermophilic, thriving best in hot springs, industrial hot-water outflows, compost heaps, and similar hot environments. *Hyperthermophilic* organisms have even higher temperature optima, typically over 80°C. These are mainly archaeans, living in boiling hot springs and around undersea thermal vents, sometimes at temperatures well above 100°C. The nucleic acids, proteins, and other cell constituents of thermophiles have various biochemical modifications that prevent their denaturation at such high temperatures, and so permit survival. The thermophilic bacterium *Thermus aquaticus* is the source of the Taq polymerase enzyme used in the polymerase chain reaction. *Compare* psychrophilic.

thiamin (vitamin B_1) One of the water-soluble B-group of vitamins. Thiamin, in the form of thiamin diphosphate, is the coenzyme for the decarboxylation of acids such as pyruvic acid. *See also* vitamin B complex.

thin-layer chromatography A chromatographic method in which a glass plate is covered with a thin layer of inert absorbent material (e.g. cellulose or silica gel) and the materials to be analyzed are spotted near the lower edge of the plate. The base of the plate is then placed in a solvent, which rises up the plate by capillary action, separating the constituents of the mixtures. Two-dimensional methods can also be employed.

thoracolumbar nervous system *See* sympathetic nervous.

threonine *See* amino acids.

threshold The value that must be reached or exceeded for an effect to happen or become apparent. For example, in embryological development, a signal molecule might need to reach a certain threshold concentration in order to induce a particular developmental response in a cell population.

thrombocyte *See* platelet.

thylakoid An elongated flattened fluid-filled sac forming the basic unit of the photosynthetic membrane system in CHLOROPLASTS and certain photosynthetic bacteria. The thylakoids are formed from a continuous thylakoid membrane, and enclose a series of interconnected spaces, the thylakoid lumen. Bound to the membrane are the light-absorbing pigments (e.g. chlorophylls), proteins, and coenzymes, including components of the electron transport chain. In bacteria, thylakoids arise from invaginations of the plasma membrane.

thymidine   The nucleoside formed when thymine is linked to D-ribose by a β-glycosidic bond.

thymine   A nitrogenous base found in DNA. It has a pyrimidine ring structure. *See illustration at* DNA.

thyroid hormone   Either of two hormones produced from thyroglobulin in the thyroid gland – thyroxine and triiodothyronine. *See* thyrotropin.

thyroid-stimulating hormone   *See* thyrotropin.

thyrotropin   (thyroid-stimulating hormone; TSH)   A hormone, produced by the anterior pituitary gland, that stimulates the thyroid gland to release thyroxine. Thyrotropin causes cells lining the thyroid follicles to take up the precursor thyroglobulin. This is processed by proteolysis within the cells, which release thyroxine into the blood. The level of thyroxine controls thyrotropin release by a negative feedback mechanism.

thyroxine (thyroid hormone; T_4)   An iodine-containing polypeptide hormone that is secreted by the thyroid gland and is essential for normal cell metabolism. Its many effects include increasing oxygen consumption and energy production.

tight junction (zonula occludens)   A type of cell junction, found especially in epithelial cells, that holds adjacent cells tightly together with no intercellular space. Tight junctions form a barrier, preventing material at the epithelial surface from penetrating between the cells, particularly at sites of absorption, such as the gut and kidney. They also stop membrane proteins and glycolipids from diffusing through the plane of the plasma membrane between the 'top' (apical) surface of the cell, and the sides and basal surfaces. The junctions consist of slender bands of proteins, particularly occludin and claudin, which form a series of interlocking ridges and grooves in the plasma membrane.

Tiselius, Arne Wilhelm Kaurin (1902–71)   Swedish chemist who worked on serum proteins. He was awarded the Nobel Prize for chemistry in 1948 for his research on electrophoresis and adsorption analysis.

tissue   A group of cells that is specialized for a particular function. Examples are connective tissue, muscular tissue, and the mesophyll of leaves. Several different tissues are often incorporated in the structure of each organ of the body. In plants, tissue cells communicate via cytoplasmic connections called plasmodesmata, whereas animal cells have a range of CELL JUNCTIONS. In both types of organisms, the cells are embedded in an EXTRACELLULAR MATRIX, which plays a crucial role in the development, organization, and cohesion of tissues. Notable features of the matrix are the CELL WALL in plants and CELL ADHESION MOLECULES in animals.

tissue culture   The growth of tissues or organs in suitable media *in vitro*. Such media must normally be sterile, correctly pH balanced, and contain all the necessary micro- and macronutrients, carbohydrates, vitamins, and growth factors required by the component cells. Studies of such cultures have shed light on physiological processes that would be difficult to follow in the living organism. For example, the cytokinins were discovered through work on tobacco-pith tissue culture.

Plant tissue-culture techniques have important practical applications. Large-scale multiplication of certain plants is done in various ways, for example by MICROPROPAGATION of excised meristems, or the culture of undifferentiated CALLUS tissue. In these cases, tissue is induced to develop roots and shoots by varying the concentrations of certain plant hormones – particularly auxins and cytokinins – in the culture medium. New plantlets can also be produced from SOMATIC EMBRYOS, formed in cell cultures under special conditions.

Tissue engineering is a relatively new area of biomedical technology concerned with producing synthetic or semisynthetic tissues that resemble natural body tissues. Implants can be constructed using different combinations of cultured cells, artificial biopolymers, and growth factors. They are designed to supplement or replace natural tissue grafts in surgery. Synthetic skin and cartilage are already in clinical use, and other tissues are being developed. In future, EMBRYONIC STEM CELLS might be cultured to create a wide range of tissues and organs.

tissue engineering   *See* tissue culture.

titin   *See* sarcomere.

T-lymphocyte   *See* T-cell.

tocopherol   *See* vitamin E.

Todd, Alexander Robertus, Lord Todd (1907–97)   British organic chemist. He was awarded the Nobel Prize for chemistry in 1957 for his work on nucleotides and nucleotide coenzymes.

tonoplast   The membrane that surrounds the large central vacuole of plant cells.

totipotent   Describing a cell with the ability to form all the types of tissues that constitute the mature organism. In plants, many, if not all, living cells are totipotent, even if the cells have completely differentiated; they can be induced to divide and produce relatively undifferentiated progeny cells if given the appropriate balance of nutrients and hormones (e.g. *see* callus). Totipotency demonstrates that each cell retains the full genetic potential of the species. Differentiated animal cells generally cannot be dedifferentiated in this way, and even embryonic stem cells may not be truly totipotent. However, the nucleus of a differentiated animal cell does retain genetic information for making all other cell types, as evidenced by the technique of NUCLEAR TRANSFER. *Compare* pluripotent.

toxin   A chemical produced by a pathogen (e.g. bacteria, fungi) that causes damage to a host cell in very low concentrations. Toxins are often similar to the enzymes of the host and interfere with the appropriate enzyme systems. *See also* endotoxins; exotoxins.

T phage   Any of various virulent phages that infect cells of *E. coli* and related bacteria. Their genome consists of a single molecule of double-stranded DNA, of variable size. The *T-even phages*, designated T2, T4, and T6, have comparatively large genomes, and are structurally more complex than the *T-odd phages*, designated T1, T3, T5, and T7. Much basic knowledge of virus replication, and protein synthesis, was obtained from studies on this group.

trace element   An element required in trace amounts (a few parts per million of food intake) by an organism for health.

tracheid   A cell that serves as one type of conducting element in the xylem tissue of plants. Tracheids are elongated cells, dead at maturity, with heavily lignified cell walls and oblique end walls. The only connection between adjacent tracheids is through paired PITS. Tracheids form the only xylem conducting tissue of ferns and most conifers. *Compare* vessel.

transcription   The process in living cells whereby RNA is synthesized according to the template embodied in the base sequence of DNA, thereby converting the cell's genetic information into a coded message (MESSENGER RNA, mRNA) for the assembly of proteins, or into the RNA components required for protein synthesis – ribosomal RNA (rRNA) and transfer RNA (tRNA). The term is also applied to the formation of single-stranded COMPLEMENTARY DNA (cDNA) from an RNA template. Details of DNA transcription differ between prokaryote and eukaryote cells, but essentially it involves three stages: initiation, elongation, and termination.

In prokaryotes, *initiation* involves binding of the enzyme RNA POLYMERASE to a promoter site in the DNA, upstream of the

IMPORTANT TRACE ELEMENTS IN PLANTS AND ANIMALS		
Trace element	*Compounds containing*	*Metabolic role*
copper	cytochrome oxidase	oxygen acceptor in respiration
	plastocyanin	electron carrier in photosynthesis
	hemocyanin	respiratory pigment in some marine invertebrates
	tyrosinase	melanin production – absence causes albinism
zinc	alcohol dehydrogenase	anaerobic respiration in plants – converts acetaldehyde to ethanol
	carbonic anhydrase	CO_2 transport in vertebrate blood
	carboxy peptidase	hydrolysis of peptide bonds
cobalt	vitamin B_{12}	red blood cell manufacture – absence causes pernicious anaemia
molybdenum	nitrate reductase	reduction of nitrate to nitrite in roots
	a nitrogen-fixing enzyme	nitrogen fixation
manganese	enzyme cofactor	oxidation of fatty acids
		bone development
fluorine	associated with calcium	tooth enamel and skeletons
boron		mobilization of food in plants?

protein-coding region of a gene cluster (*see* operon). This binding is regulated by various molecules, including repressors and activators. The polymerase unwinds the two DNA strands in the promoter region, and starts transcription. During *elongation*, RNA polymerase moves along one of the DNA strands – the template strand (or anticoding strand, since the code is carried by the complementary base sequence of the RNA) – in the $3'{\to}5'$ direction, and nucleotides are assembled to form an RNA molecule with a complementary base sequence that is antiparallel to the template strand (i.e. running in the $5'{\to}3'$ direction). The enzyme adds nucleotides in succession to the growing $3'$ end of the RNA chain, using nucleoside triphosphates as substrates. *Termination* occurs either by the intervention of a protein factor, called rho, which binds to a specific site on the RNA chain, or directly, by the formation of a hairpin loop in the RNA due to intrachain complementary base pairing. Either mechanism signals release of the RNA molecule and ends transcription.

Eukaryotes have three different types of RNA polymerase (I–III); each is responsible for transcribing different classes of genes, and forms different types of RNAs. In each case, they bind to specific start sites of genes, and are controlled by a host of different TRANSCRIPTION FACTORS, many of which bind to distant sites in the DNA, and can act variously as activators or repressors of transcription. RNA polymerase II, which transcribes protein-coding genes into mRNA, forms part of a large complex, along with 60–70 polypeptides, called the INITIATION COMPLEX. The different eukaryote RNA polymerases employ different termination mechanisms. For most protein-coding genes, termination occurs when the RNA transcript contains a sequence that specifies its own cleavage.

In eukaryotic genes the functional message is contained within discontinuous segments of the DNA strand (exons), interrupted by nonfunctional segments (introns). Initially, both exons and introns are transcribed to form a primary transcript (pre-rRNA or pre-mRNA), which constitutes HETEROGENEOUS NUCLEAR RNA (hnRNA). Subsequently the noncoding intron sequences are spliced out to form the fully functional mature mRNA transcript, which then leaves the nucleus to direct protein assembly in the cytoplasm, in the process called TRANSLATION.

transcription factor Any of numerous proteins that regulate gene activity and enable gene TRANSCRIPTION in eukaryotes. Many exert their effects by binding to specific sites on the DNA molecule. *General transcription factors* take part in the transcription of many different genes. They bind to the PROMOTER site near the start of the coding region, and guide the positioning of RNA polymerase in readiness for transcription. *Regulatory transcription factors* are usually specific for one or a few genes. They determine which genes are switched 'on' and which are turned 'off', by binding to regulatory sites, sometimes located hundreds or thousands of base pairs distant from the coding region and promoter. They are crucial to tissue-specific expression of genes, and during embryonic development they orchestrate the subtle patterns of genetic expression that lead to tissue differentiation and organ formation. Regulatory transcription factors can be divided into two broad categories: activators bind to sites called enhancers, thereby increasing transcription; repressors bind to sites called silencers, and decrease transcription. A gene can have numerous such regulatory sites, located upstream or downstream of the coding region, or within introns. For example, the *decapentaplegic* gene of the fruit fly *Drosophila*, which encodes a protein that acts as a MORPHOGEN in development, has perhaps hundreds of enhancers, each responding to a different activator.

transduction The transfer of part of the DNA of one bacterium to another by a bacteriophage. The process does occur naturally but is mainly known as a technique in recombinant DNA technology, and has been used in mapping the bacterial chromosome. There are two types: in *generalized transduction*, the transducing phage can transfer any part of the host bacterium's chromosome, whereas in *specialized transduction*, only particular parts of the host genome are transferred.

transfer cell A specialized type of plant cell in which the cell wall forms protuberances into the cell, thus increasing the surface area of the wall and plasma membrane. They are active cells, containing many mitochondria, and are concerned with short-distance transport of solutes. They are common in many situations, for example, as gland cells and epidermal cells, and in xylem and phloem parenchyma, where they are concerned with active loading and unloading of vessels and sieve tubes.

transfer RNA (tRNA) A type of RNA that participates in protein synthesis in living cells. It attaches to a particular amino acid – forming an *amino-acyl tRNA* – and imports the amino acid to the site of polypeptide assembly at the RIBOSOME when the appropriate codon is reached during TRANSLATION of messenger RNA (mRNA). Each tRNA molecule consists of 70–80 nucleotides; some regions of the molecule undergo base pairing and form a double helix, while in others the two strands separate to form loops. When flattened out the tRNA molecule has a characteristic 'cloverleaf' shape with three loops; the base of the 'leaf' carries the amino-acid binding site, and the middle loop contains the anticodon, whose base triplet pairs with the complementary codon in the mRNA molecule. Bacterial cells typically contain 30–40 different tRNAs, whereas in plant and animal cells the number is generally 50–100. Hence, some cells have fewer tRNAs than the 61 possible amino acid-specifying codons in the genetic code. This is because some tRNAs can recognize two or more different codons for their particular amino acid due to nonstandard pairing between the third bases of the codon and anticodon – a phenomenon known as 'wobble'. In eukaryotic cells, moreover, some different tRNAs will bind the same amino acid.

The correct amino acid is attached to a tRNA molecule by an enzyme called an *aminoacyl-tRNA ligase* (or aminoacyl-tRNA synthetase). There are 20 of these, one for each type of amino acid. This attachment also involves the transfer of a high-energy bond from ATP to the amino acid, which provides the energy for peptide bond formation during translation.

transformation   1. A permanent genetic recombination in a cell, in which a DNA fragment is incorporated into the chromosome of the cell. This occurs naturally among bacteria, which can take up bits of DNA through their cell wall, and integrate them into their chromosome. It can be demonstrated by growing bacteria in the presence of dead cells, culture filtrates, or extracts of related strains. The bacteria acquire genetic characters of these strains. Special treatment is usually required to cause transformation of eukaryotic cells; for example, removal of the cell wall of fungal cells. This is done for certain recombinant DNA techniques.
2. The conversion of normal cells in tissue culture to cells having properties of tumor cells. The change is permanent and transformed cells are often malignant. It may be induced by certain viruses or occur spontaneously.

transforming growth factor (TGF)   1. Transforming growth factor a (TGFa): a glycoprotein produced by macrophages and certain other tissue cells. It induces epithelial development and is a member of the epidermal growth factor family of proteins.
2. Transforming growth factor b (TGFb): any of a large family of intercellular signal proteins that play key roles in the development of vertebrates and invertebrates. They bind to cell-surface receptors, thereby activating intracellular transcription factors and leading to transcription of target genes. One member of the family is the Dpp protein, which controls dorsoventral (upper surface to lower surface) patterning in the early embryo of the fruit fly *Drosophila*. This acts as a MORPHOGEN, inducing different cell fates at different concentrations. Homologous proteins, called BMP2 and BMP4, perform a similar role in the frog embryo.

transgenic   Describing organisms that contain foreign genetic material, especially ones containing genes (*transgenes*) that are stably incorporated into the host organism's genome and transmitted to their offspring. Genetic engineering has created a wide range of transgenic animals, plants, and other organisms, for both experimental and commercial purposes. In research, transgenic organisms are used to investigate many aspects of gene function and control, for example in development and disease. Many genetically modified (GM) commercial crop plants are transgenic. Resistance of various crops (e.g. tomato, potato, green pepper) to certain viral diseases can be engineered by introducing the gene for the viral coat protein into the plant genome. Human genes for particular serum proteins, used as drugs, have been inserted in dairy cows and goats so that the transgenic animals secrete the proteins in their milk. *See* genetic engineering; recombinant DNA. *See also* gene knockout.

translation   The process whereby the genetic code of messenger RNA (mRNA) is deciphered by the machinery of a cell to make proteins. Molecules of mRNA are transcripts of the cell's genes, created by the process of TRANSCRIPTION, and relay the genetic information to the sites of protein synthesis, the RIBOSOMES. In eukaryotes these are located in the cytoplasm, so the mRNA is exported from the nucleus via NUCLEAR PORES. The first stage in translation is *initiation*, in which the two subunits of the ribosome assemble and attach to the mRNA molecule at the ribosome-binding site near the *start codon* (AUG), which signals the beginning of the message. This requires energy, supplied by the hydrolysis of GTP, and involves various proteins called INITIATION FACTORS, and the initiator transfer RNA (tRNA), which always carries the amino acid *N*-formyl methionine. In bacteria the ribosome-binding site is a sequence of eight nucleotides called the SHINE–DALGARNO SEQUENCE.

The next stage is *elongation*, in which the peptide chain is built up from its component amino acids. It requires the participation of a set of proteins called ELONGATION FACTORS. tRNA molecules successively occupy two sites – the A site and the P site – on the larger ribosome subunit in a sequence determined by consecutive codons on the mRNA. As each pair of tRNAs occupies the ribosomal sites, their amino acids are joined together by a peptide bond. As the ribosome moves along the mRNA (in the $5'{\to}3'$ direction) to the next codon, the next amino-acyl-tRNA enters the entry site (A site), while the previous occupant moves to the P site, and so on, leading to elongation of the peptide chain. Having delivered its amino acid, the depleted tRNA is released from the P site (which is then occupied by the tRNA with the growing chain) and moves to a third site on the ribosome, the exit site, or E site, from where it is discharged. This process continues until the ribosome encounters a *stop codon*. These are recognized by proteins called RELEASE FACTORS, which bind to the ribosomal A site. The polypeptide is released from the P site, and the ribosome dissociates into its two subunits, marking the *termination* of translation.

Following its release, the polypeptide undergoes various changes, such as the removal or addition of chemical groups, or even cleavage into two. This *post-translational modification* produces the fully functional protein. Folding of the protein is assisted by a class of molecules called CHAPERONES.

translocation 1. *See* chromosome mutation.
2. The movement of food materials through a plant. In vascular plants, the PHLOEM serves to translocate such substances, primarily organic carbon compounds formed by photosynthesis. These include carbohydrates (especially sugars, notably sucrose), amino acids, and other organic compounds, which are moved both upwards and downwards in the phloem.

translocon A complex of proteins within the membrane of the ENDOPLASMIC RETICULUM that permits the entry of newly synthesized polypeptide chains into the lumen. The polypeptide passes directly from a ribosome into the pore formed by the translocon.

transplantation The transfer of a tissue or organ from one part of an animal to another part or from one individual to another. *See also* graft.

transport protein A protein that enables the passage of ions or molecules through a cell membrane. Transport proteins are generally large and span the width of the membrane. So, for example, one located in the plasma membrane will have a cytosolic face, a membrane-spanning region, and an exterior face, each of which can contain several distinct functional domains. They fall into several categories. *Channel proteins* enclose a passageway through which water molecules or specific ions flow rapidly down their concentration gradients. A region of the protein commonly acts like a gate, opening and closing the channel in response to chemical or electrical signals (*see* ligand-gated ion channel, voltage-gated ion channel). Other transport proteins have specific sites to which their target substances bind. The protein then changes its conformation (three-dimensional structure) so that the bound ion or molecule is translocated across the membrane, to be released on the opposite face. Proteins that work in this way include the UNIPORTERS, which transport only one type of substance, and the COTRANSPORTERS, which move two or more different substances. The transport of substances against their concentration gradient involves energy-demanding ACTIVE TRANSPORT. Some active transport proteins are ATP-powered pumps (e.g. *see* Na^+/K^+ ATPASE), whereas others exploit existing concentration gradients.

transposition *See* transposon.

transposon A segment of an organism's DNA that can insert at various sites in the genome – a process called *transposition* – either by physically moving from place to place, or by producing a DNA copy that inserts elsewhere. Transposons represent one of the two main categories of MOBILE ELEMENTS occurring in DNA. The most rudimentary, called *insertion sequences* (IS), occur in bacteria, these are typically 1–2 kb long and comprise just one or sometimes two protein-coding genes flanked at either end by INVERTED REPEATS. The gene encodes a transposase enzyme, which catalyzes the cleavage of the insertion sequence from the chromosome and its insertion at another site. Other insertion sequences form a DNA replica of themselves, which then inserts at a new site. *Bacterial transposons* are larger than insertion sequences; besides the IS components they also contain antibiotic-resistance genes. Such transposons are used in bacterial genetics to insert within and disrupt particular genes. The antibiotic resistance conferred by the transposon serves to identify cells containing it.

The first mobile elements to be discovered were *eukaryotic transposons*, whose existence was postulated by Barbara McClintock in the 1940s, based on observations of kernel color in corn (maize). She proposed the existence of genetic entities that could enter a gene, so creating mutants in which kernel color was changed from purple to white. The entity could subsequently leave the gene, causing reversion to the wild type. These entities were later identified as being similar to bacterial insertion sequences and encoding transposase. However, transposons are rare in eukaryotes, compared to the much more common RETROTRANSPOSONS.

triacylglycerol   *See* triglyceride.

tricarboxylic acid cycle   *See* Krebs cycle.

triglyceride (**triacylglycerol**)   An ester of glycerol in which all the –OH groups are esterified; the acyl groups may be the same or different. Many LIPIDS are triglycerides in which the parent acid(s) of the acyl group(s) are long-chain fatty acids. In ani-mals the triglycerides are more frequently saturated and have higher melting points than the triglycerides of plant origin, which are generally unsaturated. *See also* carboxylic acid; glyceride.

triploblastic   Describing an animal whose body tissues derive from three embryonic GERM LAYERS: essentially, ectoderm on the outside, endoderm lining the gut and allied structures, and mesoderm between these two layers. Each of the three layers gives rise to a particular set of tissues and organs. Most animals are triploblastic; exceptions are the sponges and cnidarians.

triploid   A cell or organism containing three times the haploid number of chromosomes. Triploid organisms can arise by the fusion of a haploid gamete with a diploid gamete that has resulted from the non disjunction of chromosomes at meiosis. Alternatively, a triploid results if two sperm fertilize one egg. In humans, triploid embryos have multiple defects and an abnormal placenta; any surviving to birth show severe disability. Triploids are usually sterile because one set of chromosomes remains unpaired at meiosis, which disrupts gamete formation. In flowering plants the endosperm tissue is usually triploid, resulting from the fusion of one of the pollen nuclei with the two polar nuclei.

trisaccharide   *See* sugar.

trisomy   *See* aneuploidy.

tRNA   *See* transfer RNA.

trophoblast   The cells of the outer wall of the mammalian BLASTOCYST. It is the part of the blastocyst that is attached to the wall of the uterus and it forms the part of the early placenta that is in closest contact with the maternal tissues.

tropic movement   *See* tropism.

tropism (**tropic movement**)   A growth movement of part of a plant whose direction is related to an external stimulus. Tropisms are named according to the stimulus.

The organ is said to exhibit a positive or negative tropic response, depending on whether it grows toward or away from the stimulus respectively; for example, shoots are positively phototropic but negatively gravitropic. Growth straight toward or away from the stimulus (0° and 180° orientation respectively) is called *orthotropism*. Primary roots and shoots are orthotropic to light and gravity. By contrast, growth at any other angle to the direction of the stimulus as by branches or lateral roots is called *plagiotropism*. *See* phototropism; gravitropism. *See also* taxis.

tropomyosin A protein found in skeletal muscle that regulates contraction of the muscle filaments. It forms a rodlike molecule that, in the resting stage, blocks the myosin binding sites on the actin filaments. Another protein, *troponin*, is associated with tropomyosin. A nerve impulse triggers contraction by causing an influx of calcium ions from the sarcoplasmic reticulum. The troponin molecules bind calcium ions and in so doing cause the tropomyosin molecules to shift position slightly, so exposing the myosin binding sites and allowing contraction to proceed. Tropomyosin, but not troponin, is also found in smooth muscle. *See* sarcomere.

troponin *See* tropomyosin.

tryptophan *See* amino acids.

TSH (**thyroid-stimulating hormone**) *See* thyrotropin.

tubulin *See* microtubule. suppressor gene

tumor necrosis factor *See* lymphokine.

tumor suppressor gene Any of various genes whose products in some way regulate DNA replication and progression through the cell cycle, or promote the repair of damaged DNA. Mutations in such genes cause loss of function in cell-cycle control mechanisms, or increase the accumulation of genetic defects, which can lead to the transformation of a cell into a cancerous one. Hence, a steady accumulation of mutations in tumor suppressor genes throughout life can contribute to the eventual development of cancer. Also, individuals who inherit a mutant allele of any such gene are at increased risk of developing specific tumors early in life. For example, children who inherit a defective allele of the *RB* gene are likely to develop retinal tumors in both eyes – a condition called hereditary retinoblastoma. This is because a subsequent mutation of the other normal allele in just a single cell will give rise to a tumor. *RB* is a tumor suppressor gene whose protein normally checks passage of the cell through the restriction point in the cell cycle. Another commonly mutated tumor suppressor gene in human cancers is *p53*, whose protein acts to prevent cells with damaged DNA progressing through the cell cycle.

turgor The state, in a plant or prokaryote cell, in which the protoplast is exerting a pressure (*turgor pressure*) on the cell wall owing to the intake of water by OSMOSIS. The cell wall, being slightly elastic, bulges but is rigid enough to prevent water entering to the point of bursting. It exerts a *wall pressure* equal and opposite to the turgor pressure. The cell is then said to be turgid. Turgidity is the main means of support of herbaceous (nonwoody) plants. It also leads to elongation of growing plant cells following localized hormone-induced loosening of cell-wall components. *See* plasmolysis.

tylose An ingrowth from a parenchyma cell into an adjacent tracheid or vessel within the water-conducting tissue of a plant. Tyloses are often found in injured tissue, older wood (heartwood), and below an abscission layer, and can completely block the conducting vessel.

tyrosine *See* amino acids.

ubiquinone (coenzyme Q) A coenzyme found in bacterial membranes and the inner membrane of mitochondria that is an essential component of the ELECTRON-TRANSPORT CHAIN. In mitochondria it transfers two electrons from NADH dehydrogenase to cytochrome *c*, and simultaneously pumps two protons out of the mitochondrial matrix into the intermembrane space.

ultracentrifuge A high-speed CENTRIFUGE, operating at up to a million revolutions per second, that is used to sediment protein and nucleic acid molecules. Ultracentrifuges operate under refrigeration in a vacuum chamber and forces 50 million times gravity may be reached. The rate of sedimentation depends on the density, shape, and molecular weight of the particles, and thus the ultracentrifuge can be used to separate a mixture of large molecules, and estimate sizes. *See also* density-gradient centrifugation, differential centrifugation.

ultramicrotome *See* microtome.

ultrastructure (fine structure) The detailed structure of biological material as revealed, for example, by electron microscopy, but not by light microscopy.

undulipodium (*pl.* **undulipodia**) A eukaryotic 'flagellum'; i.e. a whiplike organelle that protrudes from a eukaryotic cell and is used chiefly for locomotion (e.g. sperm) or feeding (e.g. ciliate protists). Undulipodia include all eukaryotic cilia and flagella, which share the same essential structure, and differ markedly from bacterial flagella (see FLAGELLUM). The shaft comprises a cylindrical array of nine double MICROTUBULES surrounding a central core of two single microtubules – together forming the *axoneme*. The outer wall of the shaft is an extension of the cell membrane. Movement of the shaft is produced by projecting arms of a protein (dynein) on the microtubules, which cause adjacent microtubule doubles to slide past each other. This requires energy from ATP. One set of linker proteins join the nine doubles together in a circle, and another set project radially like spokes towards the central core. The latter are thought to play a crucial role in regulating the beating motion. At the base of the shaft near the cell surface is a basal body, or kinetosome. This consists of a cylindrical array of nine triplet microtubules, from which the shaft grows. Its structure is similar to that of a centriole. Flagella tend to be larger than cilia, and produce successive waves of bending that are propagated to the tip of the shaft, as in a sperm. Cilia characteristically beat in a different manner, with a power stroke and a recovery stroke. *See also* cilium.

unicellular Describing organisms that exist as a single cell. Such a state is characteristic of protists and bacteria and is also found in many fungi. *Compare* multicellular; acellular.

uniporter A transport protein that enables the passage of a particular substance across a cell membrane down its concentration gradient. Such movement, sometimes called *facilitated diffusion*, is powered entirely by the difference in concentration on either side of the membrane. Cells use various types of uniporter, for example, to transport amino acids, sugars, nucleosides, and other small molecules across the plasma membrane. These generally hydrophilic ('water-loving') substances would otherwise be unable to cross the hydrophobic ('water-hating') middle layer of the membrane. Uniporters are generally specific for a particular molecule or group of similar molecules. For instance, mammalian cells take in glucose using the GLUT1 uniporter. This moves glucose molecules across the plasma membrane one at a time. It can also transport the similar sugars mannose and galactose, and will work in reverse if the glucose concentration inside the cell exceeds that outside the cell.

uracil A nitrogenous base that is found in RNA, replacing the thymine of DNA. It has a pyrimidine ring structure.

urea A water-soluble nitrogen compound, $H_2N.CO.NH_2$. It is the main excretory product of catabolism of amino acids in certain animals (ureotelic animals).

uridine The nucleoside formed when uracil is linked to D-ribose by a β-glycosidic bond.

uronic acid *See* sugar acid.

Uracil

V

vacuole A spherical fluid-filled organelle of variable size found in plant and
animal cells, bounded by a single membrane and functioning as a compartment to
separate a variety of materials from the cytoplasm. Vacuoles have a variety of specialized functions, for example, as food
vacuoles or CONTRACTILE VACUOLES. Many
mature plant cells have a single large central vacuole that confines the cytoplasm to
a thin peripheral layer. It is bounded by a
membrane called the tonoplast, and serves
as a store of nutrients or a repository for
waste materials. The vacuole contains cell
sap, comprising various substances in solution, such as sugars, salts, and organic
acids, often in high concentrations resulting in a high osmotic pressure. Water
therefore moves into the vacuole by osmosis making the cell turgid.

valine *See* amino acids.

Vane, (Sir) John Robert (1927–)
British pharmacologist. He was awarded
the Nobel Prize for physiology or medicine
in 1982 jointly with S. K. Bergström and
B. I. Samuelsson for their discoveries concerning prostaglandins and related biologically active substances.

variable number tandem repeats
(VNTRs) A type of REPETITIVE DNA consisting of particular sequences, typically
15–100 bp long, that are repeated in tandem arrays of varying length (1–5 kb)
scattered throughout the genome. For example, all humans have a particular 16-
base sequence that is repeated hundreds of
times among the different chromosomes.
However, no two people are likely to have
the same number of copies at every site.
This individuality in VNTRs makes them

useful for DNA FINGERPRINTING, for example, as a means of identifying suspects, and
as genetic markers in mapping chromosomes (*see* chromosome map). *See also*
satellite DNA.

Varmus, Harold Eliot (1939–) American microbiologist and molecular biologist. He was awarded the Nobel Prize for
physiology or medicine in 1989 jointly
with M. J. Bishop for their discovery of the
cellular origin of retroviral oncogenes.

vasopressin (antidiuretic hormone) A
peptide hormone secreted by the neurohypophysis of the pituitary gland that promotes water absorption by the kidney
tubules, thereby conserving body water. It
also causes the constriction of small arteries and capillaries in the peripheral blood
circulation.

vector *See* cloning vector.

vegetal pole (vegetative pole) The end
of a spherical animal egg that is opposite
the animal pole. The vegetal pole contains
most of the yolk and is furthest from the
nucleus. *Compare* animal pole.

vegetative pole *See* vegetal pole.

vesicle Any of numerous small membrane-bound sacs that serve various functions within cells, especially in the
processing, sorting, and transport of proteins and other substances. For example,
Golgi vesicles convey material between the
different compartments of the GOLGI COM-
PLEX. Proteins destined for immediate SE-
CRETION or for delivery to organelles are
packaged in *transport vesicles*; ones for
regulated secretion are stored in *secretory*

vesicles. Many transport vesicles are coated in protein; for example, materials from the exterior taken in by endocytosis are typically packaged in CLATHRIN-COATED VESICLES. *Compare* endosome.

vessel *See* vessel member.

vessel member (vessel element) One of the cells that makes up a xylem *vessel* in the water-conducting tissue of a plant. Vessel members join end to end, and the end walls are digested away at maturity, forming a continuous vessel. In contrast to TRACHEIDS, the most advanced vessel elements are often more broad than long and have horizontal rather than slanting perforation plates.

Vigneaud *See* du Vigneaud.

villus (*pl.* **villi**) One of the microscopic finger-like projections in the lining of the small intestine. Millions of these villi enormously increase the surface area for absorption. Each villus is covered by a single layer of columnar cells through which the soluble products of digestion can readily pass into the blood or lymph. The surface area is further increased by microvilli (*see* microvillus), which cover the apical surface of each cell of the columnar epithelium. Villi also occur in the chorion (*chorionic villi*), especially in the placenta where they provide a large surface area for exchange of materials between fetal and maternal blood.

virion The extracellular inert phase of a virus. A virion consists of a protein coat (*see* capsid) surrounding one or more strands of DNA or RNA. Virions may be polyhedral or helical and vary greatly in size.

viroid A submicroscopic infectious agent, around 20 nm long and found in plants, that is smaller than a virus and consists simply of a circular strand of RNA, 300–400 nucleotides long. On entering a host cell a viroid is replicated by the cell's machinery, and can accumulate in the cell in very large numbers, eventually killing the cell. Viroids are highly infectious, and include significant plant pathogens, such as potato spindle tuber viroid, coconut cadang-cadang, and hop stunt viroid.

virulent phage A bacteriophage that infects a bacterial cell and immediately replicates, causing lysis of the host cell. *Compare* temperate phage.

virus An extremely small infectious agent, generally 30–300 nm in diameter, that requires a living cell in order to replicate itself. Viruses infect virtually all types of eukaryotic organisms as well as bacteria, and cause a variety of diseases in plants and animals, such as smallpox, the common cold, and tobacco mosaic disease. Each type of virus is usually restricted to a fairly narrow range of hosts. Viruses that infect only bacteria are called PHAGES (or bacteriophages). Outside their living host cell viruses exist as inactive particles consisting of a core of DNA or RNA surrounded by a protein coat (CAPSID). The inert extracellular form of the virus, termed a VIRION, penetrates the host membrane and liberates the viral nucleic acid into the cell. Viral growth cycles inside the cell are of two types. Most viruses undergo a *lytic cycle*, in which the viral nucleic acid is transcribed and translated by the host cell's enzymes and ribosomes to produce the proteins needed for the formation of daughter virions. The virions are released by rupture (lysis) of the host cell. In contrast, some phages enter a quiescent state inside the cell called LYSOGENY. Here the viral DNA is integrated with the host's genome. Some animal viruses, notably the RETROVIRUSES, can also integrate their genome with that of the host; these viruses can introduce ONCOGENES, which cause cancer in the host.

visual purple *See* rhodopsin.

vital stains Nontoxic coloring materials that can be used in dilute concentrations to stain living material without damaging it. Examples of vital stains include Janus green, which selectively stains mitochondria and nerve cells, and trypan blue, which has an affinity for the macrophages. *See also* staining.

vitamin A (vitamin A_1; retinol)   A fat-soluble vitamin, derived from carotenes, that is required for many body functions including normal vision. It is a precursor of retinal, a vital constituent of the visual pigment RHODOPSIN

vitamin B complex   A group of ten or more water-soluble vitamins, including THIAMIN (vitamin B_1), RIBOFLAVIN (vitamin B_2), nicotinic acid (niacin; *see* NAD), pantothenic acid (vitamin B_5), pyridoxine (vitamin B_6), CYANOCOBALAMIN (vitamin B_{12}), BIOTIN, lipoic acid, and folic acid. Many of the B vitamins act as coenzymes during respiration.

vitamin C (ascorbic acid)   A water-soluble vitamin, which is widely required in metabolism, notably as a cofactor in the formation of collagen.

vitamin D   A fat-soluble vitamin whose principal action is to increase the absorption of calcium and phosphorus from the intestine. The vitamin also has a direct effect on the calcification process in bone. The term vitamin D refers, in fact, to a group of compounds, all sterols, of very similar properties. The most important are vitamin D_2 (calciferol) and vitamin D_3 (cholecalciferol). Precursors of these are converted to the vitamins in the body by the action of ultraviolet radiation.

vitamin E (tocopherol)   A fat-soluble vitamin that serves as an antioxidant in tissues, through its ability to trap free radicals.

vitamin K (phylloquinone; menaquinone)   A fat-soluble vitamin that is required to catalyze the synthesis of prothrombin, a blood-clotting factor, in the liver.

vitamins   Organic chemical compounds that are essential in small quantities for metabolism. Many serve as cofactors or coenzymes for enzyme-catalyzed reactions. Vitamins A, D, E, and K are the fat-soluble vitamins, whereas vitamins B and C are the water-soluble vitamins.

vitelline membrane   *See* egg membrane.

VNTR   *See* variable number tandem repeats.

voltage-gated ion channel   An ION CHANNEL that opens in response to a change in membrane potential. Such channels are crucial in generating the ACTION POTENTIAL of a nerve impulse. First, SODIUM ION CHANNELS open in response to slight depolarization of the nerve cell membrane. There is an influx of sodium ions, leading to reversal of the membrane potential. This then triggers the opening of voltage-gated potassium channels, causing the outward flow of potassium ions (K^+) that restores the resting potential.

voluntary muscle   *See* skeletal muscle.

von Euler, Hans   *See* Euler-Chelpin.

von Euler, Ulf Svante (1905–83)   Son of H. von Euler. Swedish neurophysiologist and pharmacologist notable for his recognition of the importance of norepinephrine as the main neurotransmitter in mammalian adrenergic neurons. He was awarded the Nobel Prize for physiology or medicine in 1970 jointly with J. Axelrod and B. Katz.

von Hevesy   *See* Hevesy.

W

Waksman, Selman Abraham (1888–1973) Russian-born American soil microbiologist who discovered streptomycin in 1952 (with Albert Schatz). He was awarded the Nobel Prize for physiology or medicine in 1952.

Wald, George (1906–97) American physiologist. He was awarded the Nobel Prize for physiology or medicine in 1967 jointly with R. Granit and H. K. Hartline for their discoveries concerning the primary physiological and chemical visual processes in the eye.

Walker, John E. (1941–) British biochemist who worked on the enzyme mechanism underlying the synthesis of ATP. He was awarded the 1997 Nobel Prize for chemistry jointly with P. D. Boyer. The prize was shared with J. C. Skou.

wall pressure *See* turgor.

Warburg, Otto Heinrich (1883–1970) German physiologist and biochemist. He was awarded the Nobel Prize for physiology or medicine in 1931 for his discovery of the nature and mode of action of the respiratory enzyme.

water potential Symbol: ψ The chemical potential of water in a biological system compared to the chemical potential of pure water at the same temperature and pressure. It is measured in kilopascals (kPa), and pure water has the value 0 kPa; solutions with increasing concentrations of solute have more negative values of ψ, since the solute molecules interfere with the water molecules. This effect is termed the *solute potential* (or osmotic potential),

denoted by ψ_s; it is measured in kPa and always has a negative value, with increasing concentrations of solute having increasingly negative values of ψ_s. In turgid plant cells there is also a pressure exerted by the walls of the cell; this is called the *pressure potential*; it is denoted by ψ_p, and has a positive value (although in xylem cells it may be negative due to water movement in the transpiration stream). Hence for a plant cell the water potential of the cell, ψ_{cell}, is given by:

$$\psi_{cell} = \psi_p + \psi_s$$

If the water potential of a cell is less (i.e. more negative) than its surroundings the cell will take up water by osmosis, until the solute potential is just balanced by the pressure potential, when the cell is described as being fully turgid. If a cell's water potential is greater (i.e. less negative) than its surroundings, the cell will lose water to its surroundings. *See* osmosis.

Watson, James Dewey (1928–) American biologist. He was awarded the Nobel Prize for physiology or medicine in 1962 jointly with F. H. C. Crick and M. H. F. Wilkins for their discoveries concerning the molecular structure of nucleic acids and its significance for information transfer in living material.

wax One of a group of water-insoluble substances with a very high molecular weight; they are esters of long-chain alcohols with fatty acids. Waxes form protective coverings to leaves, stems, fruits, seeds, animal fur, and the cuticles of insects, serving principally as waterproofing. They may also occur in plant cell walls, for example, leaf mesophyll.

Weismannism The ideas put forward by the German biologist August Weismann (1834–1914) criticizing the theory of the inheritance of acquired bodily characteristics implicit in Lamarckism and certain aspects of Darwinism. Weismann synthesized his ideas into the 'Theory of the Continuity of the Germ Plasm', published in 1886, which emphasized the distinction between the somatic cells and the germ cells and stated that inheritance was effected only by the germ cells.

Western blotting A technique analogous to SOUTHERN BLOTTING used to separate and identify proteins (instead of nucleic acids). The protein mixture is separated by electrophoresis and blotted onto a porous membrane sheet. It is incubated with antibodies specific to the protein of interest, which bind to their target proteins. Excess antibodies are washed off, and a second antibody, specific to the first antibody, is then applied. This is linked to the enzyme alkaline phosphatase, which catalyzes a color-change reaction when a developer is added, so marking the target proteins on the sheet.

white blood cell (**white blood corpuscle**) *See* leukocyte.

white fat *See* adipose tissue.

white matter Nerve tissue that consists chiefly of the fibers (axons) of neurons and their whitish myelin sheaths. It forms the outer region of the spinal cord and occurs in many parts of the brain.

Wieland, Heinrich Otto (1877–1957) German chemist who worked on the constitution of the bile acids and related substances. He was awarded the Nobel Prize for chemistry in 1927.

Wieschaus, Eric F. (1947–) American biochemist who was jointly awarded the 1995 Nobel Prize for physiology or medicine with E. B. Lewis and C. Nüsslein-Volhard for work on the genetic control of early embryonic development.

wild type The most commonly found form of a given gene in wild populations. Wild-type alleles, often designated +, are usually dominant and produce t13he 'normal' phenotype.

Wilkins, Maurice Hugh Frederick (1916–) New Zealand-born British biophysicist. He was awarded the Nobel Prize for physiology or medicine in 1962 jointly with F. H. C. Crick and J. D. Watson for their discoveries concerning the molecular structure of nucleic acids and its significance for information transfer in living material.

Willstätter, Richard Martin (1872–1942) German organic chemist who worked on plant pigments, especially chlorophyll. He was awarded the Nobel Prize for chemistry in 1915.

Windaus, Adolf Otto Reinhold (1876–1959) German chemist. He was awarded the Nobel Prize for chemistry in 1928 for his research into the constitution of the sterols and their connection with the vitamins.

Woodward, Robert Burns (1917–79) American organic chemist. He was awarded the Nobel Prize for chemistry in 1965 for his outstanding achievements in the art of organic synthesis.

X

X chromosome The larger of the two types of sex chromosome in mammals and certain other animals. It is similar in appearance to the other chromosomes and carries many sex-linked genes. *See* sex determination.

xenograft *See* graft.

X-ray crystallography The use of *X-ray diffraction* by crystals to give information about the 3-D arrangement of the atoms in the crystal molecules. When X-rays are passed through a crystal a *diffraction pattern* is obtained because X-rays, whose wavelength is comparable with the distances between atoms, are diffracted by the atoms rather as light is diffracted by a diffraction grating. The diffracted rays are intercepted by photographic film, resulting in a complex pattern of spots, which must be interpreted. The technique is used to help construct 3-D molecular models, and has provided crucial evidence concerning the structures of some large biological molecules such as DNA, RNA, viruses, and a variety of proteins (e.g. hemoglobin, myoglobin, and lysozyme).

X-ray diffraction *See* X-ray crystallography.

xylem The water-conducting tissue in vascular plants. It consists of dead hollow cells (the tracheids and vessels), which are the conducting elements. It also contains additional supporting tissue in the form of fibers and sclereids and some living parenchyma. The secondary cell walls of xylem vessels and tracheids become thickened with lignin to give greater support. Movement of water from roots to leaves via the xylem is termed the transpiration stream.

YAC *See* yeast artificial chromosome.

Yalow, Rosalyn Sussman (1921–) American physicist. She was awarded the Nobel Prize for physiology or medicine in 1977 for the development of radioimmunoassays of peptide hormones. The prize was shared with R. Guillemin and A. V. Schally.

Y chromosome The smaller of the two types of sex chromosome in mammals and certain other animals. It is found only in the heterogametic sex (i.e. males) and typically carries few functioning genes. (about six in humans). One of these is the sex-determining gene, also called the *testis-determining factor* (*TDF*). Its presence in cells of the primordial gonads causes the development of testes, cells of which secrete the hormone testosterone, which subsequently influences the development of male features in other organs. In the absence of the *TDF* gene, the gonads develop along the default female pathway, becoming ovaries instead of testes. *See* sex determination.

yeast artificial chromosome (**YAC**) A linear plasmid that replicates efficiently inside host yeast cells and segregates into the daughter cells following cell division. YACs are used as CLONING VECTORS to clone large fragments of DNA, up to 1000 kb in length. Each contains three essential sequences: an autonomously replicating sequence (ARS) of about 100 bp, which serves as an origin of replication so that the plasmid is replicated; a centromeric region (CEN), which permits proper segregation of the replicated plasmids to daughter cells; and telomeric sequences (TEL) at each end, which again are required for successful replication.

yeasts *See Saccharomyces; Schizosaccharomyces.*

yolk A mixture of proteins, fats, and other substances that is stored in an egg to nourish the developing embryo of an animal following fertilization. The amount of yolk and its distribution within the egg vary with the type of animal. For example, most vertebrates have relatively large amounts of yolk, and this tends to be concentrated toward the lower, or vegetal pole of the egg. However, mammals, because they are nourished largely by the placenta, have evolved eggs that contain little yolk, and this is evenly distributed in the egg cytoplasm. In insects and other arthropods the yolk is typically concentrated in the center of the egg, surrounded by a thin peripheral layer of cytoplasm.

Z

Z disk *See* sarcomere.

zinc *See* trace elements.

zinc finger *See* motif.

Zinkernagel, Rolf M. (1944–) Swiss biochemist who was jointly awarded the 1996 Nobel Prize for physiology or medicine with P. C. Doherty for work on the specificity of the cell-mediated immune defense.

zona pellucida The thick clear jelly-like layer that surrounds and protects the mammalian egg. It contains species-specific sperm receptors, which permit penetration only by sperm of the same species. In the freshly ovulated egg the zona is surrounded by *cumulus (follicle) cells*, but these disperse as the sperms pass between them and penetrate the zona by enzyme action.

zonula occludens *See* tight junction.

zygote A fertilized egg. Typically it is a diploid cell resulting from the fusion of two haploid gametes (sperm and egg cell), and usually undergoes cleavage immediately to form the EMBRYO.

zygotene In meiosis, the stage in midprophase I that is characterized by the active and specific pairing (*see* synapsis) of homologous chromosomes leading to the formation of bivalents.

zymogen granule A type of secretory vesicle found in large numbers in enzyme-secreting cells. The vesicles contain an inactive precursor of the enzyme, the zymogen, such as trypsinogen in exocrine cells of the pancreas. The zymogen is activated after secretion via exocytosis at the plasma membrane.

APPENDIXES

Appendix I

Chronology

Major Events in the Development of Biochemistry and
Molecular Biology

1658	Discovery and description of red blood corpuscles by Dutch microscopist Jan Swammerdam (1637–80)
1663	English scientist Robert Hooke (1635–1703) is credited with coining the term 'cell'
1665	Hooke publishes first detailed drawings of plant cells (cork tissue) in his *Micrographia*
1673–83	Various cells (including spermatozoa, yeast cells, protozoa, and bacteria) described by Dutch microscopist Antoni van Leeuwenhoek (1632–1723)
1833	Discovery of the nucleus by the British botanist Robert Brown (1773–1858)
1838–39	The view that all organisms are composed of cells ('cell theory') is advanced by the German botanist Matthias Schleiden (1804–81) and the German physiologist Theodor Schwann (1810–82)
1858	The principle that all cells derive from pre-existing cells is proposed by the German pathologist Rudolf Virchow (1821–1902) in his *Die Cellular pathologie*
1859	British naturalist Charles Darwin (1809–82) publishes *On The Origin of Species*, describing evolution by natural selection
1860	The French chemist Louis Pasteur (1822–95) shows that fermentation is produced by substances from yeasts and bacteria (which he called 'ferments'; later (1877) designated as enzymes)
1866	The Austrian geneticist Gregor Mendel (1822–84) publishes his discoveries concerning the basic principles of inheritance, thus founding the science of genetics
1869	Nucleic acids discovered by the German biochemist Johann Friedrich Miescher (1844–95)

Chronology

1873	The Golgi complex is discovered by the Italian cytologist Camillo Golgi (1843–1926)
1882	The process of mitosis is described in detail by German cytologist Walther Flemming (1843–1905)
1888	The term 'chromosome' is introduced by the German anatomist Heinrich Waldeyer-Hartz (1836–1921)
1903	The chromosomal theory of inheritance is stated independently by American geneticist Walter Sutton and German zoologist Theodor Boveri (1862–1915)
1909	Ribose is identified in DNA by the Russian-born American biochemist Phoebus Levene (1869–1940)
1913	The first genetic map, of six genes of the fruit fly *Drosophila*, is constructed by the American geneticist Alfred H. Sturtevant (1891–1970)
1925	Cytochrome is discovered by the Russian-born British scientist David Keilin (1887–1963)
1929	The German chemist Hans Fischer (1881–1945) determines the structure of heme in hemoglobin
1932	The muscle protein myoglobin is isolated by the Swedish biochemist Hugo Theorell (1903–82)
1940	The German-born American biochemist Fritz Lipmann (1899–1986) puts forward the proposal that ATP is the carrier of chemical energy in cells
1943	The American biochemist Britton Chance (b. 1913) suggests that enzymes operate through an enzyme–substrate complex
1944	DNA is demonstrated conclusively to be the genetic material by the American bacteriologist Oswald T. Avery (1877–1955) and his associates Colin MacLeod and Maclyn McCarty
1950	The endoplasmic reticulum is discovered by the Belgian-born American cell biologist Albert Claude (1899–1983)
1953	The British biologist Francis Crick (b. 1916) and the American biologist James Watson (b. 1928) discover the structure of DNA
1955	Lysosomes are discovered by the Belgian cell biologist Christian de Duve (b. 1917)

Chronology

1956	The American biochemist Arthur Kornberg (b. 1918) discovers DNA polymerase
1956	The American molecular biologist Paul Berg (b. 1926) identifies the nucleic acid that is later called transfer DNA
1957	Details of the carbon fixation reactions of photosynthesis are published by the American biochemist Melvin Calvin (1911–97)
1960	Messenger RNA is identified by the South African-born British molecular biologist Sydney Brenner (b. 1927) and the French biochemist François Jacob (b. 1920)
1964	First cell receptors discovered by American microbiologists Keith Porter and Thomas F. Roth
1966	Deciphering of the genetic code completed by American molecular biologists Marshall W. Nirenberg (b. 1927) and Robert Holley (1922–93), and Indian biochemist Har Gobind Khorana (b. 1922)
1970	The enzyme reverse transcriptase is discovered by the American virologists Howard Temin (1934–94) and David Baltimore (b. 1938).
1970	American biologist Lyn Margulis (b. 1938) proposes the endosymbiont theory for the origin of cellular organelles.
1970	Restriction enzymes are first found by the American molecular biologist Hamilton Smith (b. 1931)
1971	The signal hypothesis to explain how cells sort proteins and deliver them to their correct destinations within the cell is proposed by German-born American cell biologist Günter Blobel (b. 1936)
1972	First recombinant DNA produced by Paul Berg
1973	First use of a plasmid to clone DNA by the American biochemists Stanley N. Cohen (b. 1935) and Herbert Boyer (b. 1936)
1975	Fusion of plant and animal cells achieved by British biologists J.A. Lucy and E.C. Cocking
1981	First expression of a foreign gene in a mouse demonstrated by Ralph Brinster and Richard Palmiter

Chronology

1983	Gene for human growth hormone inserted in mouse embryo to produce a 'supermouse'
1985	DNA fingerprinting is invented by the British biochemist Alec Jeffreys (b. 1950).
1985	American biochemist Kary Mullis (b. 1944) invents the polymerase chain reaction for amplifying DNA fragments
1986	First licence granted in the US for marketing a genetically engineered living organism
1989	The Human Genome Project is launched
1993	First cloned human embryos produced
1994	Beginnings of DNA chip technology
1997	Birth of Dolly the sheep, the first mammal to be cloned from a tissue cell taken from an adult
2001	First draft of the human genome sequence is published covering 97% of the human genome

Appendix II

Amino Acids

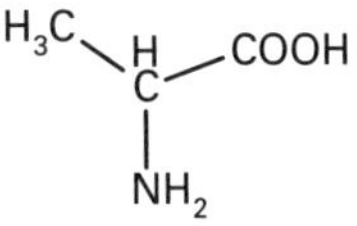

alanine

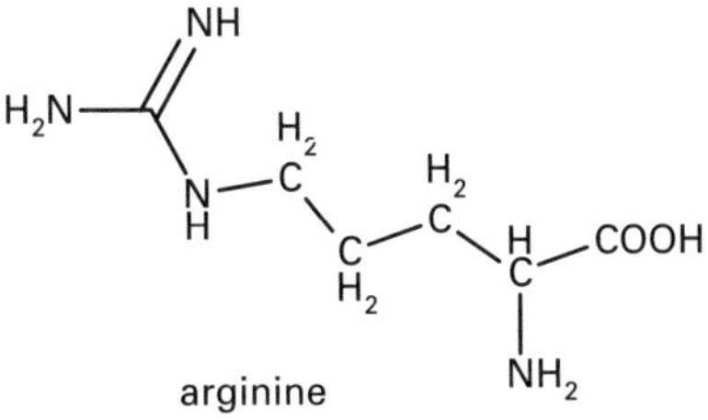

arginine

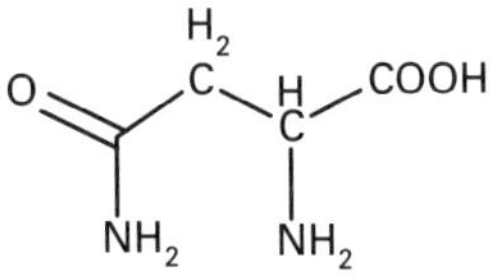

asparagine

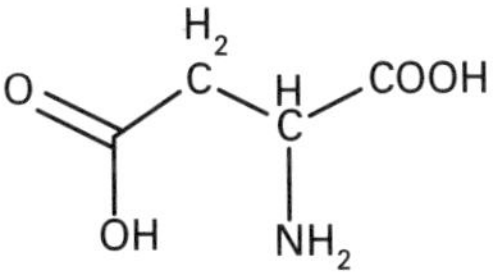

aspartine

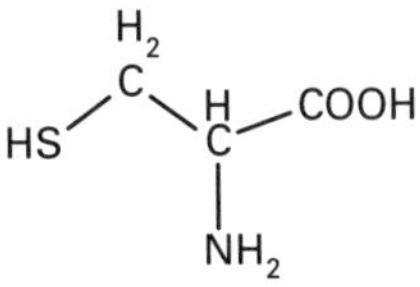

cysteine

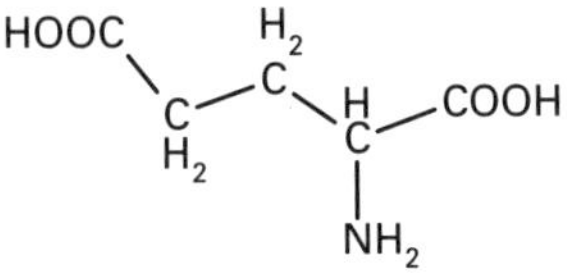

glutamic acid

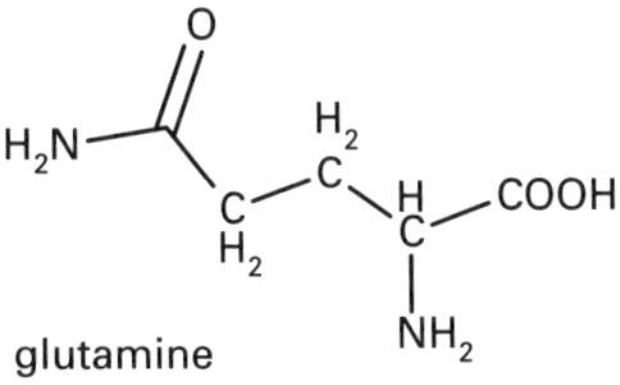

glutamine

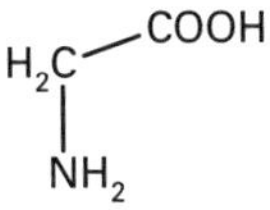

glycine

Amino Acids

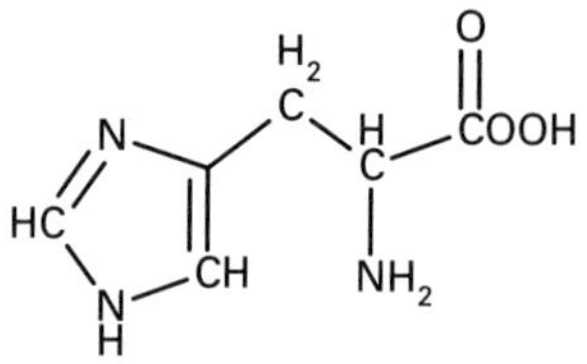

histidine

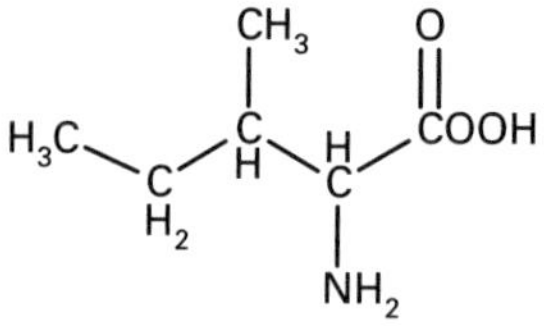

isoleucine

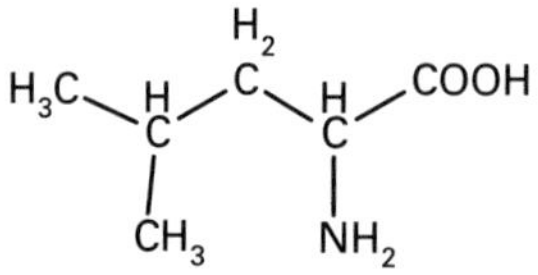

leucine

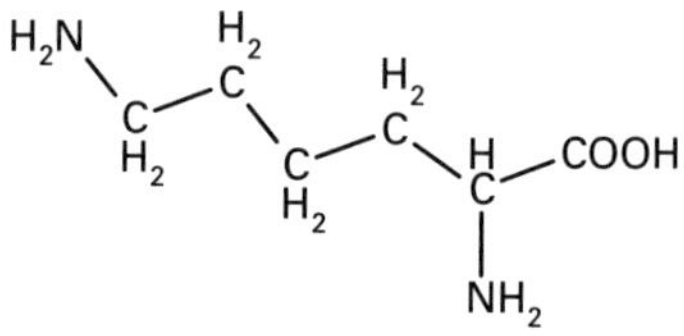

lysine

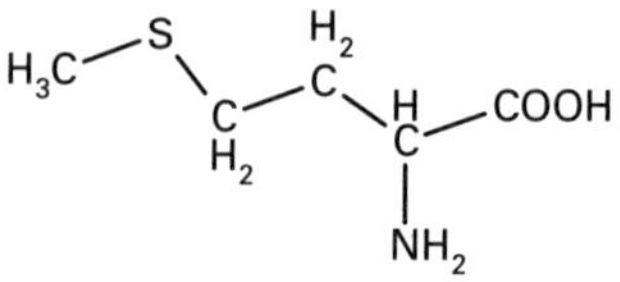

methionine

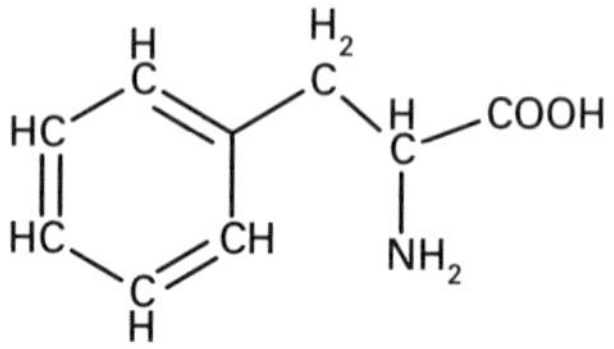

phenylalanine

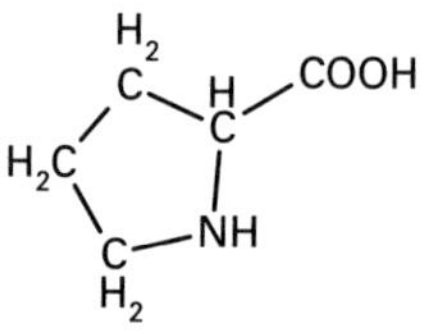

proline

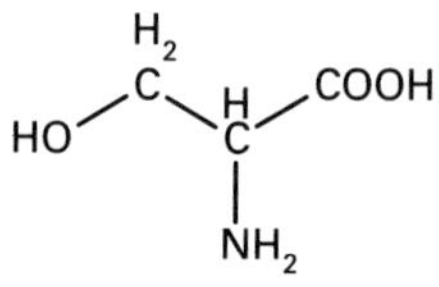

serine

Amino Acids

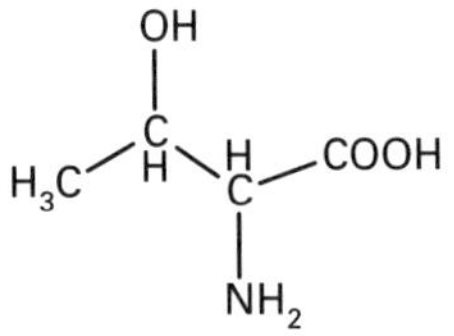

threonine

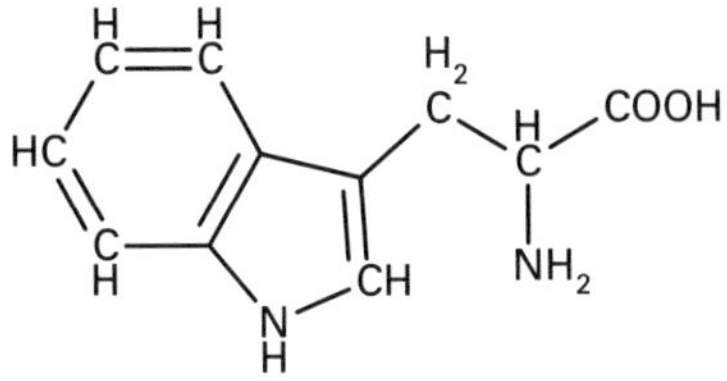

tryptophan

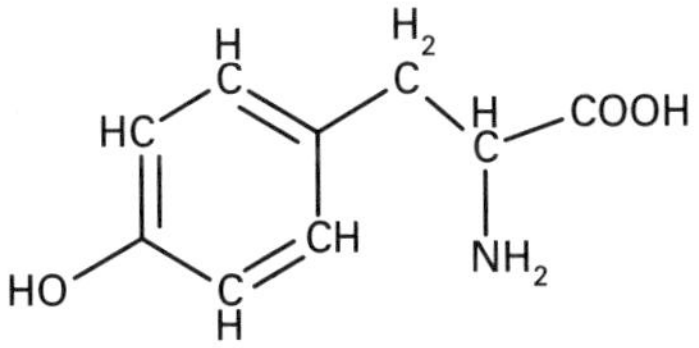

tyrosine

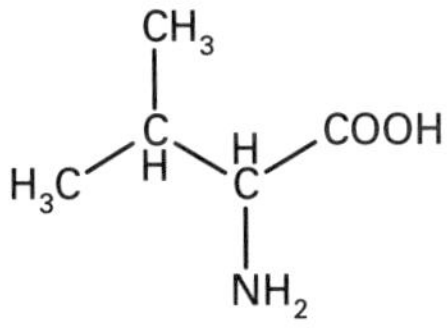

valine

Appendix III

The Genetic Code

		second base				
		U	**C**	**A**	**G**	
	U	UUU Phe	UCU Ser	UAU Tyr	UGU Cys	U
		UUC Phe	UCC Ser	UAC Tyr	UGC Cys	C
		UUA Leu	UCA Ser	UAA stop codon	UGA stop codon	A
		UUG Leu	UCG Ser	UAG stop codon	UGG Trp	G
	C	CUU Leu	CCU Pro	CAU His	CGU Arg	U
		CUC Leu	CCC Pro	CAC His	CGC Arg	C
first base		CUA Leu	CCA Pro	CAA Gln	CGA Arg	A
		CUG Leu	CCG Pro	CAG Gln	CGG Arg	G
	A	AUU Ile	ACU Thr	AAU Asn	AGU Ser	U
		AUC Ile	ACC Thr	AAC Asn	AGC Ser	C
		AUA Ile	ACA Thr	AAA Lys	AGA Arg	A
		AUG Met (start codon)	ACG Thr	AAG Lys	CUG Arg	G
	G	GUU Val	GCU Ala	GAU Asp	GGU Gly	U
		GUC Val	GCC Ala	GAC Asp	GGC Gly	C
		GUA Val	GCA Ala	GAA Glu	GGA Gly	A
		GUG Val	GCG Ala	GAG Glu	GGG Gly	G

(third base column shown on right: U, C, A, G for each group)

KEY

Ala	Alanine	Leu	Leucine
Arg	Arginine	Lys	Lysine
Asn	Asparagine	Met	Methionine
Asp	Aspartic acid	Phe	Phenylalanine
Cys	Cysteine	Pro	Proline
Gln	Glutamine	Ser	Serine
Glu	Glutamic acid	Thr	Threonine
Gly	Glycine	Trp	Tryptophan
His	Histidine	Tyr	Tyrosine
Ileu	Isoleucine	Val	Valine

Appendix IV

Webpages

The following all have information on biochemistry and on cell and molecular biology:

European Bioinformatics Centre	ebi.ac.uk
Harvard University Department of Molecular and Cellular Biology	mcb.harvard.edu/BioLinks.html
Human Genome Project	www.ornl.gov/hgmis
Internation Union of Biochemistry and Molecular Biology (Nomenclature)	www.chem.qmul.ac.uk/iubmb
Leeds University Bioinformatics Links	www.bioinf.leeds.ac.uk/bioinformatics.html
National Biological Information Infrastructure	www.nbii.gov
Oxford University Bioinformatics Centre	www.molbiol.ox.ac.uk
Southwest Biotechnology and Informatics Center	www.swbic.org
UK Human Genome Mapping Project	www.hgmp.mrc.ac.uk

Bibliography

Basic texts covering biochemistry and cell and molecular biology:

Alberts, Bruce; Johnson, Alexander; Lewis, Julian; Raff, Martin; Roberts, Keith; & Walter, Peter *Molecular Biology of the Cell.* 4th ed. New York: Garland Science, 2002

Aldridge, Susan *Biochemistry for Advanced Biology.* Cambridge, U.K.: Cambridge University Press, 1994

Boyer, Rodney F. *Modern Experimental Biochemistry.* San Francisco, Calif.: Benjamin Cummings, 2000

Campbell, Mary K. *Biochemistry.* 3rd ed. Philadelphia: Harcourt Brace, 1999

Elliott, William H. & Elliott, Daphne C. *Biochemistry and Molecular Biology.* 2nd ed. Oxford, U.K.: Oxford University Press, 2001

Horton, H. Robert; Moran, Laurence A.; Ochs, Raymond S.; Raun, David J.; & Scrimgeour, K. Gray *Principles of Biochemistry.* 3rd ed. Upper Saddle River, NJ: Prentice Hall, 2002

Lodish, Harvey; Berk, Arnold; Zipursky, S. Lawrence; & Matsudaira, Paul *Molecular Cell Biology.* 4th ed. New York: W. H. Freeman, 2000

McKee, Trudy & McKee, James R. *Biochemistry: The Molecular Basis of Life.* 3rd ed. Boston, Mass.: McGraw-Hill, 1999

Voet, Donald; Voet, Judith G.; & Pratt, Charlotte W. *Fundamentals of Biochemistry.* New York: Wiley, 1999

A more advanced text is:

Nelson, David L. & Cox, Michael M. *Lehninger Principles of Biochemistry.* 3rd. ed. New York: Worth Publishers, 2000